AF352847

Semi-Solid Processing of Alloys and Composites

Semi-Solid Processing of Alloys and Composites

Special Issue Editors

Shahrooz Nafisi
Reza Ghomashchi

MDPI • Basel • Beijing • Wuhan • Barcelona • Belgrade • Manchester • Tokyo • Cluj • Tianjin

Special Issue Editors

Shahrooz Nafisi
University of Alberta
Canada

Reza Ghomashchi
The University of Adelaide
Australia

Editorial Office
MDPI
St. Alban-Anlage 66
4052 Basel, Switzerland

This is a reprint of articles from the Special Issue published online in the open access journal *Metals* (ISSN 2075-4701) (available at: https://www.mdpi.com/journal/metals/special_issues/semi_solid_processing).

For citation purposes, cite each article independently as indicated on the article page online and as indicated below:

LastName, A.A.; LastName, B.B.; LastName, C.C. Article Title. *Journal Name* **Year**, *Article Number*, Page Range.

ISBN 978-3-03928-975-2 (Hbk)
ISBN 978-3-03928-976-9 (PDF)

Cover image courtesy of Shahrooz Nafisi.

Contents

About the Special Issue Editors . . . . **vii**

Shahrooz Nafisi and Reza Ghomashchi
Semi-Solid Processing of Alloys and Composites
Reprinted from: *Metals* **2019**, *9*, 526, doi:10.3390/met9050526 . . . . **1**

Michael Modigell, Annalisa Pola and Marialaura Tocci
Rheological Characterization of Semi-Solid Metals: A Review
Reprinted from: *Metals* **2018**, *8*, 245, doi:10.3390/met8040245 . . . . **3**

Annalisa Pola, Marialaura Tocci and Plato Kapranos
Microstructure and Properties of Semi-Solid Aluminum Alloys: A Literature Review
Reprinted from: *Metals* **2018**, *8*, 181, doi:10.3390/met8030181 . . . . **27**

Shahrooz Nafisi, Anthony Roccisano, Reza Ghomashchi and George Vander Voort
A Comparison between Anodizing and EBSD Techniques for Primary Particle Size
Measurement
Reprinted from: *Metals* **2019**, *9*, 488, doi:10.3390/met9050488 . . . . **45**

Jufu Jiang, Guanfei Xiao, Changjie Che and Ying Wang
Microstructure, Mechanical Properties and Wear Behavior of the Rheoformed 2024 Aluminum
Matrix Composite Component Reinforced by Al_2O_3 Nanoparticles
Reprinted from: *Metals* **2018**, *8*, 460, doi:10.3390/met8060460 . . . . **59**

Gabriela Lujan Brollo, Cecília Tereza Weishaupt Proni and Eugênio José Zoqui
Thixoforming of an Fe-Rich Al-Si-Cu Alloy—Thermodynamic
Characterization, Microstructural Evolution, and Rheological Behavior
Reprinted from: *Metals* **2018**, *8*, 332, doi:10.3390/met8050332 . . . . **83**

M. N. Mohammed, M. Z. Omar, Salah Al-Zubaidi, K. S. Alhawari and M. A. Abdelgnei
Microstructure and Mechanical Properties of Thixowelded AISI D2 Tool Steel
Reprinted from: *Metals* **2018**, *8*, 316, doi:10.3390/met8050316 . . . . **107**

Yongkun Li, Rongfeng Zhou, Lu Li, Han Xiao and Yehua Jiang
Microstructure and Properties of Semi-solid ZCuSn10P1 Alloy Processed with an Enclosed
Cooling Slope Channel
Reprinted from: *Metals* **2018**, *8*, 275, doi:10.3390/met8040275 . . . . **123**

Chul Kyu Jin
Microstructure of Semi-Solid Billets Produced by Electromagnetic Stirring and Behavior of
Primary Particles during the Indirect Forming Process
Reprinted from: *Metals* **2018**, *8*, 271, doi:10.3390/met8040271 . . . . **135**

Marta Ślęzak
Study of Semi-Solid Magnesium Alloys (With RE Elements) as a Non-Newtonian Fluid
Described by Rheological Models
Reprinted from: *Metals* **2018**, *8*, 222, doi:10.3390/met8040222 . . . . **151**

**Maryam Eslami, Mostafa Payandeh, Flavio Deflorian, Anders E. W. Jarfors and
Caterina Zanella**
Effect of Segregation and Surface Condition on Corrosion of Rheo-HPDC Al–Si Alloys
Reprinted from: *Metals* **2018**, *8*, 209, doi:10.3390/met8040209 . . . . **165**

Jufu Jiang, Guanfei Xiao, Ying Wang and Yingze Liu
Tribological Behavior of Nano-Sized SiCp/7075 Composite Parts Formed by Semisolid
Processing
Reprinted from: *Metals* **2018**, *8*, 148, doi:10.3390/met8030148 . . . . **183**

Ava Azadi Chegeni and Platon Kapranos
An Experimental Evaluation of Electron Beam Welded Thixoformed 7075 Aluminum Alloy
Plate Material
Reprinted from: *Metals* **2017**, *7*, 569, doi:10.3390/met7120569 . . . . **205**

About the Special Issue Editors

Shahrooz Nafisi, Dr., is an adjunct Professor at the University of Alberta, Canada. He received his B.S. and M.S. in Metallurgical Engineering from the Iran University of Science and Technology and his Ph.D. from the University of Québec. He has co-authored two books, "A New Approach to Gating Systems" (1st edition, 1997, 2nd edition, 2001) and "Semi-Solid Processing of Aluminum Alloys", ISBN 978-3-319-40333-5, Springer, Sep 2016 (republished in China, Jan 2020, ISBN: 978-7-122-34281-2), and more than 70 journal articles and conference papers. He was the 2017 profile in achievement from Professional Engineers of Canada (APEGS); the Vanadium award recipient in 2014 (Institute of Materials, Minerals and Mining "IOM3"); the 2013 best paper award of the International Metallographic Society and Metallography, Microstructures, and Analysis; and the recipient of the 2012 Association of Iron and Steel Technology (AIST) Hunt-Kelly Outstanding Paper Award.

Reza Ghomashchi, Prof., graduated from the Iran University of Science and Technology (B.Eng. Metallurgical Engineering,1978), and Cambridge (M.Phil. Materials Technology, 1979) and Sheffield (Ph.D. Metallurgy, 1983) Universities in the U.K. After a few years working at universities in the U.K., he migrated to Australia to work for BHP Steel in 1988. In early 1990, he joined the University of South Australia and worked as a Lecturer and then Senior Lecturer until 2001 when he was offered a Natural Science and Engineering Research Council of Canada (NSERC) Industrial Research Chair-Tier I (in collaboration with ALCAN, now Rio-Tinto ALCAN) and Professor Position at the University of Québec in Canada. He has also been a visiting Professor at MIT, 1994; IUST, 1999; and an adjunct Professor at the School of Mechanical Engineering at the University of Adelaide, 2007. In early 2008, he returned to Australia to take up the position of Manager, Materials Research and Development, at Sunday Solar Technologies, a start-up R&D company in Sydney and, in August 2010, he accepted his current position at the school of Mechanical Engineering of the University of Adelaide, Australia.

Editorial

Semi-Solid Processing of Alloys and Composites

Shahrooz Nafisi [1],* and Reza Ghomashchi [2],*

[1] Department of Chemical and Materials Engineering, University of Alberta, Edmonton, AB T6G 2M7, Canada
[2] The School of Mechanical Engineering, The University of Adelaide, Adelaide, SA 5005, Australia
* Correspondence: shahrooznafisi@gmail.com (S.N.); reza.ghomashchi@adelaide.edu.au (R.G.);
   Tel.: +1-503-568-0992 (S.N.); +61-8-83133360 (R.G.)

Received: 29 April 2019; Accepted: 7 May 2019; Published: 8 May 2019

A quick look through the past two centuries tells us that we may be in our third industrial revolution. The first industrial revolution (19th century) was mainly due to the introduction of steam energy, while the second (20th century) was mainly due to inexpensive oil and gas—which, by the way, brought us some unwelcome consequences; the so-called greenhouse effect and subsequent global warming and unpredictable weather patterns. We are now embracing a third industrial revolution, which could be termed the green energy and communication era. As a result, our manufacturing technologies should also follow a similar pattern: "Green manufacturing" with less energy consumption. Semi-solid metal (SSM) processing may be branded as a step forward towards green manufacturing, as it consumes less energy than its conventional counterparts. However, in spite of the many advantages of SSM processing and its viable manufacturing route, including a reduction in energy consumption, its implementation in the metal industry has been very sluggish. As strong advocates within the SSM processing community, we believe such a delay in recognizing the benefits of SSM casting of light alloys is predominantly due to the lack of proper communication between research and development (R&D) investigators and industry leaders. The Editors have tried to close the communication gap through publication of a new book [1] and the introduction of the current special issue of the Metals Journal on SSMs as an extra effort to the biannual S2P conference. We hoped an invitation of key players to highlight the latest advancements in the field would contribute towards better usage of SSM processes in industrial applications.

This special issue is focused on the recent research and findings in the field, with the aim of filling the gap between industry and academia, and to shed light on some of the fundamentals of science and technology of semi-solid processing.

This special issue provides new researches on the two main routes of semi-solid metal processing; Rheo and Thixo - casting. In addition, a variety of alloying systems and composite materials are covered in this special issue, including interesting information on welding, tribology and corrosion of SSM-processed alloys. Rheology and the correlation between structure and properties have been covered in two outstanding review articles. We would like to thank all the authors for their contribution and consideration of the reviewers' comments. Additionally, the continuous assistance of the Metals editorial staff is gratefully acknowledged.

Conflicts of Interest: The authors declare no conflict of interest.

Reference

1. Nafisi, S.; Ghomashchi, R. *Semi Solid Processing of Aluminum Alloys*; Springer International Publishing: Basel, Switzerland, 2016.

Review

Rheological Characterization of Semi-Solid Metals: A Review

Michael Modigell [1,*], Annalisa Pola [2] and Marialaura Tocci [2]

[1] Department of Engineering, German University of Technology in Oman (GUtech), PO Box 1816, Athaibah PC 130, Muscat, Oman

[2] DIMI—Mechanical and Industrial Engineering Department, University of Brescia, Via Branze, 38, 25123 Brescia, Italy; annalisa.pola@unibs.it (A.P.); marialaura.tocci@unibs.it (M.T.)

* Correspondence: michael.modigell@gutech.edu.om or michael.modigell@avt.rwth-aachen.de; Tel.: +968-2206-1110

Received: 18 February 2018; Accepted: 4 April 2018; Published: 7 April 2018

Abstract: In the present review, the main findings on the rheological characterization of semi-solid metals (SSM) are presented. Experimental results are a fundamental basis for the development of comprehensive and accurate mathematics used to design the process effectively. For this reason, the main experimental procedures for the rheological characterization of SSM are given, together with the models most widely used to fit experimental data. Subsequently, the material behavior under steady state condition is summarized. Also, non-viscous properties and transient conditions are discussed since they are especially relevant for the industrial semi-solid processing.

Keywords: rheology; semi-solid alloys; thixotropy; rheometer; compression test; viscosity

1. Introduction

In the early 1970s Flemings and coworkers [1] discovered metallic alloys in the semi-solid state with non-dendritic structure to have special rheological properties which can be exploited for a new, attractive forming process. The non-dendritic structure can be easily achieved by stirring the alloy while cooling it from the liquid state down into the semi-solid temperature range. This results in a suspension consisting of a liquid metallic phase and primary solid particles with globular or rosette-type shape [2,3], such as that shown in Figure 1, in comparison with a typical dendritic microstructure.

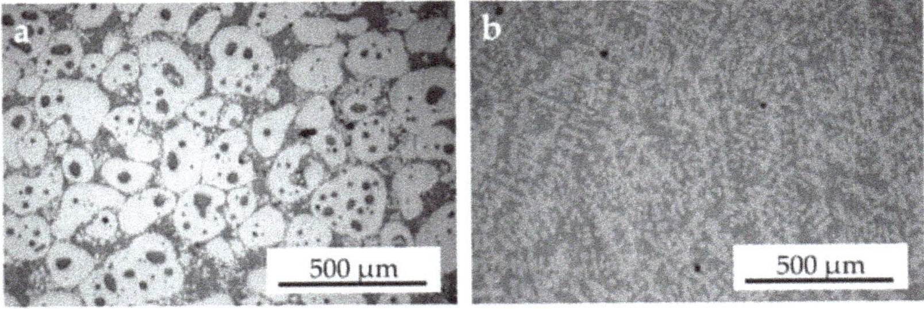

Figure 1. Typical microstructures of an A356 component obtained by (**a**) thixocasting (billet preheated to a solid fraction of 0.52–0.54) and (**b**) conventional casting (pouring temperature = 680 °C).

The rheological properties of this special type of slurry give the advantages of the semi-solid metals (SSM)-processing. In detail, the slurry can either flow like a liquid—but with non-constant viscosity—or it can behave like a solid. This is typical for suspensions with high solid fraction (*Fs*) [4], whereas fully liquid metals show Newtonian flow behavior, which is water-like [5]. The rheological properties are responsible for the die-filling behavior of SSM, which is different from fully liquid (or fully solid) materials [6,7] and which results in specific advantages in the quality of the product (low gas porosity, less shrinking, higher mechanical properties, etc.), besides those related to technological aspects (longer tool life in comparison with conventional casting processes due to the lower metal temperature, etc.) [2]. To obtain these advantages, it is necessary to fully understand the rheology of the material. This enables understanding of flow-specific phenomena, such as instabilities or segregation, and allows optimization of the process. Carefully performed experiments with respect to the mechanical, fluid dynamical, and thermal conditions lead to the development of comprehensive and accurate mathematical models, which picture the physics properly and which are used in computer simulation to design and optimize the process effectively.

Few publications provide a comprehensive discussion on the rheological behavior of SSM [8,9]. In addition, in this regard, it should be mentioned that still some aspects about the flow of semi-solid metals are not clear or contradict data are available in the literature [8]. For this reason, it appears useful to provide an overview of the current knowledge about this topic, with particular attention to the scientific innovation that took place in the past 10–15 years, while details about semi-solid metal processing can be found elsewhere [2,7,10].

In Section 2, the rheological classification of SSM is explained. In Section 3, the principles of the most frequently used devices for investing rheological properties of SSM are presented: in Section 3.1 the rotational rheometer and in Section 3.2 the compression test. Section 4 gives a summary of the applied rheological models and the corresponding constitutive equations. An overview of recent published results of the equilibrium viscosity of Al-alloys is given in Section 5. Section 6 deals with special rheological phenomena of SSM: yield stress and thixotropy. Section 7 explains the influence of Ostwald ripening in rheological experiments, which is important to consider for long-term experiments.

Before analyzing the literature, it is useful to clarify some rheological terms that are related with the rheological properties of SSM and which are frequently used improperly in the literature of SSM processing.

2. Rheological Classification of SSM

The rheological behavior of any material is found to be between two limiting, ideal cases: the ideal solid body (Hookean body), which shows deformation proportional to the stress, and the ideal viscous material (Newtonian body), which shows rate of deformation proportional to the stress. Within the viscous materials, besides the Newtonian fluids (with constant viscosity, only depending on temperature and pressure), we find the Non-Newtonian fluids. Among these, we have the non-linear pure viscous fluids, whose viscosity depends additionally on the stress and which exhibit shear thinning or shear thickening behavior. Another class of the non-linear materials are the plastic ones, which show solid behavior below a certain threshold of stress (yield stress), while they are characterized by linear (Bingham body) or non-linear behavior above the yield stress. Another class of materials shows time-dependent properties, whose rheological properties do not change immediately after change in strain but follow a specific kinetics. Viscoelastic materials have simultaneously elastic and viscous properties. Thixotropic materials show a gradual decrease of viscosity under constant stress and a recovery of the viscosity when the stress is removed; in particular, the viscosity of the initial condition will be recovered totally. As shown in Figure 2, due to thixotropy, at the beginning of shearing or after a rapid change in shear rate, the instantaneous viscosity is different from the steady state values, and it takes time for the viscosity to reach a constant value, which reflects the equilibrium condition of the structure. The opposite behavior, an increase of viscosity with time, is called rheopexy.

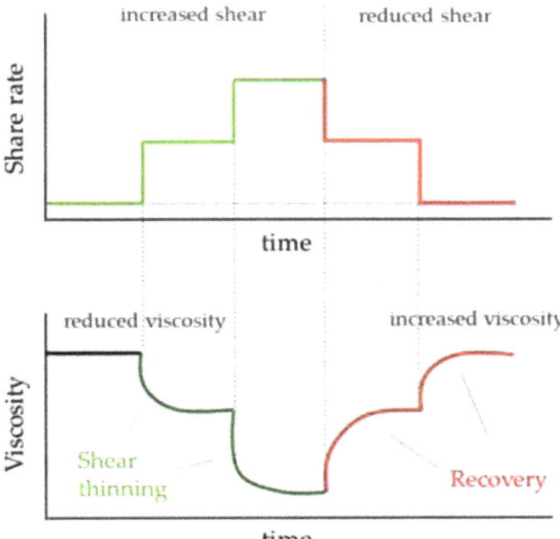

Figure 2. Schematic diagram showing the change in viscosity with time following changes in shear rate to illustrate the thixotropic behavior of semi-solid slurries [11].

Regarding the rheological classification of SSM, it is generally accepted that they are shear thinning fluids, that means that viscosity will drop with increasing shear rate (Figure 3) [12]. Additionally, they are thixotropic materials.

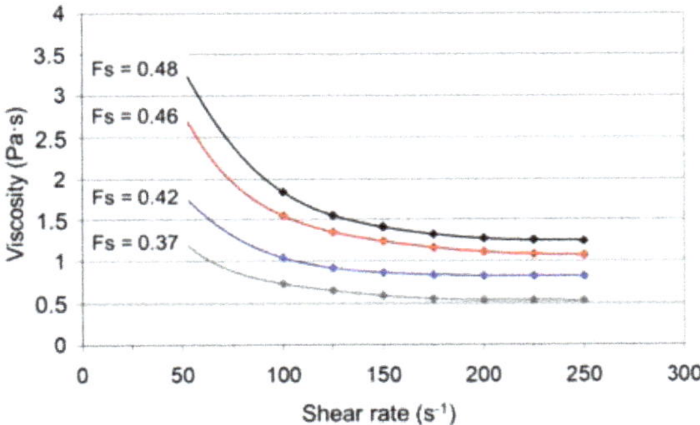

Figure 3. Example of plot of viscosity measurements in isothermal conditions from shear rate experiments for A356 alloy at various solid fractions showing the typical shear thinning behavior of SSM [12].

Finally, SSM with high solid fraction exhibit yield stress, which is nicely demonstrated with the picture of a block of Al alloy in the semi-solid state able to wear its own weight and which can be cut with an ordinary knife. A representative image of this phenomenon is shown in Figure 4; analogous pictures are available in the scientific literature, as for instance in [1].

The specific rheological properties are finally nothing more than the consequences of the changes in the internal structure of the slurry due to external forces. A simple physical model can be set up, in agreement with the general understanding of the kinematics of the SSM. It can be assumed that cohesive forces are acting between the particles of the SSM, resulting in the formation of agglomerates [13]. The particles of such agglomerates can be connected temporarily by formation of welded necks or e.g., by capillary forces. Within the agglomerates, a certain amount of liquid is immobilized, which leads to a higher apparent solid fraction [14]. Under the influence of shear forces, the agglomerates will be partially or totally disintegrated whereby the liquid phase is released and the apparent solid fraction will approach the true fraction. With this model, the steady state behavior of the material, i.e., shear thinning, can be explained since the viscosity decreases with decreasing solid fraction.

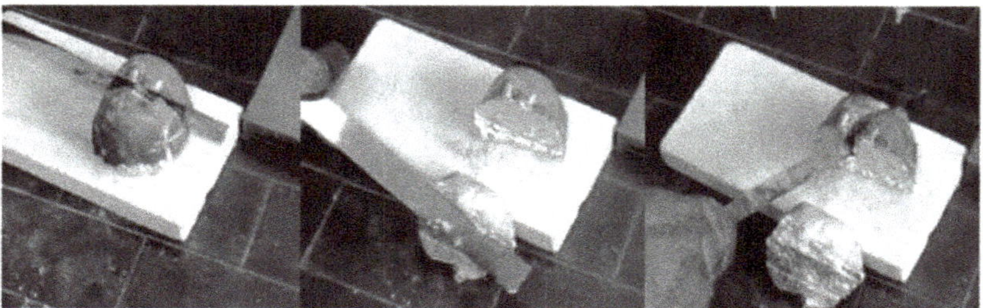

Figure 4. Billet of a semi-solid metal cut with an ordinary knife.

By reducing the shear rate, particles that meet in the shear field have the chance to agglomerate, resulting in an increase in the apparent solid fraction.

It follows that the structural change is reversible, which explains one feature of thixotropy. The deagglomeration and agglomeration processes do not happen instantaneously, but they take some time. The agglomeration is diffusion controlled and, therefore, it is much slower than the deagglomeration phenomenon.

Similar to other suspensions, the most important parameter influencing the rheological properties is the solid fraction, which depends on the temperature [2]. Experimental measurements show that an increase in solid fraction results in an increase of viscosity. In addition, also the yield stress increases with higher solid fraction, as demonstrated in scientific literature [15–18], together with the presence of the thixotropic effects. Other factors, such as particle diameter, or diameter distribution, and the shape of the particles are of minor importance. Both these parameters can be combined with the specific surface area of solid and liquid phase. The dependency on the viscosity of the liquid phase is not very strong because the liquid phase viscosity is orders of magnitude lower than of the SSM.

This rather simple structural model has been accepted widely—although there is no clear experimental evidence. Indeed, all metallographic micrographs have been produced from ordinary solidified samples with low cooling rates. It has been demonstrated [19,20] that, with cooling rates smaller than -10 K/s, diffusional processes will significantly change the size of the particles and, therefore, the appearance of the structure. To investigate the structure is difficult. There are only a few publications [20–24] that are dealing with structural experiments with X-ray tomography for SSM in rest and compression. However, up to now, no work has been published for investigations of SSM under pure shear. The evaluation of the X-ray images shows that with increasing shear rate the distribution of the particles will become more homogenous over the volume, which confirms qualitatively the physical model [24]. The compression experiment under the X-ray beam shows that for high solid fraction (0.70) the material behaves as a saturated sponge and the liquid phase is pressed out of the skeleton with the consequence that the solid fraction will change locally. This has been previously found, as well by analyses of the solid distribution in a billet after compression [25].

3. Experimental Methods for the Measurements of Rheological Properties

3.1. Shear Experiments in Rotational Rheometers

The most widely used shear rheometers for the study of semi-solid metals are the rotational rheometers with concentric arranged cylinders. The outer cylinder is a cup that contains the SSM material and in which the inner cylinder, the bob, is inserted. In the Couette-type rheometer the cup is rotating and the bob is fixed, whereas in the Searle rheometer the bob is rotating while the cup is fixed. Because of the relative movement of cup and bob, the material is sheared in the gap between them. The shear stress at the wall is related to the torque, which is measured, and the shear rate is related to the rotational speed and to the geometry. Due to inertia forces, the Searle system is sensitive for secondary flows, the Taylor vortices, which dissipate energy and cause an increase in the measured torque [26]. Depending on the geometry of the system and the properties of the sample, the vortices can occur already at rather low rotational speed. Simple criteria are available to calculate the onset of the vortices [26], which can be applied for all viscous fluids. Another effect that falsifies the measurements consists in turbulent vortices that occur for both rheometers at higher rotational speeds, which is defined by a critical Re number [27].

For Newtonian fluids, the evaluation of the viscosity from the torque and the rotational speed is rather simple. For non-Newtonian fluids, it is complicated when the rheological nature of the fluid is unknown. In this case, frequently the way of evaluation valid for Newtonian fluids is applied. This results in an apparent viscosity value and in an apparent flow curve, which does not reflect properly the physical properties of the material. For purely viscous materials, the approach of the representative shear location should be applied [28], which results in physical correct values. Alexandrou et al. [29] have shown recently that this method does work for viscoplastic materials in special cases only. In general, the processing of data collected from rotational rheometer should be evaluated with the help of computational rheology [30].

Wall slip is another phenomenon that affects rheological measurements in suspensions with any kind of shear device. The slip is caused by segregation of a thin layer of the liquid phase adjacent to the wall. This thin layer has the effect of a lubricant that reduces the friction and, consequently, the torque measured by rotational rheometers, resulting in apparently lower viscosity values. In the literature, some different geometries for the bob have been proposed to avoid slip. For instance, Modigell et al. [31] have shown that vane-type bobs are not suitable because they lead to secondary flows that influence torque measurements, whereas a grooved bob prevents slip without affecting the torque significantly. Another way to treat slip is to apply the Kiljanski method for Searle or Couette rheometers [32] (or the Mooney method for capillary systems). The idea of both methods is to evaluate the slip velocity with the help of two different geometries. Harboe et al. [33] could show that the application of the Kiljanski method results in the same flow curve as the application of a grooved rod, but it requires significantly more experimental effort.

For SSM with low solid fraction—and low corresponding viscosity—the influence of the surface tension on the experimental result must be considered. Tocci et al. [34] could demonstrate that small deviation of the symmetry of the measuring system leads to secondary forces caused by the surface tension, which is dominant for small shear rates. Consequently, the material appears to be strongly shear thinning, although it is almost Newtonian.

At the beginning of the development of SSM processes, most of the rheological investigations have been performed with low melting Sn-Pb alloys because of the lack of high temperature rheometers. Nowadays, commercial instruments for testing Al alloys are available, while, to our knowledge, only one commercial instrument is available on the market for studying steels [35]. Yekta et al. [36] and Modigell et al. [37] have used own developed instruments.

Typically, for experiments with rotational rheometers, a first important part of the procedure is the preparation of the semi-solid material by shearing it for a certain time during cooling to the desired temperature, according to the solid fraction. A proper material preparation is fundamental, especially under consideration of the Ostwald ripening (see Section 7), since the flow behavior of semi-solid metals is strongly related to the microstructure [2]. An example of the evolution of viscosity during the material preparation is presented in Figure 5 for an Al-Si alloy for a constant shear rate of $100\,\mathrm{s}^{-1}$ [12]. First, the material is sheared in the fully liquid state (630 °C) to ensure the homogeneity of the material. When the temperature decreases, a severe increase in viscosity takes place, mainly due to the formation of solid particles. Finally, when the temperature reaches the value corresponding to the desired solid fraction (0.35 at 583 °C), a first steep decrease in viscosity is observed due to the change of dendrites into globular particles because of the application of shear forces. The following less steep decrease of the viscosity is due to Ostwald ripening.

Values of viscosity for the evaluation of flow curve in steady state condition are calculated from experimental data at different shear rates.

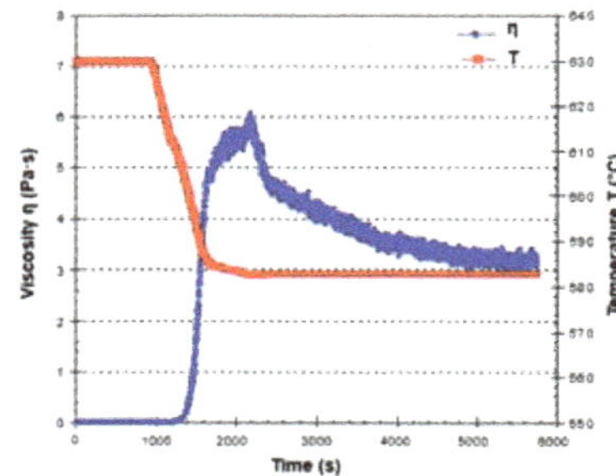

Figure 5. Plot of viscosity versus time at constant shear rate (100 s^{-1}) during cooling from liquid state (630 °C) to semi-solid condition (solid fraction of 0.35 at 583 °C) for an Al-Si alloy measured by means of a Searle rheometer [12].

3.2. Compression Tests

The compression test is a conventional testing method to acquire strain-stress curves by squeezing a sample either under a constant load between two parallel plates or with a constant speed of displacement of the plates [38]. Compression experiments are usually performed with materials characterized by a solid fraction higher than 0.5.

Various experimental configurations are possible according to the used device, an example is shown in Figure 6. Usually, the sample is first heated to the required temperature in a separate furnace or directly in the testing chamber, while, after compression, it can be quenched in water for further study of the microstructure. The applied force and the obtained displacement is monitored by a proper load cell.

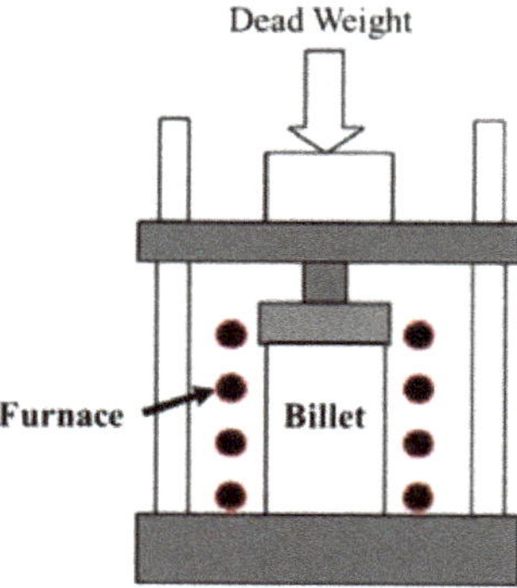

Figure 6. Scheme of parallel plate deformation set up [39].

From the stress-strain curve, it is possible to calculate rheological parameters and obtain a flow curve in terms of viscosity as a function of the shear rate, as illustrated by Laxmanan et al. [40]. It is important to mention that, with this method, the viscosity at a given shear rate is calculated under the assumption of Newtonian behavior (comparable to the simple approach with rotational rheometers).

Consequently, the calculated viscosities are apparent values only. Nevertheless, the evidence of the shear thinning behavior of SSM is obtained when the values of apparent viscosity calculated at different shear rates are compared.

Another drawback is the flow condition at the surfaces of the plates. For pure shear flow, the material should adhere to the plates. Slipping conditions result in elongation flow, which must be evaluated in a different way. In practice, none of the two conditions are completely fulfilled and the flow is of mixed mode. Additionally, the flow is non-stationary and at least 2-dimensional. Since usually the ram speed is high in order to simulate forging conditions, high accelerations are achieved, and the evaluation of thixotropic effects is difficult (as it is discussed in Section 6). At this regard, Hu et al. [40] performed compression tests with Al alloy under conditions close to forging processes and applied ram speeds up to 1000 mm/s, which results in experimental times of approximately 0.01 s. Similar values were achieved by Becker et al. [41,42] during experiments with steel.

Two practical problems arise with SSM during compression. Even at low ram speeds liquid phase is squeezed out of the sample when solid fraction is less than 0.80. This results in an inhomogeneous composition of the sample, which additionally is changing by time, although the experiments have been performed isothermally. Moreover, another problem is the cracking of the free surface with increasing deformation [43].

Temperature and compression rate can be varied to reproduce real cavity die-filling conditions, which is one advantage of this technique in comparison with shear experiments. On the other hand, the possible experimental procedures are wider for shear experiments, allowing to completely characterize the rheological behavior of the material.

Particularly, the compression rate is a key parameter since a slow compression can provide information not adequate for the understanding of the actual industrial process, which is known to take place in less than 1 s. For this reason, rapid compression tests were carried out to investigate the transient behavior of SSM [44–46]. A schematic representation of the load-displacement curve for these kinds of experiments is shown in Figure 7.

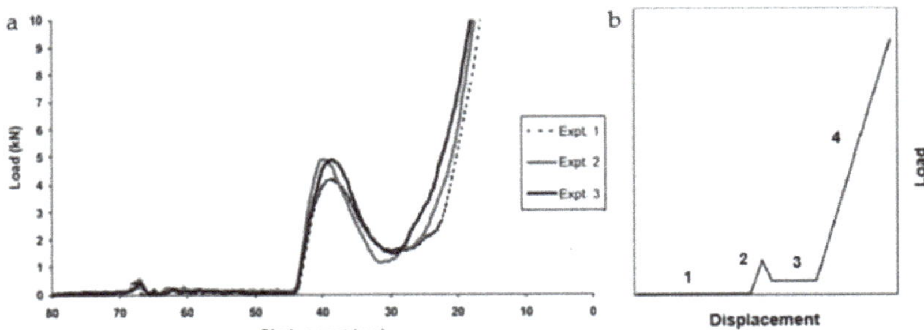

Figure 7. (a) Typical signal response to rapid compression of semi-solid A356 alloy (ram speed: 500 mm/s; soak time: 0 min; and temperature 575 °C) [44]; (b) Schematic appearance of a load vs. displacement response. Four distinct regions are present: (1) zero load prior to reaching the die; (2) an initial breakdown, or peak load (stress); (3) a relatively constant, or plateau, load after the initial breakdown; and (4) a rapid increase in the load as the die approaches complete filling [45].

In addition, the test temperature and the holding time in isothermal condition should be carefully chosen when the main aim of the experiments is to provide information adequate for the industrial process.

Finally, this experimental procedure can be applied using a close die to investigate the liquid flow during compression. This is helpful in predicting the formation of liquid segregation and surface cracks in the products obtained by SSM processing [47]. Also, drained compression tests can be performed to investigate the compressibility of the solid phase in isothermal condition [10].

4. Modelling of Rheological Properties

The evaluation of experimental rheological investigations should result in mathematical equations, called constitutive models, which should reflect the physics of the flow and the deformation process. Surely, it is difficult to include all phenomena in one model and it is generally accepted to make simplifications according to the application of the model. The simplest models are the one phase, equilibrium models, which assume the SSM to be a homogenous fluid without time-dependent properties. Under these assumptions, the Ostwald-de-Waele model—or power law—is the simplest one, assuming viscous properties only. The relationship between shear stress and shear rate can be expressed by the following equation:

$$\tau = m \, \dot\gamma^{n} \tag{1}$$

And the apparent viscosity is given by:

$$\eta = \tau/\dot\gamma = m \, \dot\gamma^{n-1} \tag{2}$$

where for $n < 1$, the fluid exhibits shear thinning properties; $n = 1$, the fluid shows Newtonian behavior; $n > 1$, the fluid shows shear-thickening behavior.

The terms m and n are two empirical parameters, the flow index and shear exponent, respectively.

Because of the simplicity of this equation, it is widely applied to process data from rheological experiments. Besides the first studies on the characterization of the rheological behavior of semi-solid slurries [38,48], mainly focused on SnPb15 alloy, also more recent papers used the Ostwald-de-Waele relationship to express the viscosity as a function of shear rate for various Al alloys and steels [49,50].

The simplicity relates to a couple of disadvantages. First, the flow index m does not have a fixed dimension because this depends on the power index. More serious is the fact that for small and large shear rates the equation results in physically non-correct values. This gives problems in its application in numerical simulation. Frequently Ostwald de Waele parameters are presented with shear exponents less than zero. This will result in physically nonsensical results, as e.g., a positive pressure gradient in a simple tube flow.

The Herschel-Bulkley equation is typically used to describe the flow of viscoplastic fluids [26]. It is a generalization of the Bingham plastic model to consider the case of a non-linear relationship between shear stress and shear rate [51].

$$\tau = \tau_y + m \, \dot\gamma^{n} \tag{3}$$

With τ_y the yield stress and m and n the flow index and the shear exponent.

It is believed τ_y to be a fundamental parameter for the modeling of semi-solid metals behavior [52]. For this reason, a Herschel-Bulkley model was applied to the numerical simulation of the semi-solid processing [30,53,54], fitting experimental results for Sn-Pb15 alloy. More recently, the same model was modified to better describe the phenomena taking place in the early stages of deformation of the solid structure [30].

To model thixotropy, three main approaches are applied. One is to describe the change in the structure with the change of the viscosity, which needs to define a rate equation for the temporal development of the viscosity [55]. Another approach is to calculate the number of existing and broken connections between the particles, which depend on the local shear rate [56].

The most promising model is the one that was originally worked out by Moore [57]. He defined a structural or coherency parameter λ, which is defined to be 1 in a fully saturated state of the structure and to be 0 when all particle bonds are broken. A rate equation for the structural parameter is set up to consider the creation and the destruction of bonds. It is frequently assumed that all parameters of the Herschel-Bulkley equation are depending on λ [58].

A more detailed model has been set up by Petera et al. [59]. The SSM is modelled as a two-phase system with a semi-fluid approach for the solid phase. A kinetical equation is introduced reflecting the change in structure. The model has been successfully applied for the simulation of die-filling experiments where the flow front was videotaped. Good agreement was achieved between simulation and experiment for the development of the flow front, the transient pressure drop in the die and the final distribution in the solid phase due to segregation [60].

The Cross model considers that at extreme boundary condition, i.e., at very low or very high shear rate, thixotropic fluids assume a Newtonian viscosity [61]. This is expressed by the following equation:

$$\eta = \eta_\infty + \frac{\eta_0 - \eta_\infty}{1 + k\,\dot{\gamma}^n} \tag{4}$$

with η_0 the viscosity for zero shear rate and η_∞ the viscosity for high shear rates and k and n parameters as in the Ostwald de Waele equation.

This model has been applied to fit the experimental results from various researches on SnPb15 alloy in a satisfactory way [62], even though consistent data about the extreme conditions are hardly available in literature and, therefore, the reliability of the model cannot be stated [7]. From the practical point of view, it does not provide an advantage compared with Ostwald de Waele model since the Cross model reduces to the Ostwald de Waele one if the extreme condition viscosities are not determined.

The above-mentioned approaches are applicable if the solid fraction is below approximately 0.65, which corresponds to the maximum packing of the solid particles. Above this value, the SSM can be treated as a "porous solid body" and the approaches of the continuum mechanics must be applied to model the relation between stress and deformation.

5. Steady State Condition: Time-Independent Properties

As aforementioned, it is fundamental to distinguish between the properties of SSM in steady state and transient conditions. In this paragraph, the main findings related to time-independent behavior will be reviewed according to the experimental procedure applied.

Comparison of data available in the literature have been done in the past for A356 and A357 [44]. It was found that the flow curves for both alloys, expressed using a power law relationship, were characterized by a slope of approximately -1, corresponding to the shear exponent. The comparison was carried out among results obtained by means of various techniques and conditions (Figure 8), which is expected to lead to discrepancies in the flow curves, even when studying the same alloy.

As additional evidence of this, Lashkari et al. [63] and Blanco et al. [49] studied a similar Al alloy containing approximately 4.5% Cu using respectively compression tests and shear rate jump experiments with a Searle rheometer. In this case, the difference is also in the range of shear rate investigated since compression tests results correspond to very low shear rates (in the order of 10^{-3}–10^{-2} s^{-1}), while in the other study a very different range was investigated (60–260 s^{-1}).

Regarding shear experiments, more recently Das et al. [64] performed various experiments on A356 alloy with a Searle-type rheometer applying the power law model to evaluate their results. Furthermore, they compared the obtained m and n parameters with the findings of previous researches. The difference between the values was mainly due to the different ranges of shear rates considered for the fitting of the flow curve since Das et al. [64] performed rheological measurements up to 1500 s^{-1} as shear rate values, while the other authors considered a narrower range (approximately up to 200 s^{-1}).

To update the available information, in the present review the attention was focused on the studies from 2003 up to now. Values of flow index and shear exponent for Al alloys were collected from various scientific publications, when available, and they are shown in Tables 1 and 2. It was chosen to organize the data according to the experimental methods used, i.e., shear (Table 1) or compression experiments (Table 2) in order to better represent the rheological behavior of these materials. At this regard, it is important to mention that experiments with rotational rheometers are useful for the investigation of the rheological behavior of semi-solid slurry with a solid fraction of 0.2–0.5, while compression experiments can provide information for materials characterized by a solid fraction higher than 0.5.

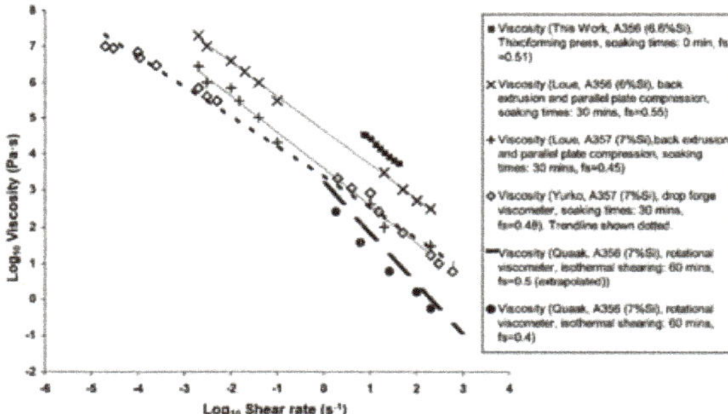

Figure 8. Comparison of apparent viscosities obtained by various experimental techniques and conditions [44].

It appears that the shear exponent assumes values between -1.5 and -1.2 for shear rates up to $300\ \mathrm{s^{-1}}$, while it decreases if a wider shear rate range is applied. Values very close to 0 are found for investigations carried out at low solid fraction. This is reasonable if it is considered that liquid metals exhibit Newtonian behavior. Consequently, it is expected that the flow of semi-solid slurries at low solid fraction would shift towards a Newtonian-like behavior.

Table 1. Power law parameters from experiments with rheometer for different alloys.

	Alloy	Solid Fraction (-)	Shear Rate Range (s⁻¹)	Flow Index m (Pa·s^n)	Shear Exponent $n-1$ (-)	Source
1	A356	0.33	3.1–124.8	269	-1.3	[65]
2	AlSi4	0.36	3.1–124.8	325	-1.3	[65]
3	A356	0.2	0–1500	166	-0.92	[64]
4	A356	0.41	0–1500	509	-1.05	[64]
5	A356	0.5	0–1500	589	-1.05	[64]
6	A356	0.4	10–300	789	-1.23	[66]
7	AlSi3	0.4	10–300	122	-1.23	[66]
-	A201 (AlCu4.5)	0.35	60–260	-	-1.35	[49]
-	A201 (AlCu4.5)	0.45	60–260	-	-1.49	[49]
8	AlSi22	0.09	10–50	2.53	-0.34	[50]
9	AlSi30	0.15	10–50	4.63	-0.39	[50]
10	AlSi30	0.20	10–50	109	-1.04	[50]

This is particularly evident if the flow curves corresponding to the parameters listed in Table 1 are plotted in a viscosity vs shear rate graph, as in Figure 9. It clearly appears that the slope of the flow curves obtained for solid fraction above 0.3 are comparable, while for lower values of solid fraction the viscosity is less dependent on the shear rate, reflecting the increasing contribution of the Newtonian liquid phase. Furthermore, also the influence of solid fraction is visible if the results for the same A356 alloys are considered. On the other hand, a scattering in the viscosity values of semi-solid AlSi alloys is observed, which underlines how these kinds of measurements can be affected by various parameters, as such as material preparation, holding time, shear history of the samples, etc. Despite the large number of experiments available in scientific literature, it is still difficult to define viscosity values for SSM in an unambiguous and systematic way.

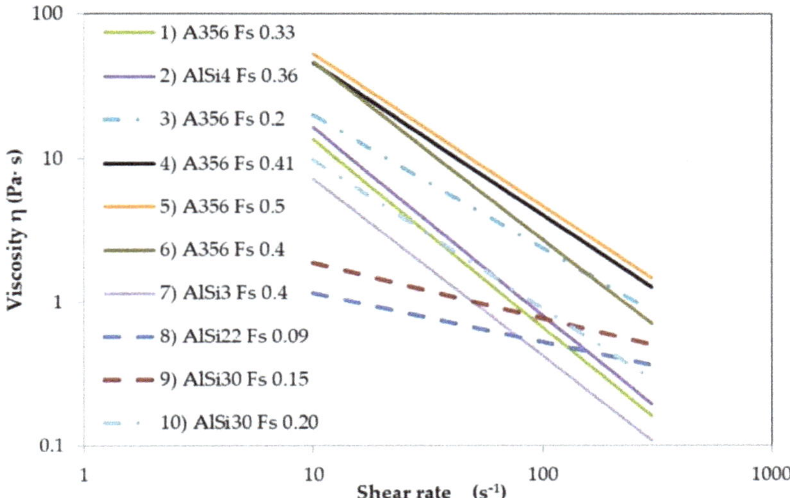

Figure 9. Comparison of flow curves according to parameters reported in the previous Table 1.

It is more complex to perform the same analysis for flow curves obtained from compression experiments, since less abundant data are available in scientific literature, as shown in Table 2.

Table 2. Power law parameters from compression experiments for different alloys.

Alloy	Solid Fraction (-)	Shear Rate Range (s^{-1})	Flow Index m (Pa·s^n)	Shear Exponent $n-1$ (-)	Source
6061	0.5–0.9	0.01–2	10^4–10^6	From -2 to -1 according to the solid fraction	[47]
7075	0.5–0.9	0.01–2	10^4–10^5	approx. -2.5	[47]
2024	0.5–0.9	0.01–2	10^3–10^5	From -2.75 to -2.5 according to the solid fraction	[47]
AlSi25	Various T (°C)	10–3000	1.78×10^7	-1.5	[67]

It is important to mention that most of the shear exponents are negative values. As mentioned above, this gives strange physical results. It must be assumed that the experiments were dominated by secondary effects that influence the experimental results, e.g., wall slippage. The application of these results seems to be doubtful. This is even more evident for the results of the compression tests, which indicate the difficulty to extract reasonable results for the shear behavior from compression tests.

On the other hand, compression experiments can provide important information about the transient and time-dependent properties of SSM close to practical application [68] even when their use for simulation might be doubtful.

6. Non-Viscous Properties

6.1. Yield Stress

It is obvious from the previous explanations that the rheological properties of SSM strongly depend on time—which is frequently forgotten. This is insofar important as any die-filling process, either casting or forging, is a non-stationary process by nature with temporally changing border conditions for the deformation of the material.

An important question is how the typical time scales of the process and the material behavior compare when discussing an appropriate approach to model the material behavior properly [51,69].

This matter becomes evident when investigating the phenomenon of yield stress in SSM. The existence of yield stress is subject of basic discussions, which seem to have philosophical character [70,71]. There is no doubt that the experimental detection of yield stress depends on the quality of the experiment. However, this holds for any mechanical property as well. To identify e.g., thixotropic properties it is necessary to have an instrument with a certain minimal temporal resolution.

In general, it is accepted that the introduction of yield stress for materials with viscoplastic behavior is at least a reasonable engineering approach to model adequately the flow behavior [11,52,72]. As with other materials, the yield stress in SSM is the consequence of the formation of a network of the particles, which is more or less mechanically stable. The direct way to investigate yield stress phenomena is the application of stress or strain ramps and the observation of the temporal behavior of strain or stress respectively [69]. Experiments [6,15] show that the yield stress depends on the solid fraction, the particle size and shape and on the resting time. With increasing resting time, the yield stress increases nearly exponentially, which leads to the concept of three different values of yield stress [15,16]:

- the static one is a constant value and it is achieved after long resting time,
- the dynamic yield stress holds for the period where the yield stress increases with time when the material is in rest after being sheared and
- the isostructural one corresponding to the value which would be measured immediately after shearing the material.

The isostructural value cannot be measured directly and must be found, for instance, by extrapolating the dynamic yield stress to resting zero time. For two Al alloys, it was found that the isostructural yield stress is less than 1/10 of the static value, which was for both alloys about 400 Pa. Comparable experiments have been performed by Solek [73] with high carbon steel, which confirms the above-mentioned findings.

Application of oscillatory shearing and comparison of the temporal development during rest of the yield stress and the loss- and storage-moduli show that the ratio of the both moduli decreases while the yield stress increases [74]. Due to the formation of a stable structure in the material the rheological nature of the material changes from viscous to elastic—fluid to solid—as indicated by the loss angle. The temporal increase of the storage modulus compares with the increase of the yield stress.

The other way to determine yield stress is an indirect one by evaluating the flow curve on the base of an appropriate visco-plastic model. A Herschel-Bulkley model was found to fit in a proper way the experimental results for semi-solid Sn-Pb15 alloy at different solid fractions [60]. The alloy was considered as a homogeneous material with thixotropic properties and it was tested under isothermal conditions. Among the parameters obtained from the application of the model to experimental results, it was possible to calculate finite values of yield stress as a function of the solid fraction. A Herschel–Bulkley approach was applied to the numerical simulation of die filling and it was validated with the results from die-filling experiments [75] using a Sn-Pb alloy. In these experiments, the evolution of the flow front as well as the pressure drop were observed during the die filling and the process was performed for different flow conditions. Similarly, a model corresponding to the Herschel-Bulkley was developed to represent the behavior of semi-solid metallic suspensions in fast transient conditions [76]. The validation of the model was carried out with short time measurements of Sn-Pb15 alloy under rapid shear rate changes, already published by other authors [77]. The model can show the increase of the shear rate in a shear rate jump and the gradual decrease after reaching a maximum. For the yield stress, the authors found interestingly a constant value of 100 Pa, independent of the time the material was in rest (which was between 0 and 5 h).

The evaluation of yield behavior of Al-Si alloys was carried out also using different experimental procedures, as such as the compression and the cone penetration method [72]. Yield stress was measured as a function of temperature, i.e., solid fraction, taking also into account different billet processing methods, as such as the addition of grain refiner and the application of magneto-hydrodynamic stirring (Figure 10).

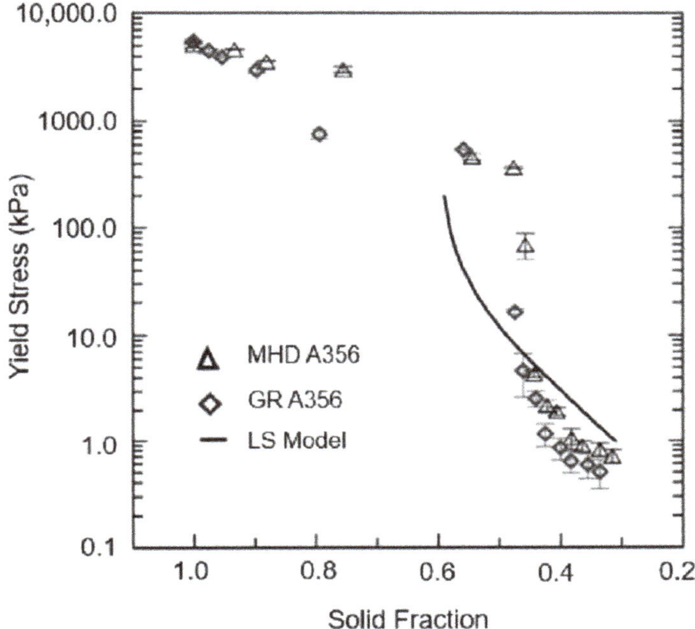

Figure 10. A comparison between measured average yield stresses of grain refined (GR) A356 and magneto-hydrodynamically stirred (MHD) A356 alloys with yield stress values predicted by the Loue–Sigworth (LS) model [72].

It was found that not only the solid fraction is affecting the yield behavior, but that also the entrapped liquid and the morphology of the solid globules can play an important role in determining the deformation resistance. A high amount of entrapped liquid can lead to an increase in yield stress because the "effective" liquid fraction is decreased, while the presence of more rounded solid particles can result in a smaller deformation resistance and, consequently, in a decrease in yield stress.

The viscoplastic behavior of A356 alloy in semi-solid state was expressed by the Herschel-Bulkley model by Simlandi et al. [78]. They also considered a time-dependent structural parameter to comprehensively describe the material behavior at different solid fractions.

An enhanced model based on the Herschel-Bulkley equation was proposed also for other metallic alloys, as such as steels [79] and Mg alloys [80]. Pouyafar et al. [79] calculated the yield stress of M2 steel from the flow curve obtained from experiments in steady-state flow by interpolation of the experimental results to evaluate the stress at zero shear rate. The enhanced model was compared to the classical one and it was found that the new model could predict in a more precise way the behavior of the material, especially at high shear rate, than the conventional one. This was confirmed also by the application of the model to the simulation of the rheometer flow. Moreover, the same model was used to simulate the material behavior during compression test [36].

6.2. Transient Behavior: Time-Dependent Properties

In the introduction, the main thixotropic properties of SSM were briefly discussed. It was mentioned that thixotropic materials show a decrease of viscosity under constant stress and a recovery of the viscosity when the loading stops. This behavior is linked to microstructural evolution of SSM, in particular to the agglomeration and deagglomeration phenomena.

Various experiments can be performed to better investigate these time-dependent properties, which focus on the response of the SSM to changes in the applied deformation rate or shear stress, as such as shear rate jump or hysteresis loop experiments [60,65,68] or rapid compression tests [45,81].

Modigell et al. [60] investigated thixotropic behavior under isothermal condition on a Sn-Pb15 alloy. Hysteresis loops experiments are particularly useful to provide qualitative information on the degree of thixotropy of the material. It was found that the faster the ramp is performed, the higher the hysteresis area and, therefore, the more evident the thixotropic behavior (Figure 11). On the other hand, Brabazon et al. [65] tested an AlSi4 alloy in a similar way but found that with slower ramp the peak viscosity increases corresponding to an increase in thixotropy, not considering the change around the hysteresis loop. They also evaluated the effect of rest time on the material thixotropy and measured an increase in the viscosity with increasing rest time.

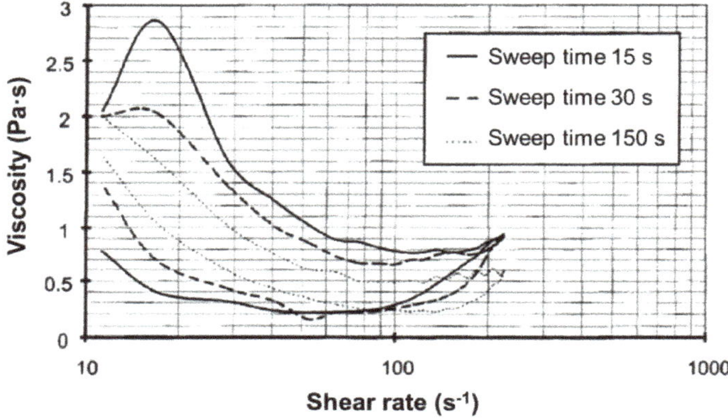

Figure 11. Hysteresis experiments at different sweep times (T = 198 °C, Fs = 0.45) [60].

It is reported [60] that, immediately after an increase in shear rate, the measured viscosity makes a sudden jump and, subsequently, a decrease follows according to the thixotropic nature of the material. The opposite happens when the shear rate is decreased. This phenomenon is said to be "isostructure behavior". It is assumed that the structure of the alloy—corresponding to the initial shear rate—will not change instantaneously to the structure corresponding to the new shear rate. Evaluation of special shear rate jump experiments in terms of an "isostructural flow curve" shows that the SSM immediately after a shear rate jump reacts in a shear thickening way [17]. Recent analyses of the literature [10] regarding this effect confirm the findings of [60]. They showed that it needs a solid fraction larger than 0.36 to be observed. It has been clarified that for shear rate jumps from shear rate zero (sample at rest) this shear thickening effect is masked by yield stress effects.

As above-mentioned, it is important to compare the time scale of the time-dependent rheological properties and the typical process times. Typical process times for die casting or forging are in the order of 1/10 s [40,82], as also confirmed by other authors [83,84] for the thixoforging of steel components. This is interestingly in the order of the response time of modern rotational rheometer. Detailed investigations about kinetics of the thixotropy in SSM are rare. The first well-known investigations of this phenomenon were done by Quaak and Peng et al. [85,86]. Quaak assumed that the process of deagglomeration and agglomeration after a change in shear rate is composed of two steps: a fast process to destroy or create agglomerates and a slow one of coarsening and sintering. This has been confirmed for Sn-Pb alloy by Koke [18], who could model the transient process assuming two different kinetics. The time scale for the fast process was in the order of 0.5 s—and seemed to be independent of shear rate—whereas the scale of the slow process was 100 times slower. Liu at al. [77] studied the effect of shear jumps on the temporal development of the shear stress for Sn-Pb15 as well (Figure 12). They only investigated the short-term behavior and the time scale they found for the thixotropic effects is comparable with the results of Quaak [85] and Koke.

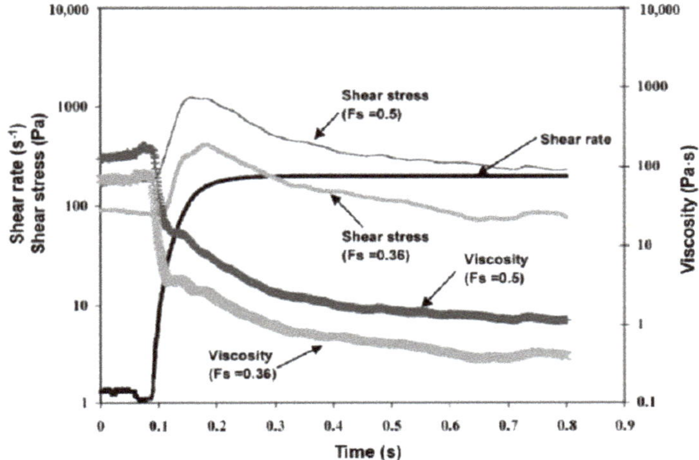

Figure 12. Shear rate jumps (1–200 s^{-1}) for Sn15Pb alloy for two different solid fractions (*Fs* = 0.5 and 0.36) in isothermal condition [77].

Rapid compression tests on AlSi alloys have been performed by Hogg et al. [45], to investigate the rheological properties under the conditions of the real manufacturing process. Time scale of the test has been around 0.065 s, so much shorter than the scale of rotational rheometer. The experiments provided data also about the effect of the holding time and the reheating temperatures. The viscosity estimated from their experiments were by around a factor of ten higher then viscosity values evaluated from shear experiments. Assuming a time scale for the thixotropy of some 1/10 s it can be concluded that in these experiments the material was not in the final state of equilibrium. Based on these studies, also numerical models were developed to describe accurately the transient behavior [30,40,46,76].

Another interesting result regarding thixotropy can be extracted from the work of Bührig-Polaczek et al. [87] who performed experiments with Al 356 with a capillary rheometer. The device, which was developed in house, was equipped with four pressure sensors along the capillary. This allowed evaluating the pressure drop along the path of flow and, consequently, it was possible to analyze the thixotropic reaction of the material. The shear rates they could realize in the capillary were up to 3200 s^{-1} and the resident time in the capillary was in the range of 1/10 s, which compares to typical process times in casting. They found shear thinning behavior in the shear rate range from 1000 to 3200 s^{-1}, whereby the viscosity for 1000 s^{-1} was approximately ten times higher than the viscosity measured in a rotational rheometer at shear rate equal to 250 s^{-1}, which compares with the findings of Hogg [45]. Evaluation of the pressure drop along the capillary showed that the viscosity close to the entrance was around 2.5 Pa·s, while it dropped towards the end of the capillary to 0.6 Pa·s (both for a shear rate of 3200 s^{-1}). The drop was almost linear and showed no tendency to become constant. This indicates that, within the residence time of 0.15 s, the thixotropic reaction is not finalized.

The consequences of this are that obviously the short term and transient behavior of the SSM is of much more interest for the typical casting and forging processes than the equilibrium flow curves.

This is in accordance with numerical analyses of the flow of a viscoplastic material in the gap of a rheometer under conditions relevant for technical processes performed by Alexandrou et al. [30]. They composed the stress function by two contributions: the "steady state" stress, evaluated from long shearing time experiments and assumed to follow a Herschel-Bulkley model, and the stress due to slurry strength, which depends on a coherency parameter, depending on time. The simulation showed that for short times the latter contribution is dominant and defines the propagation of the shear in the gap. The relevant time frame is in the order of less than 0.1 s. Evaluation of experiments in this time scale need the application of computational rheology rather than the classical evaluation of the experimental data.

7. Ostwald Ripening

Another important aspect to consider to completely describe the rheological behavior of SSM is the growth and coarsening of the globular solid particles under rest or shear condition by recrystallization. This will happen under isothermal conditions. The phenomenon is usually indicated as Ostwald ripening and it takes place in different materials, as such as emulsion systems, etc. During this process, small particles dissolve, while the larger ones coarsen with the consequence that the surface energy is minimized [88] so that in total the mean particle diameter in the material will increase.

This phenomenon was investigated in the past by several researches [89–92] mainly by means of the observation of 2D sections of samples quenched from the semi-solid state, which represented a serious limitation in the investigation of the real phenomenon. More recently, agglomeration and growth of necks between solid particles, which are different from Ostwald ripening, and the later were investigated by X-ray in situ tomography [93,94].

In systems in rest, Ostwald ripening is a diffusion-controlled process that is rather slow and has less technical importance in the frame of the subjects discussed here. The growth of the particle as a function of time can be calculated on the base of the LSW-theory (Lifshitz, Sloyozov [95] and Wagner [96]) which gives a linear relationship between the volume of the particle and the time. In agitated systems, the growth of the particle is strongly enhanced by convective transport, which results in a growth of particles in a time frame which is at least relevant for a couple of experimental investigations.

The rheological effect of the Ostwald ripening is a decrease of the viscosity because of the increase of the mean particle diameter in the sample—although solid fraction will not change because of isothermal condition. This effect is demonstrated in Figure 13.

The Sn-Pb melt, initially fully liquid, is cooled down into the two-phase condition. First, the viscosity will rise, subsequently it will reach a maximum and will start to drop. The first decrease after the maximum is due to the formation of non-dendritic particles. At the first small peak, this process is more or less finalized. The further decrease of the viscosity is mainly due to convective Ostwald ripening. The increase in particle size is clearly demonstrated in the metallographic pictures. The time scale of this process is in the order of 60 min, depending on the shear rate. The higher the shear rate, the faster the growth.

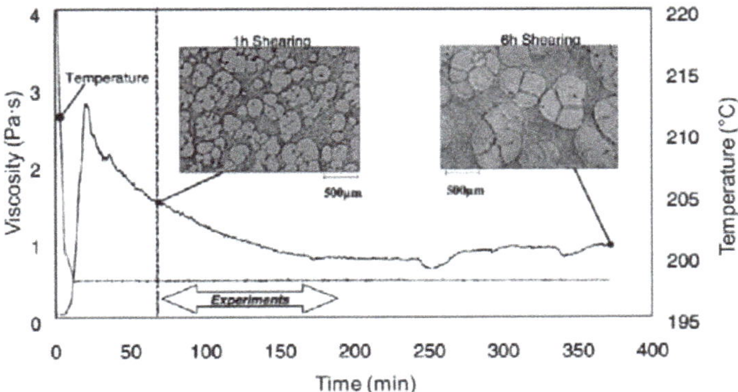

Figure 13. Viscosity vs. time at constant shear rate (110 s^{-1}). Initial cooling rate is 1 °C/min. Steady-state temperature is 198 °C, corresponding to a solid fraction Fs = 0.45. The metallographic pictures show samples quenched after 1 and 6 h of constant shearing (Sn–15% Pb) [17].

It is obvious that the Ostwald ripening falsifies the experimental results when performing long-term investigations, such as, for example, making shear rate jump experiments. For instance, in the case of A356 alloy tested in a rotational rheometer at 0.40 solid fraction, it can be seen in Figure 14 that at the end of the experiment the viscosity for a shear rate of 80 s^{-1} is significantly lower than at the beginning. Evaluation of these data will result in an overestimation of the shear thinning effect. Modigell et al. [97] have developed a simple model considering convective transport in the system, which allows correcting the data for any shear, as shown in Figure 14. For sufficient large shear rates, the temporal change of the diameter is linear with time.

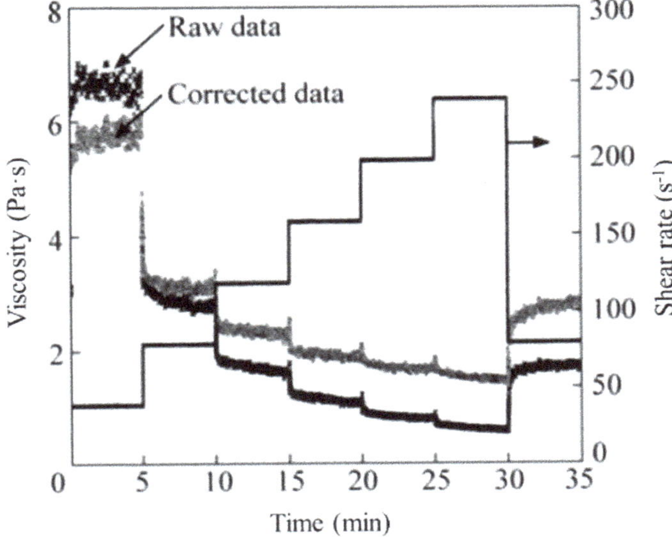

Figure 14. Shear rate experiment results with A356 under solid fraction of 0.40 [97].

The effect of isothermal and non-isothermal stirring on particle size was also evaluated for Al-Si alloys [98–101] and Mg alloys [102]. Sukumaran et al. [99] investigated the evolution of particle diameter with shearing time for an Al-Si alloy with and without the addition of grain refiner (Figure 15). In this case, it was found that first the particle size decreases due to the fragmentation of the dendritic structure. Subsequently, particle size reaches a minimum value before starting to increase due to the discussed Ostwald ripening mechanism, which is considered the main mechanism for coarsening of solid globules, while agglomeration phenomena is believed to contribute in a minor way. Due to the presence of the grain refiner, the growth of dendrites during semi-solid processing is inhibited, promoting the formation of an equiaxed microstructure, in comparison with the base material, and accelerating the formation of a globular microstructure, as well as the coarsening and ripening phenomena. Therefore, the coarsening of particles takes place earlier for the alloy with grain refiner in comparison with the base alloy, as visible in Figure 15.

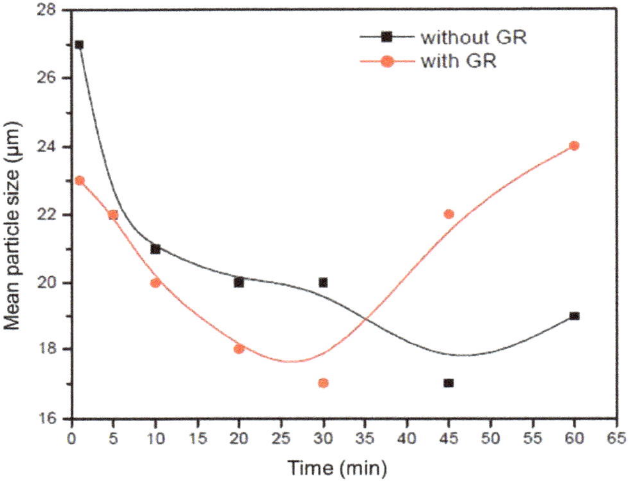

Figure 15. Plot of nominal particle diameter vs. time of isothermal stirring (shear rate 210 s^{-1}) at 615 °C [99].

The same mechanism is illustrated also by Chen et al. [102], who provided the following schematic diagram (Figure 16) and correlated the evolution of the particle size to the apparent viscosity for a Mg alloy.

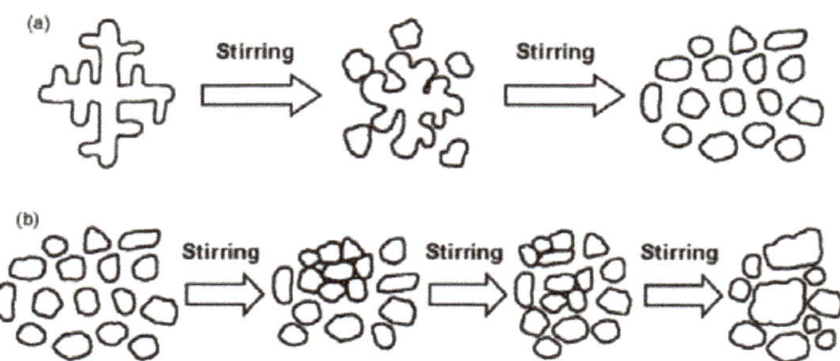

Figure 16. Schematic diagram of the variation of microstructure for: (**a**) decreasing apparent viscosity and (**b**) increasing apparent viscosity [102].

8. Summary

In the present review, a summary of the basic aspects of the rheology of semi-solid metals are presented. These materials exhibit a complex behavior and several parameters must be considered to correctly characterize it. For this reason, the distinction between steady state and time-dependent properties is fundamental. Furthermore, additional important aspects as the evaluation of yield stress and Ostwald ripening mechanism are discussed.

In general, the analyses of the recent literature show that a lot of experimental and theoretical work has been done and that the initial findings of the early researches are confirmed. A database has been provided for higher melting alloys due to technical development in the field of rheometers. Even for steels, basic data are now available provided by shear experiments.

Most of the data give information about rheological behavior under equilibrium conditions. In technical practice, the processes are far from equilibrium due to the high process speeds. For better understanding of the process phenomena and for reliable simulation of the forming process, the short-term behavior of the SSM must be investigated more deeply than it has been done up to now. This means that the thixotropic properties, as well as the yield phenomena, must be studied to clearly understand the process phenomena under technical conditions. Additionally, there is a lack of knowledge in the relationship between the composition of the alloy and the rheological properties.

Conflicts of Interest: The authors declare no conflict of interest.

References

1. Flemings, M.C. Behavior of metal alloys in the semisolid state. *Metall. Trans. A* **1991**, *22*, 957–981. [CrossRef]
2. Hirt, G.; Kopp, R. *Thixoforming: Semi-Solid Metal Processing*; Wiley-VCH Verlag GmbH & Co. KGaA: Weinheim, Germany, 2009.
3. Pola, A.; Montesano, L.; Tocci, M.; La Vecchia, G.M. Influence of Ultrasound Treatment on Cavitation Erosion Resistance of AlSi7 Alloy. *Materials* **2017**, *10*, 256. [CrossRef] [PubMed]
4. Barnes, H.A.; Hutton, J.F.; Walters, K. *An Introduction to Rheology*; Elsevier Science: Amsterdam, The Netherlands, 1993.
5. Bakhtiyarov, S.I.; Overfelt, R.A. Measurement of liquid metal viscosity by rotational technique. *Acta Mater.* **1999**, *47*, 4311–4319. [CrossRef]
6. Modigell, M.; Koke, J. Rheological modelling on semi-solid metal alloys and simulation of thixocasting processes. *J. Mater. Process. Tech.* **2001**, *111*, 53–58. [CrossRef]
7. Atkinson, H.V. Modelling the semisolid processing of metallic alloys. *Prog. Mater. Sci.* **2005**, *50*, 341–412. [CrossRef]
8. Atkinson, H.; Favier, V. Does Shear Thickening Occur in Semisolid Metals? *Metall. Mater. Trans. A* **2016**, *47*, 1740–1750. [CrossRef]
9. Lashkari, O.; Ghomashchi, R. The implication of rheology in semi-solid metal processes: An overview. *J. Mater. Process. Tech.* **2007**, *182*, 229–240. [CrossRef]
10. Kirkwood, D.H.; Suery, M.; Kapranos, P.; Atkinson, H.V.; Young, K.P. *Semi-Solid Processing of Alloys*; Springer: Berlin, Germany, 2009.
11. McLelland, A.R.A.; Henderson, N.G.; Atkinson, H.V.; Kirkwood, D.H. Anomalous rheological behaviour of semi-solid alloy slurries at low shear rates. *Mater. Sci. Eng. A* **1997**, *232*, 110–118. [CrossRef]
12. Hirt, G.; Uggowitzer, P.J.; Bleck, W.; Friedrich, B.; Schneider, J.M.; Modigell, M.; Kopp, R.; Bobzin, K.; Telle, R.; Bührig-Polaczek, A.; et al. *Final Report of the Joint Research Program SFB289 Forming of Metals in the Semi-Solid State and Their Properties*; RWTH-Aachen University: Aachen, Germany, 2007.
13. Quaak, C.J.; Katgermann, L.; Kool, W.H. Viscosity Evaluation of Partially Solidified Aluminium Slurries after a Shear Rate Jump. In Proceedings of the 4th International Conference on Semi-Solid Processing of Alloys and Composites, Sheffield, UK, 19–21 June 1996.
14. Windhab, E.J. Process–Structure–Rheology Relationships of Multiphase Food Systems. In Proceedings of the 1st International Symposium on Food Rheology and Structure, Zurich, Switzerland, 16–21 March 1997.

15. Modigell, M.; Hufschmidt, M. Dynamic and Static Yield Stress of Metallic Suspensions. *Solid State Phenom.* **2006**, *116–117*, 587–590. [CrossRef]

16. Moll, A.; Modigell, M. Yield stress phenomena in semi-solid alloys. *Int. J. Mater. Form.* **2010**, *3*, 779–782. [CrossRef]

17. Koke, J.; Modigell, M. Flow behaviour of semi-solid metal alloys. *J. Non-Newton. Fluid. Mech.* **2003**, *112*, 141–160. [CrossRef]

18. Koke, J. *Rheologie Teilerstarrter Metalllegierungen*; Fortschr.-Ber. VDI Reihe 5 Nr. 620; VDI Verlag: Düsseldorf, Germany, 2001; Volume 620.

19. Harboe, S.J. *Investigation of Rheological and Microstructural Properties of Semi-Solid Aluminium Copper Alloy during Isothermal Shear*; Shaker Verlag GmbH: Aachen, Germany, 2017.

20. Zabler, S.; Rueda, A.; Rack, A.; Riesemeier, H.; Zaslansky, P.; Manke, I.; Garcia-Moreno, F.; Benhart, J. Coarsening of grain-refined semi-solid Al-Ge32 alloy: X-ray microtomography and in situ radiography. *Acta Mater.* **2007**, *55*, 5045–5055. [CrossRef]

21. Cai, B.; Karagadde, S.; Rowley, D.; Marrow, T.J.; Connolley, T.; Lee, P.D. Time-resolved synchrotron tomographic quantification of deformation-induced flow in a semi-solid equiaxed dendritic Al-Cu alloy. *Scr. Mater.* **2015**, *103*, 69–72. [CrossRef]

22. Kareh, K.M.; Lee, P.D.; Atwood, R.C.; Connolley, T.; Gourlay, C.M. Revealing the micromechanisms behind semi-solid metal deformation with time-resolved X-ray tomography. *Nat. Commun.* **2014**, *5*, 1–7. [CrossRef] [PubMed]

23. Terzi, S.; Salvo, L.; Suery, M.; Dahle, A.; Boller, E. In situ microtomography investigation of microstructural evolution in Al-Cu alloys during holding in semi-solid state. *Trans. Nonferr. Metals Soc. China* **2010**, *20*, s734–s738. [CrossRef]

24. Modigell, M.; Pola, A.; Suery, M.; Zang, C. Investigation of correlations between shear history and microstructure of semi-solid alloys. *Solid State Phenom.* **2013**, *192–193*, 251–256. [CrossRef]

25. Kang, C.G.; Choi, J.S.; Kim, K.H. The effect of strain rate on macroscopic behavior in the compression forming of semi-solid aluminum alloy. *J. Mater. Process. Tech.* **1999**, *88*, 159–168. [CrossRef]

26. Macosko, C.W. *Rheology—Principles, Measurements, and Applications*; Wiley-VCH, Inc.: Hoboken, NJ, USA, 1994.

27. Van Wazer, J.R.; Lyons, J.; Lim, K.; Colwell, R.E. *Viscosity and Flow Measurements*; Wiley: New York, NY, USA, 1963.

28. Schummer, P.; Worthoff, R.H. An elementary method for the evaluation of a flow curve. *Chem. Eng. Sci.* **1978**, *33*, 759–763. [CrossRef]

29. Alexandrou, A.N.; Georgiou, G.C.; Economides, E.-A.; Modigell, M. Determining true material constants of semisolid slurries from rotational rheometer data. *Solid State Phenom.* **2016**, *256*, 153–172. [CrossRef]

30. Alexandrou, A.N.; Georgiou, G. On the early breakdown of semisolid suspensions. *J. Non-Newton. Fluid Mech.* **2007**, *142*, 199–206. [CrossRef]

31. Modigell, M.; Pape, L. A comparison of measuring devices used to prevent wall slip in viscosity measurements of metallic suspensions. *Solid State Phenom.* **2008**, *141–143*, 307–312. [CrossRef]

32. Kiljański, T. A method for correction of the wall-slip effect in a Couette rheometer. *Rheol. Acta* **1989**, *28*, 61–64. [CrossRef]

33. Harboe, S.; Modigell, M. Wall slip of semi-solid A356 in couette rheometers. *AIP Conf. Proc.* **2011**, *1353*, 1075–1080. [CrossRef]

34. Tocci, M.; Zang, C.; Cadórniga Zueco, I.; Pola, A.; Modigell, M. Rheological properties of liquid metals and semisolid materials at low solid fraction. *Solid State Phenom.* **2016**, *256*, 133–138. [CrossRef]

35. Solek, K.; Rogal, L.; Kapranos, P.; Solek, K.P.; Rogal, L.; Kapranos, P. Evolution of Globular Microstructure and Rheological Properties of Stellite 21 Alloy after Heating to Semisolid State. *J. Mater. Eng. Perform.* **2017**, *26*, 115–122. [CrossRef]

36. Yekta, F.H.; Vanini, A.S. Simulation the Flow of Semi-Solid Steel Alloy Using an Enhanced Model. *Metals Mater. Int.* **2015**, *21*, 913–922. [CrossRef]

37. Modigell, M.; Volkmann, T.; Zang, C. A High -Precision Rotational Rheometer for Temperatures up to 1700 °C. *Solid State Phenom.* **2013**, *192–193*, 359–364. [CrossRef]

38. Laxmanan, V.; Flemings, M.C. Deformation of Semi-Solid Sn-15 Pct Pb Alloy. *Metall. Trans. A* **1980**, *11*, 1927–1937. [CrossRef]

39. Nafisi, S.; Lashkari, O.; Ghomashchi, R.; Ajersch, F.; Charette, A. Microstructure and rheological behavior of grain refined and modified semi-solid A356 Al–Si slurries. *Acta Mater.* **2006**, *54*, 3503–3511. [CrossRef]

40. Hu, X.G.; Zhu, Q.; Atkinson, H.V.; Lu, H.X.; Zhang, F.; Dong, H.B.; Kang, Y.L. A time-dependent power law viscosity model and its application in modelling semi-solid die casting of 319s alloy. *Acta Mater.* **2017**, *124*, 410–420. [CrossRef]

41. Becker, E.; Favier, V.; Bigot, R.; Cezard, P.; Langlois, L. Impact of experimental conditions on material response during forming of steel in semi-solid state. *J. Mater. Process. Tech.* **2010**, *210*, 1482–1492. [CrossRef]

42. Becker, E.; Langlois, L.; Favier, V.; Bigot, R. Thermomechanical modelling and simulation of C38 thixoextrusion steel. *Solid State Phenom.* **2015**, *217–218*, 130–137. [CrossRef]

43. Shimahara, H.; Baadjou, R.; Kopp, R.; Hirt, G. Investigation of flow behavior and microstructure on X210CrW12 steel in semi-solid state. *Solid State Phenom.* **2006**, *116–117*, 189–192. [CrossRef]

44. Liu, T.Y.; Atkinson, H.V.; Kapranos, P.; Kirkwood, D.H.; Hogg, S.C. Rapid Compression of Aluminum Alloys and Its Relationship to Thixoformability. *Metall. Mater. Trans. A* **2003**, *34*, 1545–1554. [CrossRef]

45. Hogg, S.C.; Atkinson, H.V.; Kapranos, P. Semi-Solid Rapid Compression Testing of Spray-Formed Hypereutectic Al-Si Alloys. *Metall. Mater. Trans. A* **2004**, *35*, 899–910. [CrossRef]

46. Favier, V.; Atkinson, H.V. Micromechanical modelling of the elastic–viscoplastic response of metallic alloys under rapid compression in the semi-solid state. *Acta Mater.* **2011**, *59*, 1271–1280. [CrossRef]

47. Kim, W.Y.; Kang, C.G.; Kim, B.M. The effect of the solid fraction on rheological behavior of wrought aluminum alloys in incremental compression experiments with a closed die. *Mat. Sci. Eng. A* **2007**, *447*, 1–10. [CrossRef]

48. Joly, P.A.; Mehrabian, R. The rheology of a partially solid alloy. *J. Mater. Sci.* **1976**, *11*, 1393–1418. [CrossRef]

49. Blanco, A.; Azpilgain, Z.; Lozares, J.; Kapranos, P.; Hurtado, I. Rheological characterization of A201 aluminum alloy. *Trans. Nonferr. Metals Soc. China* **2010**, *20*, 1638–1642. [CrossRef]

50. Heidary, D.S.B.; Akhlaghi, F. Experimental Investigation on the Rheological Behavior of Hypereutectic Al-Si Alloys by a Precise Rotational Viscometer. *Metall. Mater. Trans. A* **2010**, *41*, 3435–3442. [CrossRef]

51. Chhabra, R.P.; Richardson, J.F. *Non-Newtonian Flow in the Process Industries*; Butterworth Heinemann: Oxford, UK, 2004.

52. Burgos, G.R.; Alexandrou, A.N.; Entov, V. Thixotropic rheology of semisolid metal suspensions. *J. Mater. Process. Tech.* **2001**, *110*, 164–176. [CrossRef]

53. Burgos, G.R.; Alexandrou, A.N. Flow development of Herschel-Bulkley fluids in a sudden three-dimensional square expansion. *J. Rheol.* **1999**, *43*, 485–498. [CrossRef]

54. Ahmed, A.; Alexandrou, A.N. Processing of semi-solid materials using a shear-thickening Bingham fluid model. *Am. Soc. Mech. Eng. Fluids Eng. Div.* **1994**, *179*, 83–87.

55. Barnes, H.A. Thixotropy—A review. *J. Non-Newton. Fluid Mech.* **1997**, *70*, 1–33. [CrossRef]

56. Denny, D.A.; Brodkey, R.S. Kinetic Interpretation of Non-Newtonian Flow. *J. Appl. Phys.* **1962**, *33*, 2269–2274. [CrossRef]

57. Moore, F. The rheology of ceramic slips and bodies. *Trans. Brit. Ceram. Soc.* **1959**, *58*, 470–494.

58. Tonmukayakul, N.; Pan, Q.Y.; Alexandrou, A.N.; Apelian, D. Transient Flow Characteristics and Properties of Semi Solid Aluminium Alloy A356. In Proceedings of the 8th International Conference on Semi-Solid Processing of Alloys and Composites, Limassol, Cyprus, 21–23 September 2004; pp. 167–172.

59. Petera, J.; Kotynia, M. The numerical simulation of the thixoforming. *Inzynieria Chem. Proces.* **2001**, *22*, 1103–1108.

60. Modigell, M.; Koke, J. Time-Dependent Rheological Properties of Semi-Solid Metal Alloys. *Mech. Time-Depend. Mater.* **1999**, *3*, 15–30. [CrossRef]

61. Cross, M.M. Rheology of Non-Newtonian Fluids: A New Flow Equation for Pseudoplastic Systems. *J. Colloidal Sci.* **1965**, *20*, 417–437. [CrossRef]

62. Liu, T.Y. Rheology of Semisolid Alloys under Rapid Change in Shear Rate. Ph.D. Thesis, University of Sheffield, Sheffield, UK, 2002.

63. Lashkari, O.; Ajersch, F.; Charette, A.; Chen, X.-G. Microstructure and rheological behavior of hypereutectic semi-solid Al–Si alloy under low shear rates compression test. *Mater. Sci. Eng. A* **2008**, *492*, 377–382. [CrossRef]

64. Das, P.; Samanta, S.K.; Dutta, P. Rheological Behavior of Al-7Si-0.3Mg Alloy at Mushy State. *Metall. Mater. Trans. B* **2015**, *46*, 1302–1313. [CrossRef]

65. Brabazon, D.; Browne, D.J.; Carr, A.J. Experimental investigation of the transient and steady state rheological behaviour of Al-Si alloys in the mushy state. *Mater. Sci. Eng. A* **2003**, *356*, 69–80. [CrossRef]

66. Tocci, M.; Pola, A.; La Vecchia, G.M.; Modigell, M. Characterization of a New Aluminium Alloy for the Production of Wheels by Hybrid Aluminium Forging. *Procedia Eng.* **2015**, *109*, 303–311. [CrossRef]
67. Fukui, Y.; Nara, D.; Kumazawa, N. Evaluation of the Deformation Behavior of a Semi-solid Hypereutectic Al-Si Alloy Compressed in a Drop-Forge Viscometer. *Metall. Mater. Trans. A* **2015**, *46*, 1908–1916. [CrossRef]
68. Azzi, L.; Ajersch, F. Analytical Modeling of the Rheological Behavior of Semisolid Metals and Composites. *Metall. Mater. Trans. B* **2006**, *37*, 1067–1074. [CrossRef]
69. Cheng, D.C.-H. Yield stress: A time-dependent property and how to measure it. *Rheol. Acta* **1986**, *25*, 542–554. [CrossRef]
70. Barnes, H.A.; Walters, K. The yield stress myth? *Rheol. Acta* **1985**, *24*, 323–326. [CrossRef]
71. Barnes, H.A. The yield stress—A review or 'παντα ρει'—Everything flows? *J. Non-Newton. Fluid Mech.* **1999**, *81*, 133–178. [CrossRef]
72. Pan, Q.Y.; Apelian, D.; Alexandrou, A.N. Yield Behavior of Commercial Al-Si Alloys in the Semisolid State. *Metall. Mater. Trans. B* **2004**, *35*, 1187–1202. [CrossRef]
73. Solek, K. Identification of the steel viscosity and dynamic yield stress for the numerical modelling of casting simulations in the semi-solid state. *Arch. Metall. Mater.* **2017**, *62*, 195–200. [CrossRef]
74. Harboe, S.; Modigell, M. Yield stress in semi-solid alloys—The dependency on time and deformation history. *Key Eng. Mater.* **2013**, *554–557*, 523–535. [CrossRef]
75. Hufschmidt, M.; Modigell, M.; Petera, J. Modelling and simulation of forming processes of metallic suspensions under non-isothermal conditions. *J. Non-Newton. Fluid Mech.* **2006**, *134*, 16–26. [CrossRef]
76. Gautham, B.P.; Kapur, P.C. Rheological model for short duration response of semi-solid metals. *Mater. Sci. Eng. A* **2005**, *393*, 223–228. [CrossRef]
77. Liu, T.Y.; Atkinson, H.V.; Ward, P.J.; Kirkwood, D.H. Response of Semi-solid Sn-15 Pct Pb to Rapid Shear-Rate Changes. *Metall. Mater. Trans. A* **2003**, *34*, 409–417. [CrossRef]
78. Simlandi, S.; Barman, N.; Chattopadhyay, H. Study on Rheological Behavior of Semisolid A356 Alloy during Solidification. *Trans. Indian Inst. Metals* **2012**, *65*, 809–814. [CrossRef]
79. Pouyafar, V.; Sadough, S.A. An Enhanced Herschel–Bulkley Model for Thixotropic Flow Behavior of Semisolid Steel Alloys. *Metall. Mater. Trans. B* **2013**, *44*, 1304–1310. [CrossRef]
80. Liang, L.; Mian, Z. Theoretical research on rheological behavior of semisolid slurry of magnesium alloy AZ91D. *Comput. Mater. Sci.* **2015**, *102*, 202–207. [CrossRef]
81. Omar, M.Z.; Atkinson, H.V.; Kapranos, P. Thixotropy in Semisolid Steel Slurries under Rapid Compression. *Metall. Mater. Trans. A* **2011**, *42*, 2807–2819. [CrossRef]
82. Kirkwood, D.H.; Ward, P.J. Numerical Modelling of Semi-Solid Flow under Processing Conditions. *Fundam. Thixoforming Process.* **2004**, *75*, 519–524. [CrossRef]
83. Becker, E.; Bigot, R.; Rivoirard, S.; Faverolle, P. Experimental investigation of the thixoforging of tubes of low-carbon steel. *J. Mater. Process. Tech.* **2018**, *252*, 485–497. [CrossRef]
84. Lozares, J.; Azpilgain, Z.; Hurtado, I.; Ortubay, R.; Berrocal, S. Thixo Lateral Forging of a Commercial Automotive Spindle From LTT45 Steel Grade. *Key Eng. Mater.* **2012**, *504–506*, 357–360. [CrossRef]
85. Quaak, C.J. Rheology of Partially Solidified Aluminium Alloys and Composites. Ph.D. Thesis, Technische Univesiteit Delft, Delft, The Netherlands, 1996.
86. Peng, H.; Wang, K.K. Steady State and Transient Rheological Behaviour of a Semi-Solid Tin-Lead Alloy in Simple Shear Flow. In Proceedings of the 4th International Conference Semi-Solid Processing of Alloys and Composites, Sheffield, UK, 19–21 June 1996; pp. 2–9.
87. Bührig-Polaczek, A.; Afrath, C.; Modigell, M.; Pape, L. Comparison of rheological measurement techniques for semi-solid aluminium alloys. *Solid State Phenom.* **2006**, *116–117*, 610–613. [CrossRef]
88. Tadros, T. (Ed.) Ostwald Ripening. In *Encyclopedia of Colloid and Interface Science*, 1st ed.; Springer: Berlin/Heidelberg, Germany, 2013.
89. Courtney, T.H. Microstructural evolution during liquid phase sintering: Part II. Microstructural coarsening. *Metall. Trans. A* **1976**, *8*, 685–689. [CrossRef]
90. Courtney, T.H. A reanalysis of the kinetics of neck growth during liquid phase sintering. *Metall. Trans. A* **1977**, *8*, 671–677. [CrossRef]
91. Poirier, D.R.; Ganesan, S.; Andrews, M.; Ocansey, P. Isothermal coarsening of dendritic equiaxial grains in Al–15.6 wt %Cu alloy. *Mater. Sci. Eng. A* **1991**, *148*, 289–297. [CrossRef]

92. Bender, W.; Ratke, L. Ostwald ripening of liquid phase sintered Cu Co dispersions at high volume fractions. *Acta Mater.* **1998**, *46*, 1125–1133. [CrossRef]

93. Limodin, N.; Salvo, L.; Suery, M.; DiMichiel, M. In situ investigation by X-ray tomography of the overall and local microstructural changes occurring during partial remelting of an Al–15.8 wt % Cu alloy. *Acta Mater.* **2007**, *55*, 3177–3191. [CrossRef]

94. Terzi, S.; Salvo, L.; Suery, M.; Dahle, A.K.; Boller, E. Coarsening mechanisms in a dendritic Al–10% Cu alloy. *Acta Mater.* **2010**, *58*, 20–30. [CrossRef]

95. Lifshitz, I.M.; Sloyozov, V.V. The kinetics of precipitation from supersaturated solid solutions. *J. Phys. Chem. Solids* **1961**, *19*, 35–50. [CrossRef]

96. Wagner, C. Theorie der alterung von niederschlägen durch umlösung. *Z. Elektrochem.* **1961**, *65*, 581–591.

97. Modigell, M.; Pola, A. Modeling of shear induced coarsening effects in semi-solid alloys. *Trans. Nonferr. Metals Soc. China* **2010**, *20*, 1696–1701. [CrossRef]

98. Yang, Y.S.; Tsao, C.-Y.A. Viscosity and structure variations of Al–Si alloy in the semi-solid state. *J. Mater. Sci.* **1997**, *32*, 2087–2092. [CrossRef]

99. Sukumaran, K.; Pai, B.C.; Chakraborty, M. The effect of isothermal mechanical stirring on an Al–Si alloy in the semisolid condition. *Mater. Sci. Eng. A* **2004**, *369*, 275–283. [CrossRef]

100. Barman, N.; Dutta, P. Rheology of A356 Alloy during Solidification under Stirring. *Trans. Indian Inst. Metals* **2014**, *67*, 101–104. [CrossRef]

101. Rogal, L.; Dutkiewicz, J.; Atkinson, H.V.; Lityńska-Dobrzyńska, L.; Czeppe, T.; Modigell, M. Characterization of semi-solid processing of aluminium alloy 7075 with Sc and Zr additions. *Mater. Sci. Eng. A* **2013**, *580*, 362–373. [CrossRef]

102. Chen, H.I.; Chen, J.C.; Liao, J.J. The influence of shearing conditions on the rheology of semi-solid magnesium alloy. *Mater. Sci. Eng. A* **2008**, *487*, 114–119. [CrossRef]

Review

Microstructure and Properties of Semi-Solid Aluminum Alloys: A Literature Review

Annalisa Pola [1,*] , **Marialaura Tocci [1]** and **Plato Kapranos [2]**

[1]   DIMI—Mechanical and Industrial Engineering Department, Via Branze, 38-25123 Brescia, Italy; m.tocci@unibs.it

[2]   Department of Materials Science & Engineering, The University of Sheffield, Sir Robert Hadfield Building, Mappin Street, Sheffield S1 3JD, UK; p.kapranos@sheffield.ac.uk

*   Correspondence: annalisa.pola@unibs.it; Tel.: +39-030-371-5576

Received: 4 February 2018; Accepted: 6 March 2018; Published: 13 March 2018

Abstract: Semi-solid processing of aluminum alloys is a well-known manufacturing technique able to combine high production rates with parts quality, resulting in high performance and reasonable component costs. The advantages offered by semi-solid processing are due to the shear thinning behavior of the thixotropic slurries during the mold filling. This is related to the microstructure of these slurries consisting of solid, nondendritic, near-globular primary particles surrounded by a liquid matrix. This paper presents a review on the formation of this nondendritic microstructure, reports on the different proposed mechanisms that might be responsible, and illustrates the relationship between microstructure and properties, in particular, tensility, fatigue, wear, and corrosion resistance.

Keywords: semi-solid; microstructure; mechanical properties; wear; corrosion

1. Introduction

Semi-solid metal (SSM) processing is a manufacturing technique where an alloy, in the form of a slurry of near-globular primary particles in a liquid matrix, is injected into a die, allowing the production of near-net-shape components. The main advantage of this technology is related to the flow properties of the metal in the form of a slurry, which, in the semi-solid state, is non-Newtonian and exhibits shear thinning behavior [1,2]. The viscosity of the SSM slurry is higher than when fully liquid, reducing the risk of turbulent [3] or spray flow, which is more typical of conventional pressure die-casting [3,4]. However, thanks to the shear thinning behavior of the slurry, under the influence of a shear force acting on it when it flows into the die, the viscosity decreases and the metal slurry is able to fill the cavity completely in a nonturbulent manner. As a consequence, semi-solid cast parts are almost free of gas porosity.

Injecting a partially solidified alloy slurry has the added benefit that shrinkage porosity is virtually absent [5]. The low or even absent porosity allows the production of structural parts with good mechanical properties that can also undergo subsequent heat treatments or welding operations.

Semi-solid processing guarantees higher performance than die-casting, while maintaining a number of the advantages of die-casting, such as good dimensional tolerances, high production rates, high surface quality, complex near-net-shape parts, and thin sections with very limited need of any finishing operations [6]. In addition, when compared to conventional die-casting, SSM processing increases die life because of the lower stress associated with the lower injection temperatures and speeds (i.e., lower mold attack and erosion, lower thermal shock), reduced cycle times and risk of hot tearing due to the lower temperature of the metal slurry (no over-heating), and associated lower energy consumption [4,6,7].

On the other hand, SSM manufacturing requires specialized equipment for alloy preparation, combined with strict control of process parameters, particularly the alloy temperature, i.e., the solid/liquid fraction. Unfortunately, all these tend to increase the production costs [8], even though recent investigations have shown that SSM processing is only slightly more expensive than conventional die-casting and cheaper than some other competing foundry processes [9].

Since its development in the early 1970s at MIT [10], much research has been performed worldwide, aimed mainly at developing new and alternative routes of feedstock production for different alloys suitable for SSM processing. At present, there are a number of different techniques used to produce semi-solid castings, differentiated by the percentage of liquid/solid fraction they employ and the way they produce the alloy in the semi-solid state. SSM methods can be divided in two main categories according to their processing route, known as rheocasting and thixocasting [11]. In rheocasting, the semi-solid slurry is prepared in situ from the liquid state down to a certain percentage of solid fraction (usually between 10 and 30% [12]), and then directly transferred into the shot sleeve for being injected into the die. In thixocasting, a billet, characterized by an almost globular or rosette-like microstructure developed through some specific route, is reheated in the mushy zone (semi-solid region) to an appropriately chosen solid fraction (usually between 50% and 60% [13]), placed in a modified shot sleeve and finally injected into the die [14]. Another classification methodology distinguishes the SSM methods according to the initial step used for obtaining the semi-solid feedstock, i.e., from the liquid state, by controlled solidification or from the solid state, via heavy plastic deformation and recrystallization [11].

Some of the most common routes for feedstock material preparation are: mechanical stirring, such as the SSR [15,16] or GISS [17] processes, electromagnetic stirring (EMS or MHD) [18,19], ultrasonic stirring (UTS) [20,21], New Rheocasting (NRC or UBE) [22,23], cooling slope [24], twin screw [25], Rheometal [26], liquid mixing method [27], SEED [28], thermomechanical [29], and SIMA [14,30].

A number of interesting reviews about the different technologies available for obtaining nondendritic slurries can be found in the literature [11,13,14,31–33]. Nevertheless, independently from the chosen technique, the fundamental concept is based in developing feedstock with a microstructure of a solid, near-globular phase surrounded by a liquid matrix when in the semi-solid state. As already mentioned, when a shear stress is applied, the near-globular solid particles move easily between and over each other, reducing the viscosity and making the material behave like a liquid. On the contrary, when a shear stress is applied on a dendritic microstructure, the liquid remains entrapped between dendrite arms and prevents them from moving freely, thus increasing the viscosity of the alloy [32].

This paper presents a review on the formation of nondendritic microstructures, microstructures that have a key role in semi-solid processing, and discusses the different proposed mechanisms together with ways to analyze SSM microstructures. In addition, the review provides information on the variation of mechanical properties and corrosion behavior through modification of microstructures that are typical of SSM processing.

2. Formation of Nondendritic Microstructures in SSM Processing

During the early experiments performed by Spencer et al. on a Sn–Pb15 alloy [10], it was found that the microstructure of the material was strongly affected by the constant shearing of the alloy when in semi-solid state. Particularly, it was shown that shearing action causes the formation of a nondendritic grain structure, which is the distinctive characteristic of semi-solid alloys. Moreover, with further shearing during cooling, it is also possible to obtain spheroidal particles, typically with some entrapped liquid [4]. The authors also reported that high shear rates and slow cooling rates can promote the formation of spherical particles instead of rosette-like ones [10].

The steps for the formation of nondendritic microstructures have been extensively studied over the years and one of the first proposed mechanisms is shown in the schematic illustration of Figure 1.

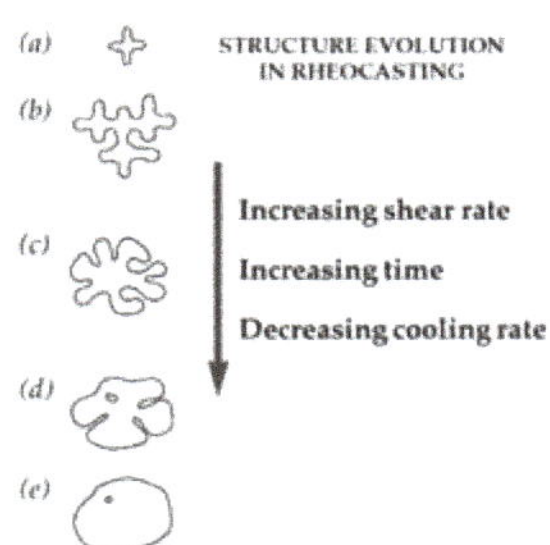

Figure 1. Globule formation during stirring in the semi-solid range: (**a**) initial dendritic fragment, (**b**) dendritic growth, (**c**) rosette, (**d**) ripened rosette, and (**e**) spheroid [4].

According to Flemings and co-workers [4], in the early stages of solidification, as it happens for all metallic materials, dendrites form in the liquid. However, unlike conventional solidification, the shearing action affects the dendritic morphology, which changes into that of a "rosette" due to different phenomena. Various explanations about the conversion mechanisms from dendritic to globular morphology can be found in the literature like ripening, shear, bending and abrasion with other growing crystals, dendrite fragmentation, remelting of dendrite arms, and growth control mechanisms [2]. Vogel et al. [34], for instance, proposed that under a shear force the dendrite arms bend plastically, thus introducing large misorientations into the arms and forming dislocations. At high temperatures these dislocations rearrange themselves inducing, under specific conditions, the detachment of dendrite arms [35] as shown in Figure 2. These dendrite fragments act as nuclei, coarsening and leading to the presence of globules of the primary phase [4,35,36].

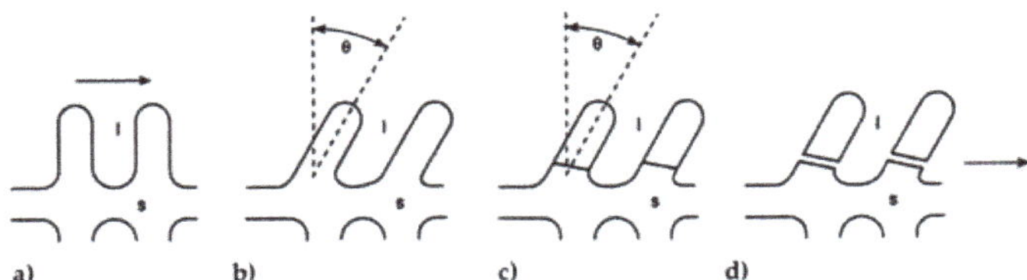

Figure 2. Schematic model of fragmentation mechanism: (**a**) undeformed dendrite, (**b**) after bending, (**c**) dislocation rearrangement to give grain boundary, and (**d**) grain boundary wetting [35].

In contrast, Molennar et al. [37] proposed that rosette-like particles are the result of cellular growth. Mullis [38] reported that bending could bring about rosette formation without any need of mechanical effects due to shearing. According to Hellawell [39], the secondary dendrite arms can separate at their roots because of solute enrichment and thermosolutal convection that determines their remelting rather than breaking off for simple mechanical interactions (Figure 3). He suggested that in the solidification range, the solid is completely ductile and dendrites can be bent but not broken. Hence, the detachment of the secondary arms can be explained by a local remelting phenomenon. In particular, the remelting can occur either by recalescence of the whole system or by local recalescence due to fluctuations caused by convective phenomena or stirring [14,40].

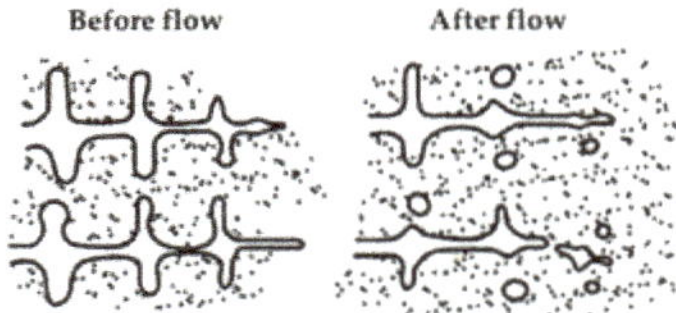

Figure 3. Schematic model of dendrite arms remelting [4].

The effect of the fluid flow characteristics on the morphology of solidification structures was also studied, by means of Monte Carlo simulation, by Das et al. [41]. They found that a rotational motion under laminar flow promotes rosette-like morphology due to a periodic stabilizing and destabilizing of the solid–liquid interface, while a turbulent flow hinders dendritic growth resulting in a compact morphology due to a stable solid–liquid interface. The presence of a concurrent mechanism was also proposed.

An interesting review of the various proposed mechanisms is reported in [2].

Figure 4 shows a typical microstructure of semi-solid castings, which consists of rosette-like or even globular grains (Figure 4a) and a dendritic structure typical of conventional casting processes (Figure 4b) [42].

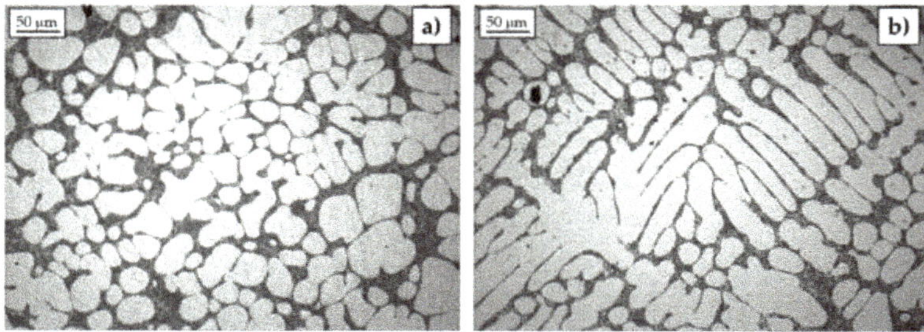

Figure 4. Typical microstructures of AlSi7 component (**a**) semi-solid cast and (**b**) conventionally cast.

3. Microstructural Analysis of Semi-Solid Alloys

In the analysis of microstructures of SSM components, dendrite arm spacing cannot be measured as in the case of conventional castings [43]; instead, the microstructural parameters taken into account are the size of the globules, their shape factor, i.e., their roundness, and the amount of entrapped liquid [44].

The globule size should be large enough to build an almost rigid solid phase network and, at the same time, small enough that the slurry can flow into the die cavity similarly to a liquid. The dimensions of the globules are usually defined as their mean diameters. It is typically assumed that the minimum thickness that can be filled by SSM should not be lower than 20–30 times the grain radius [11]. In particular, according to some findings, the optimum primary particle size for SSM alloys is lower than 100 µm [45]. However, the grain size distribution measured by 2D analysis is difficult to determine and sometimes an extensive analysis of serial sections is needed to guarantee reliable data. A more accurate and consistent analysis of the grain size distribution and evolution can be performed via 3D examination methods, like X-ray microtomography [46–50], as shown in Figure 5.

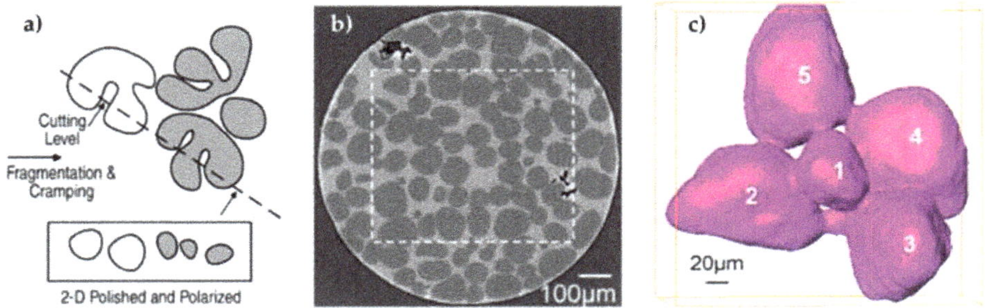

Figure 5. (**a**) Schematic representation of 2D analysis [50], (**b**) 3D image processing with 2D slice, and (**c**) volume representation of the acquired tomography data [51].

The shape factor is known to strongly affect the slurry viscosity. For laminar die filling in particular, the solid particles should preferably be round and separated from each other [11]. The shape factor (F) is defined as:

$$F = 4\pi A / P^2$$

where A is the area and P is the perimeter of the particles.

A shape factor equal to 1 represents the case of a perfect circle and it reduces to zero with an increasing amount of irregularity, i.e., highly branched or elongated microstructures. In semi-solid processing, a shape factor above 0.6 is considered as appropriate [44]. Because the SSM particles can have complex morphologies, care should be taken when interpreting metallographic sections, as errors can arise if isolated secondary branches are taken into account as real single particles [52]. Thus, in order to obtain more reliable results, a high number of particles have to be measured.

The entrapped liquid is a distinct feature of the thixocasting route [11,53]. This liquid, as shown in Figure 6, does not contribute to the sliding of the globules during processing. Therefore, it follows that the liquid fraction is lower than the theoretical one and that the viscosity increases during die filling due to a sponge effect that can be induced [11]. The amount of entrapped liquid can be estimated by image analysis of 2D polished section areas, however, as in the case of the shape factor and globule size, errors can be introduced when some of the entrapped liquid islands appear to be isolated even though they are connected to the liquid phase at deeper levels. A more thorough investigation by means of 3D analysis allows more reliable data to be obtained.

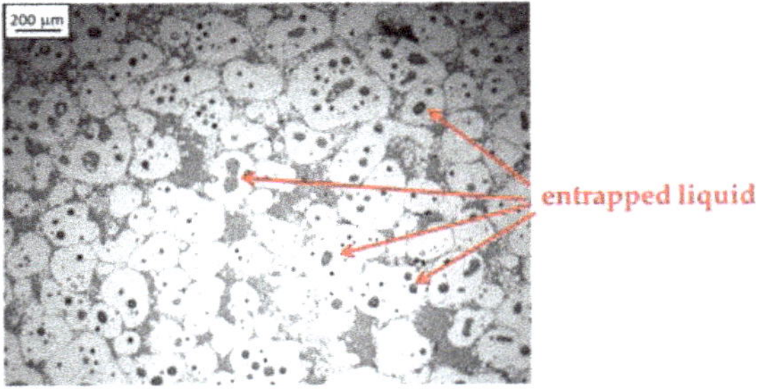

Figure 6. Thixocasting microstructure with entrapped liquid in the globules.

Clearly, different production routes of semi-solid parts can result in different microstructures, from rosette-like to near globular and details about them can be easily found in the literature [2,14].

4. Performance of Semi-Solid Aluminum Alloys

In recent years, several studies have been focused on the mechanical properties of parts fabricated by semi-solid/thixoforming processes, often in comparison to conventional routes.

4.1. Mechanical Properties

Al alloys are widely used in semi-solid processing as discussed in the previous section. The proven better quality of components obtained by SSM processes, and associated better properties, are often cited in textbooks about the topic [11,13], in particular highlighting the possibility to perform T6 heat-treatment to further increase their characteristics. There are also many scientific studies on the mechanical properties of Al SSM alloys (both casting and wrought ones), even though significantly less than those on microstructural modification in comparison with conventional casting alloys.

4.1.1. Tensile Behavior

Since the very first production attempts, it was evident that semi-solid parts have high mechanical properties comparable to those of the forged material and better than permanent mold castings [4]. Over the years, these findings have been confirmed by many authors. In fact, the enhancement in performance of parts manufactured by SSM processing compared to traditionally cast parts is supported by various studies mainly on Al–Si [54–59], Al–Cu [60,61], and Al–Zn alloy families [62,63].

Bergsma et al. [64], for instance, reported that the tensile strength of 357 and modified 319 semi-solid formed aluminum alloys are superior to conventionally cast alloys due to the reduction in porosity and the spherical microstructure, when an effective optimization of heat treatment parameters is achieved. Cerri et al. [65] showed excellent ultimate tensile strength and yield strength of 319 alloy after appropriate heat treatment, in the order of 350 MPa and 280 MPa, respectively, thus almost 100 MPa higher than the conventionally cast counterparts. Similarly, Zhu et al. [66] analyzed different casting and forging alloys used industrially for the production of compressor wheels, finding that strength and ductility approach those of the forged components after T61 heat treatment, as shown in Figure 7. Nevertheless, in their work, the influence of microstructural features is not thoroughly investigated.

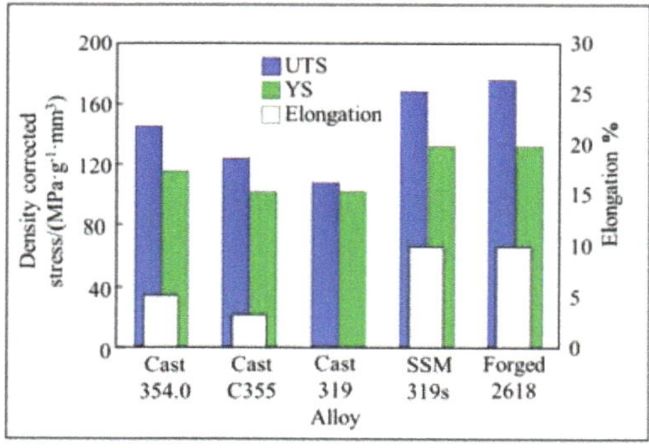

Figure 7. SSM 319 vs. cast C355, 354.0, 319 alloys and forged 2618 alloy after T61 heat treatment [66].

On the contrary, Haga et al. [67] discussed the influence of the size of primary globules on tensile properties of semi-solid castings. They underlined that the remarkable tensile properties are due not only to the nondendritic microstructure, but also to the small size of the primary α-Al, which is especially effective in enhancing the elongation to fracture. In particular, for the A356, the elongation can reach up to 18% by using the cooling slope method.

Another study on an Al–Si–Mg–Fe alloy demonstrated the influence of the shape of primary α-Al globules on ultimate tensile strength and elongation [68], a fact that is shown in Figure 8, i.e., when the shape factor increases (the more rounded the primary globules), tensile and elongation properties also increase.

Lü et al., investigating the behavior of rheocast 5052 alloy in comparison with gravity (GC) and high pressure (HPDC) die casting [69], noticed that fine and uniform microstructure throughout the entire SSM sample would effectively reduce stress concentrations at the grain boundaries under applied stress. They concluded that the globular shape is effective in enhancing the tensile strength and ductility, as detectable by fractographic analyses that show an almost ductile fracture mode for the rheocast alloy instead of the mixed ductile brittle fracture experienced by the gravity cast samples and with smaller dimples than those on the conventional casting sample (Figure 9).

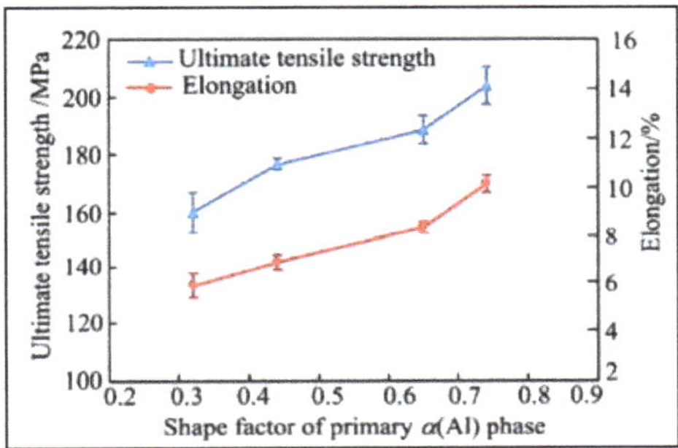

Figure 8. Effect of shape factor of primary α-Al phase on ultimate tensile strength and elongation of semi-solid slurry for Al–Si–Mg–Fe alloy in as-cast condition [68].

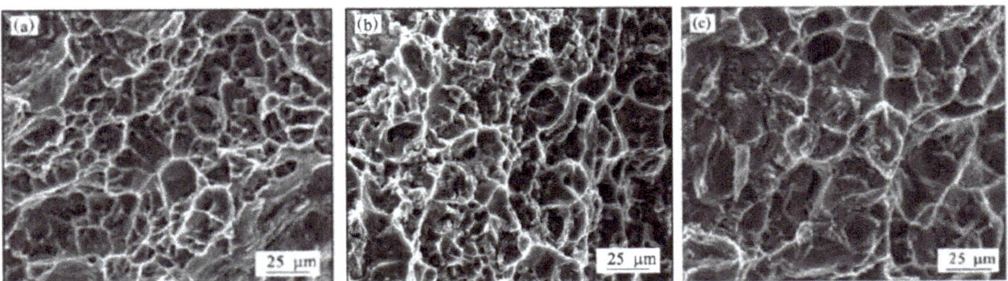

Figure 9. SEM micrographs of tensile fracture for samples produced under different processing conditions: (**a**) Rheo-HPDC; (**b**) conventional HPDC, and (**c**) conventional GC [69].

In Figure 10a, the comparison between the properties of semi-solid, casting, and forging Al–Si alloys summarized by Brochu et al. [70] is shown, further supporting the above-mentioned advantages. Clearly, different SSM process routes can result in different mechanical properties, as shown in Figure 10b [71]. However, the improved trend, as compared to traditional casting process, appears to hold in all cases as a consequence of the higher soundness of the components and the enhanced microstructure of semi-solid alloys, as discussed above.

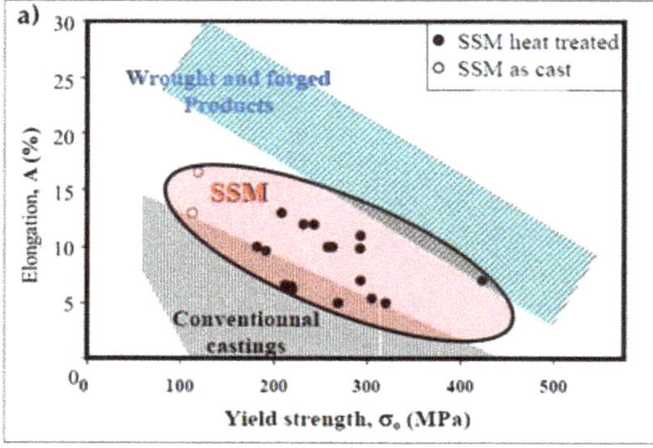
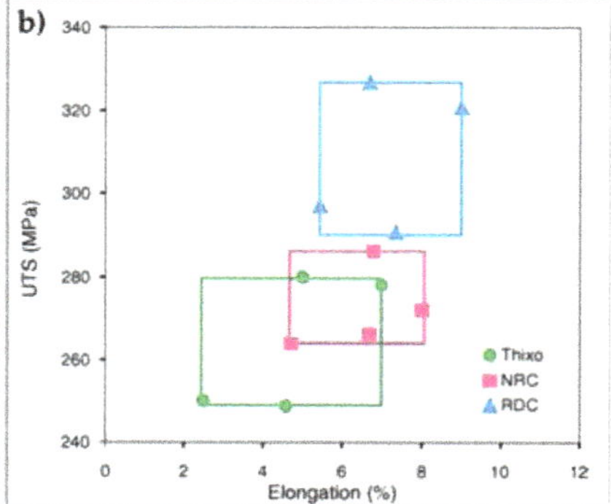

Figure 10. (**a**) Mechanical properties of SSM, cast and forged aluminum alloys [70] and (**b**) comparison of mechanical properties of A357 alloy under T5 condition obtained from different processing technologies [71].

Apart from the influence of primary α-Al grains, the effect of different alloying elements, as well as the influence of secondary phases in SSM alloys, were examined. For example, the addition of Si and Fe in a 206 aluminum alloy for an automotive application was investigated by Lemieux et al. [72], who found remarkable performance of the tested rheocast components.

The influence of a significantly higher Si content (approximately 20 wt. %) was examined in some studies, like that on a rheocast Al–Si–Cu alloy [73]. Recently, hypereutectic alloys processed by SSM methods have attracted the interest of researchers because of their heat-resistant properties. In hypereutectic Al–Si alloys, primary Si grains solidify as coarse plate-like particles, which can be refined by SSM processing, thus improving tensile properties, as demonstrated by Zheng et al. [74] studying the properties of AlSi30 rheo-diecast (RDC) compared to conventional die-casting. They pointed out that the UTS, elongation, and hardness of the SSM samples are approximately 57.9%, 42.9%, and 20.6% higher than those of the die-casting ones, respectively. They attributed this to the finer compact primary Si grains, which can reduce or even eliminate crack initiation, combined with reduced porosity. The development of a series of hypereutectic alloys based on the A390 composition (17% Si, 5% Cu, 0.5% Mg) and their thixoforming, have already been described by Kapranos et al. [75], together with their resulting microstructures and mechanical properties. Again, the main advantages of thixoforming were related to the improved size and morphology of brittle Si particles in comparison with conventionally cast parts, in addition to the expected spheroidisation of the Al matrix. The successful thixoforming of an automotive brake drum was reported and represented an interesting example of substitution of a conventional cast iron part with an aluminum one.

Concerning the effect of secondary phases (i.e., iron intermetallic compounds), interesting analyses can be found in the work of Shabestari et al. [56]. This is particularly interesting since numerous studies have reported the correlation between amount and morphology of secondary phases, for instance, intermetallic ones containing Fe, Mn, Cr, Ni, and mechanical properties [76–80] for casting Al–Si alloys. In particular, these authors found that the peculiar microstructural characteristics of the thixoformed alloy, such as the extremely low porosity, fine and equiaxed morphology of the α-Al grains and uniform distribution of intermetallic compounds fragmented by the process route, enhance strength and elongation, in comparison with the as-cast condition. A similar topic was also discussed by Möller et al. [81] in a study of the microstructural and tensile properties of semi-solid metal high pressure die cast F357 alloy with various additions of Fe, Ni, and Cr. In this case, the formation of intermetallic phases containing Fe and Ni lead to a decrease in strength and ductility due to their microcracking during tensile tests.

Nowadays, the attention of the researchers is focused on the investigation of the properties of less conventional alloys for SSM processing, like AlSi8 [82], Al5Fe4Cu [83], Al–Zn–Mg–Cu [84], AlZnMg alloys with Sc addition [85], etc., in order to evaluate the advantages in their application as semi-solid manufactured products.

4.1.2. Fatigue Behavior

Different researchers have compared high cycle fatigue resistance of SSM alloys (mainly based on Al–Si) with that of conventional casting ones [64,70,86–97]. There is a general consensus in these papers that the SSM samples show higher fatigue resistance, even though the data about the high cycles fatigue strength results is quite scattered (ranging from 60 to 180 MPa) depending on the process route used, casting and heat treatment parameters, as well as the possible presence of defects. The authors agree that the superior fatigue performance of SSM parts is related to the fact that they are almost free from defects like gas and shrinkage porosity as well as oxide inclusions. Thus, in defect-free parts, the microstructural constituents play a key role in fatigue crack nucleation and propagation [98]. As reported in [70,99], for instance, a high volume fraction of primary α-phase significantly increases the resistance to crack initiation. In particular, Park et al. found that fatigue cracks propagate mainly by cutting through the primary α-phase when a low volume fraction of α-phase is present; on the contrary, when the volume fraction is high, they mainly bypass the primary α-phase following the phase boundaries [87].

Other reasons for fatigue improvements seem to be the smaller globule size as well as the finer size and distribution of eutectic Si particles of SSM castings [100]. Relating to the former, it is reported that fatigue strength increases with decreasing primary α-phase size and that also, the globular morphology plays a positive role [70,86]. Additionally, the level of α globules agglomeration, which determines the size and distribution of Al–Si eutectic regions, influences the fatigue crack threshold. Concerning the Si particles, their interface with the α-phase in the eutectic can act as nucleation point of fatigue cracks because of the mismatch of plastic deformation between each other that causes fracture and/or decohesion of Si lamellae. It follows that their more uniform distribution and fine size improve fatigue crack initiation resistance, as documented by different authors [70,87].

Ragab et al. [89] showed that grain boundaries in SSM microstructures act as a barrier to propagation of short cracks. Additionally, they provide evidence that fatigue failure is also associated with the presence of oxides as well as slip bands and intermetallic phases. Clearly, as in the case of conventional casting, with the presence of platelet-like and needle-like shapes, Fe-rich intermetallic compounds reduce the fatigue properties, as their morphology is conducive to high stress concentrations, thus making them a source of cracks able to cause failure [95]. Figure 11 shows an example of a fatigue fracture surface [95]. A mixed fracture mode can be clearly seen with both cleavage cracks, which induce brittle fracture, and dimples in the α-phase, which denote a ductile fracture.

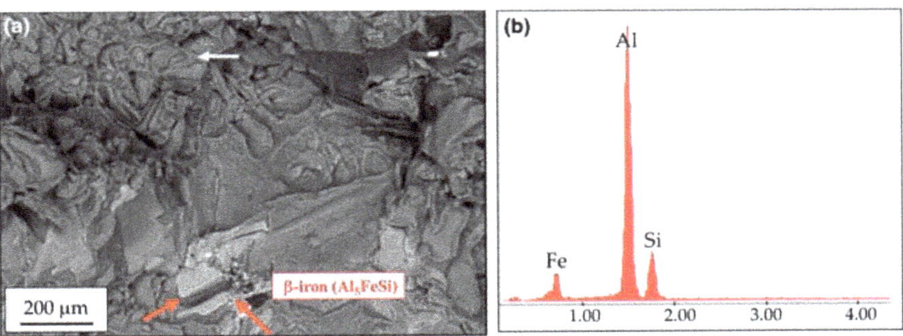

Figure 11. Scanning electron microscope analysis of the fatigue fracture surface of SSM A357 samples after T6 treatment [95].

It is known that iron up to 1.20% is needed in conventional die-casting to prevent alloy sticking onto die surfaces at high temperatures due to chemical, metallurgical, and mechanical interactions [101]. Interestingly, Al alloys for SSM contain lower amounts of Fe than conventional alloys for die-casting due to the reduced risk of die soldering related to the lower injection temperatures and speeds used during the process.

4.2. Wear Resistance

Regarding wear resistance, the dry sliding behavior of some SSM alloys has been evaluated in comparison with conventionally cast parts. Dey et al. show that the globular microstructures of A356 alloy castings appeared beneficial for dry wear resistance, as compared to dendritic ones [102]. In particular, the results of friction coefficients of SSM samples were lower than that of conventional cast specimens for all loads. The same holds for the wear loss.

A remarkable improvement in wear resistance of semi-solid specimens in comparison with conventional castings is also reported by Bayoumi et al. [103], who measured a lower wear rate for semi-solid processed A356 alloy. On the other hand, a comparable erosion mechanism was observed. Concurring results were found by other authors. The results from another study on the tribological properties of A356 alloy [104] showed that, for all applied loads, heat treatment enhanced the alloy behavior, while improvements caused by thixocasting were not systematic, i.e., a lower friction coefficient was noticed just for lower specific loads. Overall better wear resistance of thixocast materials, as compared to the original alloy, was attributed mainly to the improved distribution and smaller size of Si particles.

Several hypereutectic Al–Si compositions have also been investigated. It is a known fact that the presence of hard Si particles distributed in the matrix induces an outstanding wear resistance. However, the presence of casting defects like porosity, typical of traditional foundry processes, reduces their performance. In this regard, SSM processes make these alloys promising candidates for heavy wear applications. Very interesting results were obtained with A390 SSM alloy [105].

Birol et al. stated that the enhanced wear performances of hypereutectic Al–Si alloys are linked to the uniform distribution of fine primary Si particles [106]. Thus, the combination of a favorable silicon dispersion and the better soundness induced by the semi-solid processing gives superior wear performance.

Likewise, other chemical compositions showed comparable results in terms of improved wear behavior [63,107]. Recent studies on A319 confirm that the uniform distribution of Si, the reduction in porosity level, and the different morphology, size, and distribution of intermetallic phases obtained by SSM processing are responsible for better wear behavior than conventionally cast samples, even though the predominant wear mechanism remains the same for both the alloys [108,109].

Other damaging mechanisms of SSM components have been investigated over the years, such as cavitation resistance [110,111]. It was reported that the globular microstructure, as obtained by ultrasound treatment methods (UTS), increases the cavitation erosion resistance of the alloy because of the higher chemical and microstructural homogeneity, the morphology of the primary particles, and the refined structure of the eutectic due to the treatment itself. A comparison of the eroded area, at macroscopic scale, of conventionally cast (NUST) and SSM (UST) Al–Si7 samples is shown in Figure 12. It can be clearly seen that the highest damage was experienced by the conventionally cast alloy in the as-cast condition (Figure 12a), whereas the highest erosion resistance was exhibited by the heat-treated semi-solid sample (Figure 12d).

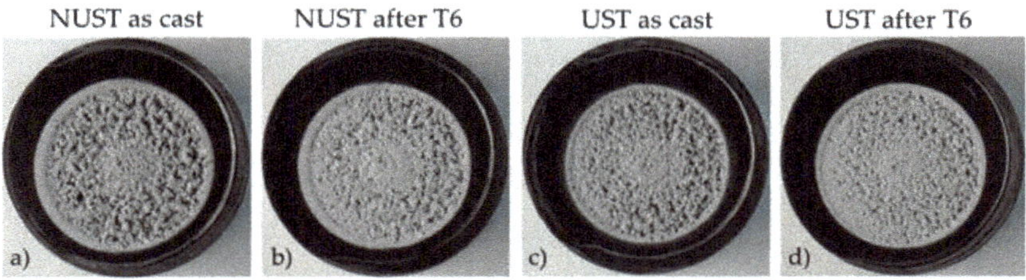

Figure 12. Surface topographies of the damaged areas of conventionally as-cast (**a**) before and (**b**) after T6, semi-solid in (**c**) as-cast and (**d**) after T6 [111].

4.3. Corrosion Resistance

Corrosion behavior of SSM aluminum alloys has not been explored to the same extent as mechanical performance and investigations on this property are more recent. For instance, Bastidas et al. [112] studied the pitting corrosion of A357 rheocast alloy, showing that it preferentially moves through the eutectic regions due to the Si particles that play a remarkable role in the corrosion process and to the cathodic properties of the intermetallic compounds.

A comparison between SSM and permanent mold cast A357 alloy was studied by Yu et al., who showed that both the resistance to corrosion and stress corrosion cracking is higher for semi-solid microstructures [97]. Similar results were obtained by other authors when comparing A356 semi-solid and low-pressure die-cast components. They highlighted that the number of pits and the degree of corrosion is higher for the conventionally cast products than in rheocast ones [99]. Similarly, Arrabal et al. [113] provide evidence about the better resistance of SSM as compared to gravity cast A356 alloy, as detectable by its lower cathodic current densities (Figure 13).

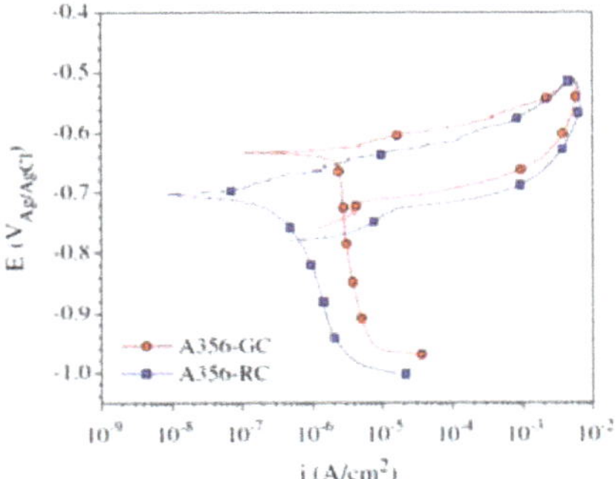

Figure 13. Cathodic polarisation curves of (**a**) gravity cast (GC) and (**b**) SSM (RC) A356 alloy after 1 h immersion in 3.5 wt. % NaCl naturally-aerated solution [113].

The improved corrosion resistance of the SSM Al–Si alloys is related to the reduced area ratio between Si particles and α-phase lamellae in the eutectic compared to that of their traditionally cast counterparts [99,114,115]. In particular, the differences in Si particle size and shape and, consequently, in the area ratio between silicon and α-phase in the eutectic, are related to the different solidification rates combined with the different applied pressures during the casting processes [114].

Analyzing the damage by scanning electron microscopy, it is noted that the α globules appear almost not to have been attacked, while the Si particles and the intermetallic compounds in the eutectic have a fundamental role in the corrosion damage [113,116], acting as local cathodes (Figure 14).

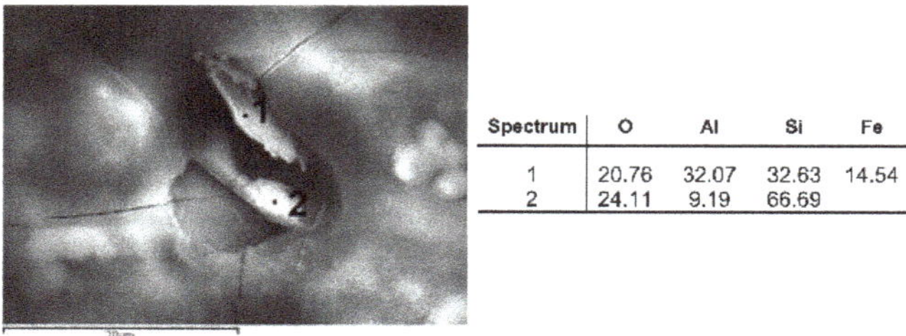

Spectrum	O	Al	Si	Fe
1	20.76	32.07	32.63	14.54
2	24.11	9.19	66.69	

Figure 14. Cathodic effect of silicon and intermetallic particles [116].

Interestingly, some authors have also investigated the effect of surface eutectic segregation on corrosion resistance [117,118]. This thin layer of liquid phase can be found in semi-solid products due to the "sponge" effect [11], when the slurry is injected under pressure into the die, and it contains a higher amount of alloying elements than the bulk. Because, as reported above, the pitting corrosion in Al–Si alloys occurs preferentially in the eutectic area, the semi-solid parts characterized by the presence of this segregation layer are more prone to the phenomenon. Recently, 3D SEM tomography showed that corrosion takes place more at the interface between α-phase and Fe-rich intermetallics than at the eutectic Si ones [119].

Some investigations are also available in the literature about other Cu-containing aluminum alloys for SSM processing. Again, the intermetallic compounds were shown to exhibit a cathodic behavior as compared to the α-phase [120].

5. Summary

In the present review paper, the microstructural characteristics of various semi-solid Al alloys are thoroughly summarized, together with the description of the evolution of their typical nondendritic microstructure during solidification. Furthermore, the influence of microstructural features on mechanical properties is systematically analyzed. This is fundamental in order to understand the different performance of SSM parts in comparison with components obtained by conventional production routes. Apart from tensile properties, other important characteristics are also discussed in order to provide a complete overview of the performance of semi-solid Al alloys, such as fatigue behavior, wear, and corrosion resistance.

Conflicts of Interest: The authors declare no conflict of interest.

References

1. Koke, J.; Modigell, M. Flow behaviour of semi-solid metal alloys. *J. Non-Newton. Fluid Mech.* **2003**, *112*, 141–160. [CrossRef]
2. Fan, Z. Semisolid metal processing. *Int. Mater. Rev.* **2002**, *47*, 49–85. [CrossRef]
3. Modigell, M.; Koke, J. Rheological modelling on semi-solid metal alloys and simulation of thixocasting processes. *J. Mater. Process. Technol.* **2001**, *111*, 53–58. [CrossRef]
4. Flemings, M.C. Behavior of metal alloys in the semisolid state. *Metall. Trans. A* **1991**, *22*, 957–981. [CrossRef]
5. Kirkwood, D.H. Semisolid metal processing. *Int. Mater. Rev.* **1994**, *39*, 173–189. [CrossRef]
6. Kapranos, P.; Ward, P.J.; Atkinson, H.V.; Kirkwood, D.H. Near net shaping by semi-solid metal processing. *Mater. Des.* **2000**, *21*, 387–394. [CrossRef]
7. Kiuchi, M.; Kopp, R. Mushy/Semi-Solid Metal Forming Technology—Present and Future. *CIRP Ann.-Manuf. Technol.* **2002**, *51*, 653–670. [CrossRef]
8. Chiarmetta, G. Thixoforming of automobile components. In Proceedings of the Fourth International Conference on Semi-Solid Processing of Alloys and Composites, Sheffield, UK, 19–21 June 1996.
9. Midson, S.P. Industrial applications for aluminum semi-solid castings. *Solid State Phenom.* **2014**, *217–218*, 487–495. [CrossRef]

10. Spencer, D.B.; Mehrabian, R.; Flemings, M.C. Rheological behavior of Sn-15 pct Pb in the crystallization range. *Metall. Trans.* **1972**, *3*, 1925–1932. [CrossRef]

11. Hirt, G.; Kopp, R. *Thixoforming: Semi-Solid Metal Processing*; Wiley-VCH, Verlag GmbH & Co. KGaA: Weinheim, Germany, 2008.

12. Midson, S. Rheocasting processes for semi-solid casting of aluminum alloys. *Die Cast. Eng.* **2006**, *50*, 48–51.

13. Kirkwood, D.H.; Suéry, M.; Kapranos, P.; Atkinson, H.; Young, K.P. *Semi-Solid Processing of Alloys*; Springer: Berlin, Germany, 2010.

14. Nafisi, S.; Ghomashchi, R. *Semi-Solid Processing of Aluminum Alloys*; Springer: Berlin, Germany, 2016.

15. Yurko, J.A.; Martinez, R.A.; Flemings, M.C. Commercial development of the semi-solid rheocasting. In Proceedings of the 22nd International Congress on The North American Die Casting Association (NADCA '03), Indianapolis, IN, USA, 15–18 September 2003.

16. Yurko, J.; Boni, R. Semi-solid rheocasting [SSR™ semi-solid rheocasting]. *Metall. Ital.* **2006**, *98*, 35–41.

17. Wannasin, J.; Martinez, R.A.; Flemings, M.C. A novel technique to produce metal slurries for semi-solid metal processing. *Solid State Phenom.* **2006**, *116–117*, 366–369. [CrossRef]

18. Niedermaier, F.; Langgartner, J.; Hirt, G.; Niedick, I. Horizontal continuous casting of SSM billets. In Proceedings of the Fifth International Conference on Semi-Solid Processing of Alloys and Composites, Golden, CO, USA, 23–25 June 1998.

19. Kenney, M.P.; Evans, G.M.; Farrior, C.P.; Kyonka, A.A.; Koch, K.P. *Semisolid Metal Casting and Forging, Metal Handbook, Vol. 15, "Casting"*; ASM Publication: Des Plaines, IL, USA, 2002.

20. Abramov, V.O.; Abramov, O.V.; Straumal, B.B.; Gust, W. Hypereutectic Al–Si based alloys with a thixotropic microstructure produced by ultrasonic treatment. *Mater. Des.* **1997**, *18*, 323–326. [CrossRef]

21. Pola, A.; Arrighini, A.; Roberti, R. Effect of ultrasounds treatment on alloys for semisolid application. *Solid State Phenom.* **2008**, *141–143*, 481–486. [CrossRef]

22. Giordano, P.; Chiarmetta, G. New rheocasting: A valid alternative to the traditional technologies for the production of automotive suspension parts. In Proceedings of the 8th International Conference on Semi-Solid Processing of Alloys and Composites, S2P 2004, Limassol, Cyprus, 21–23 September 2004.

23. Cardoso, E.; Atkinson, H.V.; Jones, H. Microstructural evolution of A356 during NRC processing. In Proceedings of the 8th International Conference on Semi-Solid Processing of Alloys and Composites, S2P 2004, Limassol, Cyprus, 21–23 September 2004.

24. Haga, T.; Suzuki, S. Casting of aluminum alloy ingots for thixoforming using a cooling slope. *J. Mater. Process. Technol.* **2001**, *118*, 169–172. [CrossRef]

25. Fan, Z.; Bevis, M.J. Semi-solid processing of engineering alloys by a twin-screw rheomoulding process. *Mater. Sci. Eng. A* **2001**, *299*, 210–217. [CrossRef]

26. Granath, O.; Wessén, M.; Cao, H. Porosity reduction possibilities in commercial aluminium A380 and magnesium AM60 alloy components using the rheometal[™] process. In Proceedings of the 4th International Conference High tech Die Casting, Brescia, Italy, 9–10 April 2008.

27. Findon, M.; de Figueredo, A.M.; Apelian, D.; Makhlouf, M.M. Melt mixing approaches for the formation of thixotropic semisolid metal structure. In Proceedings of the 7th International Conference on Semi-Solid Processing of Alloys and Composites, Tsukuba, Japan, 25–27 September 2002.

28. Langlais, J.; Lemieux, A. The SEED technology for semi-solid processing of aluminum alloys: A metallurgical and process overview. *Solid State Phenom.* **2006**, *16*, 472–477. [CrossRef]

29. Vieira, E.A.; Kliauga, A.M.; Ferrante, M. On the formation of spheroidal microstructures in a semi-solid Al–Si alloy by thermomechanical processing. *Scr. Mater.* **2007**, *57*, 1165–1168. [CrossRef]

30. Tzimas, E.; Zavaliangos, A. Evolution of near-equiaxed microstructure in the semisolid state. *Mater. Sci. Eng. A* **2000**, *289*, 228–240. [CrossRef]

31. Atkinson, H.V. Modelling the semisolid processing of metallic alloys. *Prog. Mater. Sci.* **2005**, *50*, 341–412. [CrossRef]

32. Mohammed, M.N.; Omar, M.Z.; Salleh, M.S.; Alhawari, K.S.; Kapranos, P. Semisolid metal processing techniques for nondendritic feedstock production. *Sci. World J.* **2013**, *2013*, 752175. [CrossRef] [PubMed]

33. Nafisi, S.; Ghomashchi, R. Semi-solid metal processing routes: An overview. *Can. Metall. Q.* **2005**, *44*, 289–304. [CrossRef]

34. Vogel, A.; Doherty, R.D.; Cantor, B. Stir-cast microstructure and slow crack growth. In *Solidification and Casting of Metals: Proceedings of an International Conference on Solidification*; Metals Society: London, UK, 1979; p. 518.

35. Doherty, R.D.; Lee, H.-I.; Feest, E.A. Microstructure of stir-cast metals. *Mater. Sci. Eng.* **1984**, *65*, 181–189. [CrossRef]

36. Loué, W.R.; Suéry, M. Microstructural evolution during partial remelting of Al-Si7Mg alloys. *Mater. Sci. Eng. A* **1995**, *203*, 1–13. [CrossRef]

37. Molenaar, J.M.M.; Katgerman, L.; Kool, W.H.; Smeulders, R.J. On the formation of the stircast structure. *J. Mater. Sci.* **1986**, *21*, 389–394. [CrossRef]

38. Mullis, A.M. Growth induced dendritic bending and rosette formation during solidification in a shearing flow. *Acta Mater.* **1999**, *47*, 1783–1789. [CrossRef]

39. Hellawell, A. Grain evoluation in conventional rheocasting. In Proceedings of the 4th International Conference on Semi-Solid Processing of Alloys and Composites, Sheffield, UK, 19–21 June 1996.

40. Flemings, M.C.; Yurko, J.A.; Martinez, R.A. Solidification processes and microstructures. In Proceedings of the TMS Annual Meeting, Charlotte, NC, USA, 14–18 March 2004.

41. Das, A.; Ji, S.; Fan, Z. Morphological development of solidification structures under forced fluid flow: A Monte-Carlo simulation. *Acta Mater.* **2002**, *50*, 4571–4585. [CrossRef]

42. Pola, A.; Arrighini, A.; Roberti, R. Ultrasounds: A new technology for alloys degassing, grain refinement and obtainment of a thixotropic structure. In Proceedings of the International Conference on Aluminium Alloys (ICAA), Aachen, Germany, 22–26 September 2008.

43. Campbell, J. *Castings*; Butterworth-Heinemann: Oxford, UK, 2003; Chapter 9.

44. Kapranos, P.; Haga, T.; Bertoli, E.; Pola, A.; Azpilgain, Z.; Hurtado, I. Thixo-extrusion of 5182 aluminium alloy. *Solid State Phenom.* **2008**, *141–143*, 115–120. [CrossRef]

45. Apelian, D. Semi-Solid Processing Routes and Microstructure Evolution. In Proceedings of the Seventh International Conference titled Advanced Semi-Solid Processing of Alloys and Composites, Tsukuba, Japan, 24–28 September 2002; pp. 25–30.

46. Suery, M. Microstructure of semi-solid alloys and properties. In Proceedings of the 8th International Conference on Semi-Solid Processing of Alloys and Composites, S2P 2004, Limassol, Cyprus, 21–23 September 2004.

47. Suéry, M. *Mise en Forme des Alliages Métalliques a l'État Semi Solide*; Hermes Science Publications: Paris, France, 2002.

48. Terzi, S.; Salvo, L.; Suery, M.; Dahle, A.; Boller, E. In situ microtomography investigation of microstructural evolution in Al-Cu alloys during holding in semi-solid state. *Trans. Nonfeer. Met. Soc. China* **2010**, *20*, s734–s738. [CrossRef]

49. Modigell, M.; Pola, A.; Suéry, M.; Zang, C. Investigation of correlations between shear history and microstructure of semi-solid alloys. *Solid State Phenom.* **2013**, *192–193*, 251–256. [CrossRef]

50. Nafisi, S.; Ghomashchi, R. Combined grain refining and modification of conventional and rheo-cast A356 Al-Si alloy. *Mater. Charact.* **2006**, *57*, 371–385. [CrossRef]

51. Limodin, N.; Salvo, L.; Suéry, M.; DiMichiel, M. In situ investigation by X-ray tomography of the overall and local microstructural changes occurring during partial remelting of an Al–15.8 wt.% Cu alloy. *Acta Mater.* **2007**, *55*, 3177–3191. [CrossRef]

52. Nafisi, S.; Ghomashchi, R. The microstructural characterization of semi-solid slurries. *JOM* **2006**, *58*, 24–30. [CrossRef]

53. Pola, A.; Roberti, R.; Frerini, F. Microstructure and mechanical behaviour of cast aluminium components obtained by thixocasting and traditional processes. In Proceedings of the 8th International Conference on Semi-Solid Processing of Alloys and Composites, S2P 2004, Limassol, Cyprus, 21–23 September 2004.

54. Chen, Z.; Mao, W.; Wu, Z. Mechanical properties and microstructures of Al alloy tensile samples produced by serpentine channel pouring rheo-diecasting process. *Trans. Nonfeer. Met. Soc.* **2011**, *21*, 1473–1479. [CrossRef]

55. Jamaati, R.; Amirkhanlou, S.; Toroghinejad, M.R.; Niroumand, B. Significant improvement of semi-solid microstructure and mechanical properties of A356 alloy by ARB process. *Mater. Sci. Eng. A* **2011**, *528*, 2495–2501. [CrossRef]

56. Shabestari, S.G.; Parshizfard, E. Effect of semi-solid forming on the microstructure and mechanical properties of the iron containing Al–Si alloys. *J. Alloy. Compd.* **2011**, *509*, 7973–7978. [CrossRef]

57. Wu, S.; Lu, S.; An, P.; Nakae, H. Microstructure and property of rheocasting aluminum-alloy made with indirect ultrasonic vibration process. *Mater. Lett.* **2012**, *73*, 150–153. [CrossRef]

58. Zhao, J.-W.; Wu, S.S. Microstructure and mechanical properties of rheo-diecasted A390 alloy. *Trans. Nonfeer. Met. Soc.* **2010**, *20*, 754–757. [CrossRef]

59. Cerri, E.; Evangelista, E.; Spigarelli, S.; Cavaliere, P.; DeRiccardis, F. Effects of thermal treatments on microstructure and mechanical properties in a thixocast 319 aluminum alloy. *Mater. Sci. Eng. A* **2000**, *284*, 254–260. [CrossRef]

60. Dai, W.; Wu, S.; Lu, S.; Lin, C. Effects of rheo-squeeze casting parameters on microstructure and mechanical properties of AlCuMnTi alloy. *Mater. Sci. Eng. A* **2012**, *538*, 320–326. [CrossRef]

61. Jiang, H.; Lu, Y.; Huang, H.; Li, X.; Li, M. Microstructural evolution and mechanical properties of the semisolid Al-4Cu-Mg alloy. *Mater. Charact.* **2003**, *51*, 1–10.

62. Xu, C.; Zhao, J.; Guo, A.; Li, H.; Dai, G.; Zhang, X. Effects of injection velocity on microstructure, porosity and mechanical properties of a rheo-diecast Al-Zn-Mg-Cu aluminum alloy. *J. Mater. Process. Technol.* **2017**, *249*, 167–171. [CrossRef]

63. Alipour, M.; Aghdam, B.G.; Rahnoma, H.E.; Emamy, M. Investigation of the effect of Al–5Ti–1B grain refiner on dry sliding wear behavior of an Al–Zn–Mg–Cu alloy formed by strain-induced melt activation process. *Mater. Des.* **2013**, *46*, 766–775. [CrossRef]

64. Bergsma, S.C.; Li, X.; Kassner, M.E. Semi-solid thermal transformations in Al–Si alloys: II. The optimized tensile and fatigue properties of semi-solid 357 and modified 319 aluminum alloys. *Mater. Sci. Eng. A* **2001**, *297*, 69–77. [CrossRef]

65. Cerri, E.; Cabibbo, M.; Cavaliere, P.; Evangelista, E. Mechanical behaviour of 319 heat treated thixo cast bars. *Mater. Sci. Forum* **2000**, *331*, 259–264. [CrossRef]

66. Zhu, Q.; Midson, S.P. Semi-solid moulding: Competition to cast and machine from forging in making automotive complex components. *Trans. Nonfeer. Met. Soc. China* **2010**, *20*, s1042–s1047. [CrossRef]

67. Haga, T.; Kapranos, P. Simple rheocasting processes. *J. Mater. Process. Technol.* **2002**, *130–131*, 594–598. [CrossRef]

68. Burapa, R.; Janudom, S.; Chucheep, T.; Canyook, R.; Wannasin, J. Effects of primary phase morphology on mechanical properties of Al-Si-Mg-Fe alloy in semi-solid slurry casting process. *Trans. Nonfeer. Met. Soc. China* **2010**, *20*, 857–861. [CrossRef]

69. Lu, S.; Wu, S.; Zhu, Z.; An, P.; Mao, Y. Effect of semi-solid processing on microstructure and mechanical properties of 5052 aluminum alloy. *Trans. Nonfeer. Met. Soc. China* **2010**, *20*, 758–762. [CrossRef]

70. Brochu, M.; Verreman, Y.; Ajersch, F.; Bucher, L. Fatigue Behavior of Semi-Solid Cast Aluminum: A Critical Review. *Solid State Phenom.* **2008**, *141*, 725–730. [CrossRef]

71. Fan, Z.; Fang, X.; Ji, S. Microstructure and mechanical properties of rheo-diecast (RDC) aluminium alloys. *Mater. Sci. Eng. A* **2005**, *412*, 298–306. [CrossRef]

72. Lemieux, A.; Langlais, J.; Bouchard, D.; Grant Chen, X. Effect of Si, Cu and Fe on mechanical properties of cast semi-solid 206 alloys. *Trans. Nonfeer. Met. Soc. China* **2010**, *20*, 1555–1560. [CrossRef]

73. Wu, S.; Zhong, G.; Wan, L.; An, P.; Mao, Y. Microstructure and properties of rheo-diecast Al-20Si-2Cu-1Ni-0.4Mg alloy with direct ultrasonic vibration process. *Trans. Nonfeer. Met. Soc. China* **2010**, *20*, 763–767. [CrossRef]

74. Zheng, Z.-K.; Ji, Y.-J.; Mao, W.-M.; Yue, R.; Liu, Z.-Y. Influence of rheo-diecasting processing parameters on microstructure and mechanical properties of hypereutectic Al–30%Si alloy. *Trans. Nonfeer. Met. Soc. China* **2017**, *27*, 1264–1272. [CrossRef]

75. Kapranos, P.; Kirkwood, D.H.; Atkinson, H.V.; Rheinlander, J.T.; Bentzen, J.J.; Toft, P.T.; Debel, C.P.; Laslaz, G.; Maenner, L.; Blais, S.; et al. Thixoforming of an automotive part in A390 hypereutectic Al–Si alloy. *J. Mater. Process. Technol.* **2003**, *135*, 271–277. [CrossRef]

76. Taylor, J.A. Iron-containing intermetallic phases in Al-Si based casting alloys. *Procedia Mater. Sci.* **2012**, *1*, 19–33. [CrossRef]

77. Di Giovanni, M.T.; Cerri, E.; Casari, D.; Merlin, M.; Arnberg, L.; Garagnani, G.L. The Influence of Ni and V Trace Elements on High-Temperature Tensile Properties and Aging of A356 Aluminum Foundry Alloy. *Metall. Mater. Trans. A* **2016**, *47*, 2049–2057. [CrossRef]

78. Shabestari, S.G. The effect of iron and manganese on the formation of intermetallic compounds in aluminum–silicon alloys. *Mater. Sci. Eng. A* **2004**, *383*, 289–298. [CrossRef]

79. Tocci, M.; Pola, A.; Raza, L.; Armellin, L.; Afeltra, U. Optimization of heat treatment parameters for a nonconventional Al-Si-Mg alloy with cr addition by DoE method. *Metall. Ital.* **2016**, *108*, 141–144.
80. Ceschini, L.; Boromei, I.; Morri, A.; Seifeddine, S.; Svensson, I.L. Effect of Fe content and microstructural features on the tensile and fatigue properties of the Al-Si10-Cu2 alloy. *Mater. Des.* **2012**, *36*, 522–528. [CrossRef]
81. Moller, H.; Stumpf, W.E.; Pistorius, P.C. Influence of elevated Fe, Ni and Cr levels on tensile properties of SSM-HPDC Al-Si-Mg alloy F357. *Trans. Nonfeer. Met. Soc. China* **2010**, *20*, 842–846. [CrossRef]
82. Qi, M.F.; Kang, Y.L.; Zhu, G.M. Microstructure and properties of rheo-HPDC Al-8Si alloy prepared by air-cooled stirring rod process. *Trans. Nonfeer. Met. Soc. China* **2017**, *27*, 1939–1946. [CrossRef]
83. Bo, L.; Jun, H.H.; Guang, Y.X. The heat treatment response of semisolid formed Al–5Fe–4Cu alloy. *Philos. Mag.* **2017**, *98*, 605–622. [CrossRef]
84. Chen, G.; Chen, Q.; Wang, B.; Du, Z.-M. Microstructure evolution and tensile mechanical properties of thixoformed high performance Al-Zn-Mg-Cu alloy. *Met. Mater. Int.* **2015**, *21*, 897–906. [CrossRef]
85. Zhao, J.; Xu, C.; Dai, G.; Wu, S.; Han, J. Microstructure and properties of rheo-diecasting wrought aluminum alloy with Sc additions. *Mater. Lett.* **2016**, *173*, 22–25. [CrossRef]
86. Hayat, N.; Toda, H.; Kobayashi, T.; Wade, N. Experimental investigations of fatigue characteristics of AC4CH cast aluminum alloys fabricated through rheocast and squeeze cast methods. *Mater. Sci. Forum* **2002**, *396*, 1353–1358. [CrossRef]
87. Park, C.; Kim, S.; Kwon, Y.; Lee, Y.; Lee, J. Fracture behavior of thixoformed 357-T5 Al alloys. *Metall. Mater. Trans. A* **2004**, *35*, 1017–1027. [CrossRef]
88. Davidson, C.J.; Griffiths, J.R.; Badiali, M.; Zanada, A. Fatigue properties of a semi-solid cast Al–7Si–0.3 Mg–T6 alloy. *Metall. Sci. Technol.* **2000**, *18*, 27–31.
89. Ragab, K.A.; Bouazara, M.; Bouaicha, A.; Mrad, H. Materials performance and design analysis of suspension lower-arm fabricated from Al-Si-Mg castings. *Key Eng. Mater.* **2016**, *710*, 315–320. [CrossRef]
90. Blad, M.; Johannesson, B.; Nordberg, P.; Winklhofer, J. Manufacturing and fatigue verification of two different components made by semi-solid processing of aluminium TX630 alloy. *Solid State Phenom.* **2016**, *256*, 328–333. [CrossRef]
91. Alain, A.A.; Myriam, B.; Heinrich, M. Effect of the rheocasting process and of the SLS layer on the fatigue behavior of 357 aluminum alloy. *Solid State Phenom.* **2014**, *217*, 227–234.
92. Gan, Y.X.; Overfelt, R.A. Fatigue property of semisolid A357 aluminum alloy under different heat treatment conditions. *J. Mater. Sci.* **2006**, *41*, 7537–7544. [CrossRef]
93. Rosso, M.; Peter, I.; Villa, R. Effects of T5 and T6 heat treatments applied to rheocast A356 parts for automotive applications. *Solid State Phenom.* **2008**, *141*, 237–242. [CrossRef]
94. Bergsma, S.C.; Kassner, M.E.; Evangelista, E.; Cerri, E. The optimized tensile and fatigue properties of electromagnetically stirred and thermally transformed semi-solid 357 and modified 319 aliminum alloys. In Proceedings of the 6th International Conference on Semi-Solid Processing of Alloys and Composites, S2P 2000, Turin, Italy, 27–29 September 2000; pp. 319–324.
95. Bouazara, M.; Bouaicha, A.; Ragab, K.A. Fatigue Characteristics and Quality Index of A357 Type Semi-Solid Aluminum Castings Used for Automotive Application. *J. Mater. Eng. Perform.* **2015**, *24*, 3084–3092. [CrossRef]
96. Ragab, K.A.; Bouazara, M.; Bouaicha, A.; Allaoui, O. Microstructural and mechanical features of aluminium semi-solid alloys made by rheocasting technique. *Mater. Sci. Technol. (UK)* **2017**, *33*, 646–655. [CrossRef]
97. Yu, Y.; Kim, S.; Lee, Y.; Lee, J. Phenomenological Observations on Mechanical and Corrosion Properties of Thixoformed 357 Alloys: A Comparison with Permanent Mold Cast 357 Alloys. *Metall. Mater. Trans. A* **2002**, *33*, 1339–1412. [CrossRef]
98. Brochu, M.; Verreman, Y.; Ajersch, F.; Bouchard, D. High cycle fatigue strength of permanent mold and rheocast aluminum 357 alloy. *Int. J. Fatigue* **2010**, *32*, 1233–1242. [CrossRef]
99. Park, C.; Kim, S.; Kwon, Y.; Lee, Y.; Lee, J. Mechanical and corrosion properties of rheocast and low-pressure cast A356-T6 alloy. *Mater. Sci. Eng. A* **2005**, *391*, 86–94. [CrossRef]
100. Lados, D.A.; Apelian, D. Fatigue crack growth mechanisms during dynamic loading of conventionally and SSM cast aluminum components. In Proceedings of the 8th International Conference on Semi-Solid Processing of Alloys and Composites, S2P 2004, Limassol, Cyprus, 21–23 September 2004; pp. 833–842.
101. Kaufman, G.; Rooy, E.L. *Aluminum Alloy Castings Properties, Processes, and Applications*; ASM International: Materials Park, OH, USA, 2004; p. 15.

102. Dey, A.K.; Poddar, P.; Singh, K.K.; Sahoo, K.L. Mechanical and wear properties of rheocast and conventional. *Mater. Sci. Eng. A* **2006**, *435–436*, 521–529. [CrossRef]

103. Bayoumi, M.A.; Negm, M.I.; El-Gohry, A.M. Microstructure and mechanical properties of extruded Al-Si alloy (A356) in the semi-solid state. *Mater. Des.* **2009**, *30*, 4469–4477. [CrossRef]

104. Vencl, A.; Bobic, I.; Miskovic, Z. Effect of thixocasting and heat treatment on the tribological properties of hypoeutectic Al–Si alloy. *Wear* **2008**, *264*, 616–623. [CrossRef]

105. Midson, S.; Keist, J.; Svare, J. Semi-Solid Metal Processing of Aluminum Alloy A390. In Proceedings of the SAE 2002 World Congress, Detroit, MI, USA, 4–7 March 2002.

106. Birol, Y.; Birol, F. Wear properties of high-pressure die cast and thixoformed aluminium alloys for connecting rod applications in compressors. *Wear* **2008**, *265*, 590–597. [CrossRef]

107. Pola, A.; Montesano, L.; Gelfi, M.; Roberti, R. Semisolid processing of Al-Sn-Cu alloys for bearing applications. *Solid State Phenom.* **2012**, *192*, 562–568. [CrossRef]

108. Alhawari, K.S.; Omar, M.Z.; Ghazali, M.J.; Salleh, M.S.; Mohammed, M.N. Evaluation of the microstructure and dry sliding wear behaviour of thixoformed A319 aluminium alloy. *Mater. Des.* **2015**, *76*, 169–180. [CrossRef]

109. Alhawari, K.S.; Omar, M.Z.; Ghazali, M.J.; Salleh, M.S.; Mohammed, M.N. Dry sliding wear behaviour of thixoformed hypoeutectic Al-Si-Cu alloy with different amounts of magnesium. *Compos. Interfaces* **2016**, *23*, 519–531. [CrossRef]

110. Pola, A.; Montesano, L.; Sinagra, C.; Gelfi, M.; La Vecchia, G.M. Influence of globular microstructure on cavitation erosion resistance of aluminium alloys. *Solid State Phenom.* **2016**, *256*, 51–57. [CrossRef]

111. Pola, A.; Montesano, L.; Tocci, M.; La Vecchia, G.M. Influence of Ultrasound Treatment on Cavitation Erosion Resistance of AlSi7 Alloy. *Materials* **2017**, *10*, 256. [CrossRef] [PubMed]

112. Bastidas, J.M.; Forn, A.; Baile, M.T.; Polo, J.L. Pitting corrosion of A357 aluminium alloy. *Mater. Corros.* **2001**, *52*, 691–696. [CrossRef]

113. Arrabal, R.; Mingo, B.; Pardo, A.; Mohedano, M.; Matykina, E.; Rodríguez, I. Pitting corrosion of rheocast A356 aluminium alloy in 3.5wt.% NaCl solution. *Corros. Sci.* **2013**, *73*, 342–355. [CrossRef]

114. Tahamtan, S.; Boostani, A.F. Quantitative analysis of pitting corrosion behavior of thixoformed A356 alloy in chloride medium using electrochemical techniques. *Mater. Des.* **2009**, *30*, 2483–2489. [CrossRef]

115. Tahamtan, A.; Tahamtan, S.; Boostani, A.F.; Nazemi, H. Pitting corrosion of thixoformed, rheocast and gravity cast A356-T6 alloy in chloride media. *Corros. Eng. Sci. Technol.* **2009**, *44*, 384–388. [CrossRef]

116. Arrighini, A.; Gelfi, M.; Pola, A.; Roberti, R. Effect of ultrasound treatment of AlSi5 liquid alloy on corrosion resistance. *Mater. Corros.* **2010**, *61*, 218–221. [CrossRef]

117. Masuku, E.P.; Möller, H.; Curle, U.A.; Pistorius, P.C.; Li, W. Influence of surface liquid segregation on corrosion behavior of semi-solid metal high pressure die cast aluminium alloys. *Trans. Nonfeer. Met. Soc. China* **2010**, *20*, s837–s841. [CrossRef]

118. Möller, H.; Masuku, E.P. The Influence of Liquid Surface Segregation on the Pitting Corrosion Behavior of SemiSolid Metal High Pressure Die Cast Alloy F357. *Open Corros. J.* **2009**, *2*, 216–220. [CrossRef]

119. Mingo, B.; Arrabal, R.; Pardo, A.; Matykina, E.; Skeldon, P. 3D study of intermetallics and their effect on the corrosion morphology of rheocast aluminium alloy. *Mater. Charact.* **2016**, *112*, 122–128. [CrossRef]

120. Forn, A.; Rupérezu, E.; Baile, M.T.; Campillo, M.; Menargues, S.; Espinosa, I. Corrosion behaviour of A380 aluminiun alloy by semi-solid rheocasting. In Proceedings of the 10 ESAFORM Conference on Material Forming, Zaragoza, Spain, 18–20 April 2007.

Article

A Comparison between Anodizing and EBSD Techniques for Primary Particle Size Measurement

Shahrooz Nafisi [1,2,3,*], **Anthony Roccisano** [1], **Reza Ghomashchi** [1] **and George Vander Voort** [4]

[1] School of Mechanical Engineering, The University of Adelaide, Adelaide, SA 5005, Australia;
   a.roccisano@adelaide.edu.au (A.R.); reza.ghomashchi@adelaide.edu.au (R.G.)
[2] Department of Chemical and Materials Engineering, University of Alberta, Edmonton, AB T6G 1H9, Canada
[3] Consolidated Metco, Vancouver, WA 98661, USA
[4] Consultant—Struers Inc., Wadsworth, IL 60083, USA; george@georgevandervoort.com
* Correspondence: shahrooznafisi@gmail.com

Received: 26 March 2019; Accepted: 24 April 2019; Published: 27 April 2019

Abstract: Proper understanding and knowledge of primary particle or grain size is of paramount importance in manufacturing processes as it directly affects various properties including mechanical behavior. Application of optical microscopy coupled with etching techniques has been used conventionally and in conjunction with color metallography (polarized microscopy) has been the preferred method for grain size measurement. An advanced technique as an alternative to light microscopy is using electron backscatter diffraction (EBSD). A comparison is made between these two techniques using Al-7Si alloy produced with various casting techniques to highlight the cost and time of the sample preparation and analysis for both techniques. Results showed that color metallography is certainly a faster technique with great accuracy and a much cheaper alternative in comparison with EBSD.

Keywords: polarized light microscopy; anodic etching; EBSD; grain; globule; Al-Si alloy; semi-solid metal processing; EMS; thixocasting

---

1. Introduction

Grain size plays an important role on the properties of metallic materials. Normally, the structure is observed on the plane of the polished surface using optical microscopy which is a two-dimensional (2D) measurement. The 2D analysis is not a complete representative of the 3D structure and in some instances may lead to a biased conclusion. However, it is the current practice for microstructural characterization and therefore there is an urge to ensure the procedure is as effective and accurate as possible. An example is shown in Figure 1 highlighting the deficiency of 2D analysis. The secondary dendritic branches are treated as individual and isolated particles during automatic image analysis processing [1].

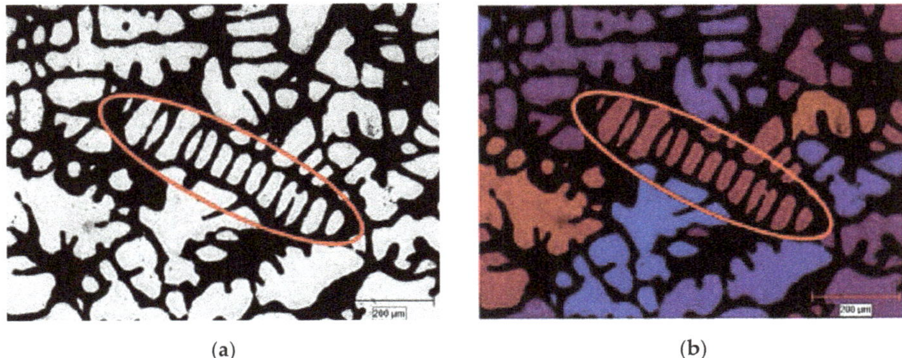

Figure 1. Branches of dendrites, A356 alloy, quenched from 598 °C [1]: (**a**) bright field, and (**b**) polarized light image with sensitive tint plate, polished and anodized with Barker's reagent.

Various techniques have been used to overcome this inadequacy including advanced serial sectioning [2,3], X-ray tomography [4,5], and analysis of crystallographic orientations of the particles by electron backscatter diffraction (EBSD) technique (either 2D or 3D) [6,7]. The competing issues in using any of the aforementioned methods are the required equipment, the operator's skill and the analysis time which are eventually interpreted in terms of the cost of these methods. This is the theme of the current report; the accuracy, the cost (equipment), and time of two widely used methods of "anodic etching-polarized light microscopy" and EBSD in the analysis of semi solid metal (SSM) processed Al-Si alloy structure. The distinction between the individual grains and particles is important for SSM processes as it not only verifies the effectiveness of SSM process but also greatly affects the finished SSM parts mechanical properties.

Anodic oxidation, or anodizing, is an electrolytic process for depositing an oxide film on the metal surface epitaxial to the underlying grain structure. It is similar in nature to heat tinting or tint etching where an interference film is produced on the surface of metals. In this method, the sample is placed in the anodizing solution as the anode connected to a stainless steel or graphite plate or bar acting as the cathode. The resulting interference film colors when viewed under a polarizing light are a function of the anodic film thickness. The thickness, however, depends on the anodizing voltage, the anodizing solution composition, and the composition and/or structures of the phases present in the specimens and anodizing time. There are plenty of reports in open literature noting that certain etchants for Body Centered Cubic (BCC) and Face Centered Cubic (FCC) metals produce an interference film on the grain which results in grain contrast effect in bright field light microscopy. When viewed in polarized light, it yields color images that can be further enhanced by adding a sensitive tint plate [8,9].

EBSD is a technique that provides crystallographic information by analyzing crystalline samples in the scanning electron microscope (SEM). In EBSD, a stationary electron beam strikes a 70 degrees tilted sample and the diffracted electrons form a pattern on a fluorescent screen. The diffraction pattern is unique to the crystal structure and crystal orientation of the sample region from which it was generated. Diffraction patterns are used to measure the crystal orientation, grain boundary misorientations, discriminate between different materials, and to provide useful information about local crystalline perfection. By scanning in a grid across a polycrystalline sample and measuring the crystal orientation at each point, the resulting map will reveal the constituent grain morphology, orientations, and boundaries. In addition, the data shows the preferred crystal orientations (texture) present within the material [10,11].

Color metallography techniques have been developed for many metals and alloys, but they have not been utilized widely by metallographers, despite their obvious benefits. Color etchants are usually phase-specific and they will fully reveal the grain structure or specific second-phase constituents. Some, like Barker's anodizing solution, produce results that can only be observed using cross polarized light. Coloration in all cases, if weak, can be enhance by adding a sensitive tint filter (sometimes called a lambda plate). The coloration is due to variations in the crystallographic orientation of the grains or particles. Black and white etchants cannot reveal such differences. EBSD is commonly used to reveal crystallographic orientations between grains and phases, but the process requires an SEM equipped with an EBSD system. Coarser step sizes between diffraction patterns reduce the precision of the method, although it shortens the scan time. Specimen preparation for EBSD is much more challenging for most laboratories, and more time consuming. All preparation-induced damage must be removed to get high-quality EBSD grain maps, and the specimen surfaces must remain perfectly flat [12,13]. So, the use of color metallography by anodizing Al and its alloys, followed by examination using polarized light and a sensitive tint filter, is much faster, easier, and less expensive than EBSD.

This article attempts to highlight the effectiveness of employing polarized light microscopy in conjunction with anodizing, in revealing the individual grains or primary particles in SSM processed Al-Si alloy. The distribution of grains and particles are then compared with the grain size obtained from EBSD technique to emphasize the simplicity and low cost of color metallography. As a result of such practice, it was managed to identify the critical degrees of misorientation as the required criterion to differentiate between grains and subgrains in SSM processed alloys when using EBSD, which in its own right is an important finding by itself.

2. Materials and Methods

Binary Al-7% Si alloys (6.7–6.9% Si, 0.8–0.81% Fe) were prepared in an electric resistance furnace. Two different molds were used; for higher cooling rate, a copper mold with a water-cooled jacket was used and for the lower cooling rate, a CO_2 bonded silica-sand mold was employed. Ingots were 76-mm in diameter and 300-mm long. The entire configuration was placed in an electromagnetic stirring machine (EMS). The frequency was set to 50 Hz and the current was 100 and 30 A for copper and sand molds, respectively (the application of magnetic field was stopped when the alloy temperature in the molds reached 400 °C on cooling). The application of different currents was intended to ensure that the applied magnetic force to stir the molten alloy is almost the same for both the sand and copper molds. Experimental details are explained in further detail elsewhere [14]. Pouring temperature was changed between 630 and 690 °C. The cooling rate in the copper and sand molds for the conventional ingot (with no stirring) was about 4.8 and 3.3 °C/s, respectively (the cooling rates were calculated in the liquid state above the liquidus temperature).

For thixocast trials (reheating to semi-solid region), samples were cut from the transverse sections (200 mm from the bottom of EMS billets), in areas between the billet center and wall, and were reheated in a single coil 5 kW induction furnace operating at 80 kHz. Samples were placed vertically on an insulator plate. Temperature variation during the tests was monitored by attaching thermocouples to both the billet center and the wall. The induction furnace was controlled by the central thermocouple and the wall thermocouple was used to establish if there is any transverse temperature gradient within the billet. The reheating cycle included 2–3 min of heating up to 583 ± 3 °C and 10 min holding time at this temperature before water quenching (at the selected holding temperature of 583 ± 3 °C, there is about 38–40% fraction of solid alloy according to ThermoCalc calculations). For EMS cast samples, the metallographic specimens were cut transversely at 200-mm from the bottom of the billets, mounted in Bakelite, ground, and polished conventionally down to 0.05 μm colloidal silica to develop high quality EBSD (Struers, Willich, Germany) maps. Thixocast samples were prepared simply by cutting a transverse section of the quenched samples.

EBSD analysis was undertaken with an FEI Inspect F50 Field Emission Gun (FEG) Scanning Electron Microscope (SEM) with an EDAX Hikari detector. In each specimen at least four regions were located using a microhardness indent and were scanned through EBSD. At least four regions per specimen were scanned at 200X magnification, with a scan frame of 800×800 μm^2 and a step size of 1.5 μm taking approximately 45 min to scan. In addition, each specimen had a region scanned at 100×

magnification with a scan frame of 1600×1600 μm^2 and a step size of 2 μm^2 taking approximately 90 min to scan. After the EBSD analysis, the samples were anodized. Samples were immersed in a solution of Barkers reagent (1.8% fluoroboric acid in water) and anodized until a color shift was observed under polarized light microscopy (20 V direct current, 4 min). Polarized light microscopy was undertaken on a Zeiss AXIO Imager.M2m (Carl Zeiss Microscopy, Jena, Germany) equipped with reflected light polarizer and, rotatable analyzer with lambda plate. Regions that were scanned with EBSD were subsequently viewed under polarized light microscopy and the grain boundary properties determined under each were compared.

The acquisition time for color metallography and EBSD is provided in Table 1 indicating the approximate time spent acquiring grain boundary data. Sample preparation time is typically 30 min longer for EBSD samples due to the increased colloidal silica polishing time required to produce strong patterns and the stricter cleaning regime required to produce good scans. The EBSD acquisition time is heavily dependent upon the available equipment and the resolution of the scan with the newer detectors offering shorter scan times for the same resolution, however, there are extended waiting periods in all models from pumping and venting the chamber as well as locating the scan area. In addition to the above time allocation to acquire the desired outcome, it is important to point out that the level of skill to run the EBSD is far more demanding than the one for anodizing and optical microscopy. On top of that, the capital investment is nowhere comparable.

Table 1. Time consumption acquiring grain boundary data through color metallography and electron backscatter diffraction (EBSD).

Process Parameter	Color Metallography (minutes)	EBSD (minutes)
Sample Preparation	20	50
Anodizing	10 (per sample)	-
Optical Microscopy	10 (per sample)	-
EBSD	-	75 (per scan)
Post-processing of Results	-	15 (per scan)
Total	40	140

3. Results and Discussions

Generally, aluminum hypoeutectic alloys consist of two main constituents, primary α-Al particles and eutectic mixture of α-Al and Si (in this specific alloy, due to the extra addition of iron, some β-iron intermetallics were formed). Effects of different cooling rates on the silicon and iron flake sizes and distribution in the as-cast EMS samples are shown in Figure 2. Lower cooling rate in the sand mold shows larger dendrites comparing to that of the copper mold specimens. Dendrite size had a great difference since the cooling rate is higher in the copper mold (Figure 2c,d). At higher magnifications, a great difference could be seen in the size of the silicon eutectic flakes and β–iron intermetallics. By pouring the alloy in the copper mold, i.e., with a higher cooling rate, silicon and β–iron intermetallics got thinner and shorter (compare Figure 2b with Figure 2d).

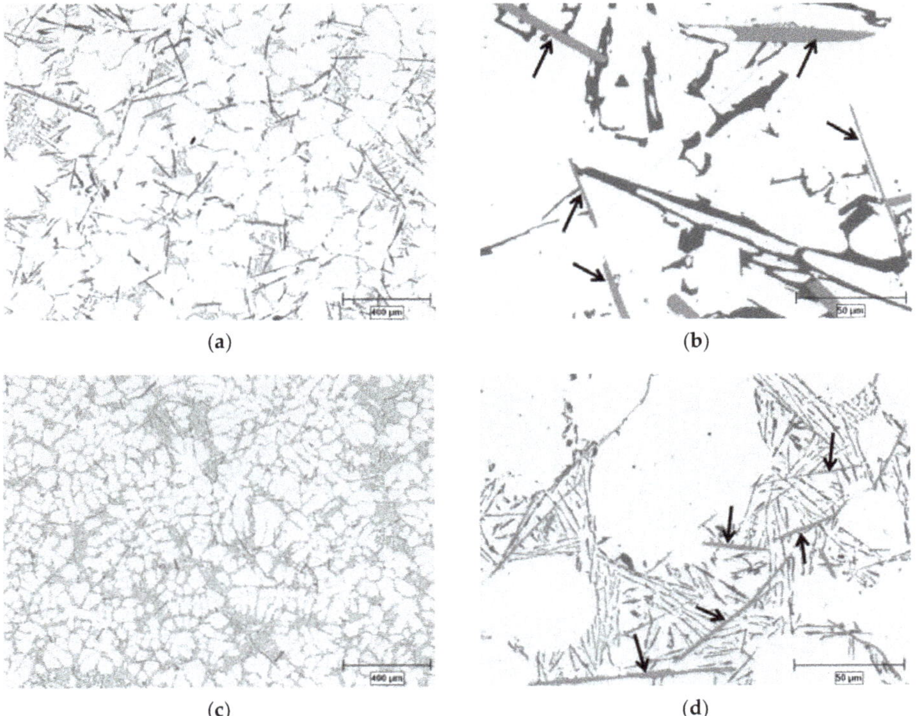

(a)   (b)

(c)   (d)

Figure 2. Optical micrographs showing the effect of cooling rate on the microstructure of as-cast magnetic stirring machine (EMS) billets: (**a**,**b**) sand mold and (**c**,**d**) copper mold, pouring temperature 690 °C, etched with 0.5% Hydrofluoric acid (HF) (arrows show some of the iron-intermetallics). Scale bar is 400 μm for (**a**) and (**c**), 50 μm for (**b**) and (**d**).

During isothermal holding, the eutectic mixture was re-melted while the primary α-Al phase was coarsened. There was also a driving force towards reduction of interfacial area which results in globularization of the primary α-Al particles (Figure 3).

In order to discuss the correlation between EBSD and anodizing, it is better to examine the thixocast structure first which, due to the better clarity of individual particles, makes comparison much easier and convenient than for the EMS which the particles sometimes are intertwined.

Figure 4 shows an example of color metallography and EBSD of the same area of the sample cast in copper mold at 630 °C and isothermally held at ~583 °C for 10 min. As discussed earlier, by isothermal holding, primary α–Al particles became spherical and some of the primary particles possibly sintered together. The colors in the polarized light micrograph, Figure 4a, were indications of different orientations of the primary particles while a similar color indicates the same nucleation and growth pattern. In this specific example, particles K–L, J–M, M–N, O–P, T–X, X–Y, and Z–O possibly sintered, i.e., two or more isolated solid particles joined and formed a pseudo cluster that is associated mainly to the stabilized contact during post heat treatment processing [1]. In the semi-solid science, this phenomenon is defined as "agglomeration" [15–17] where primary particles come into contact and possibly sinter together to form agglomerates.

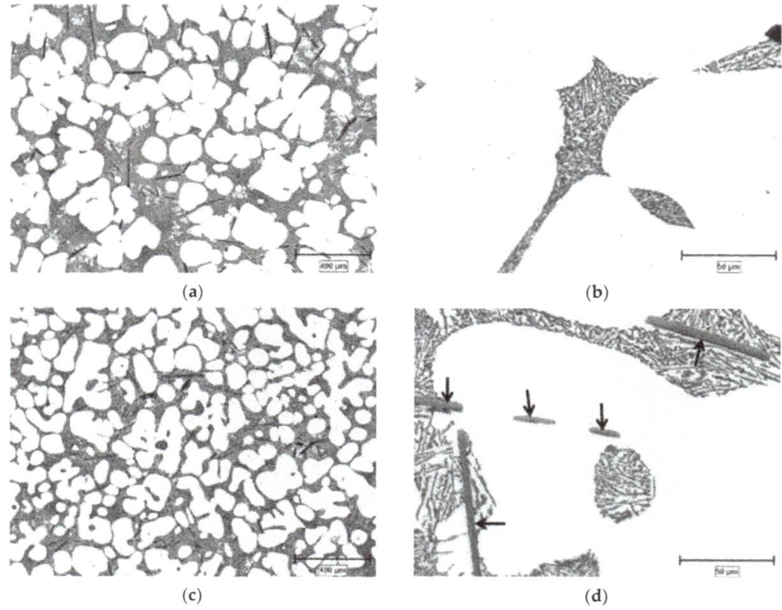

Figure 3. Optical micrographs showing the effect of isothermal holding on the microstructure of thixocast samples: (**a**,**b**) sand mold and (**c**,**d**) copper mold, pouring temperature 690 °C, etched with 0.5% HF (arrows show some of the iron-intermetallics). Scale bar is 400 µm for (**a**,**c**), 50 µm for (**b**,**d**).

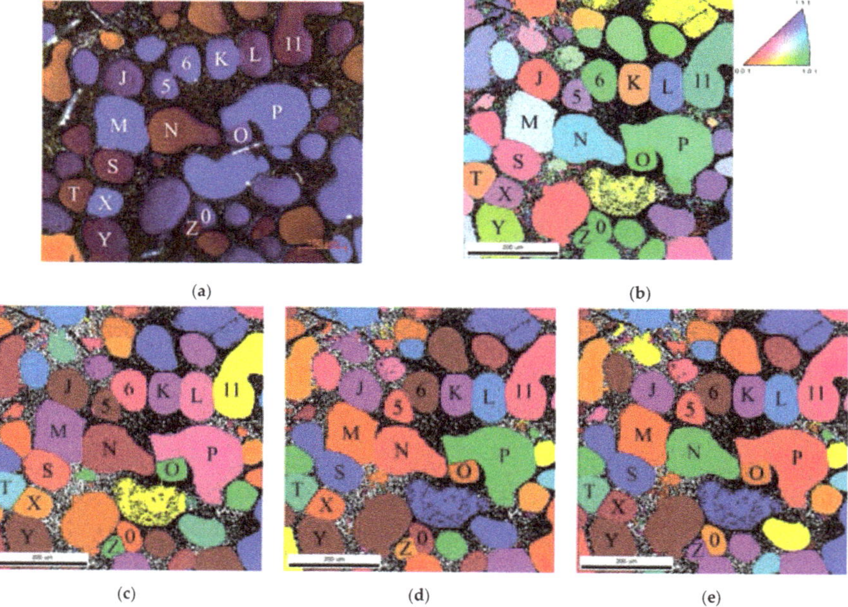

Figure 4. Thixocast sample, copper mold, poured at 630 °C; (**a**) polarized light image with sensitive tint plate, polished and anodized with Barker's reagent, (**b**) inverse pole figure (IPF) map, (**c**) 5° grain map, (**d**) 10° grain map, (**e**) 15° grain map.

Table 2 presents information such as misorientation between particles, and whether they are detectable as individual grains when the grain boundary misorientation criteria were set at five, 10, and 15 degrees. For example, particles K and L had a misorientation of 55.6 degree specifying that these two are indeed two separate grains. Changing the critical misorientation angle from five to 10 and then 15 (Figure 4c–e) did not change the grain map and proved that these two particles are two individual grains and polarized image confirmed the fact. The investigated areas in all samples were analyzed using the same methodology.

Table 2. EBSD data for thixocast sample, copper mold, poured at 630 °C (Y, Yes; N, No).

Grain Boundary	Misorientation	EBSD 5° Misorientation	EBSD 10° Misorientation	EBSD 15° Misorientation	Color Metallography
KL	55.6	Y	Y	Y	Y
JM	32.7	Y	Y	Y	Y
MN	55.6	Y	Y	Y	Y
MS	42.3	Y	Y	Y	Y
TX	29.8	Y	Y	Y	Y
XY	43	Y	Y	Y	Y
OP	59.3	Y	Y	Y	Y
Z0	58.7	Y	Y	Y	Y
L 11	38.3	Y	Y	Y	Y

Figure 5 presents another example of color metallography and EBSD of the sample cast in copper mold at 690 °C and isothermally held at ~583 °C for 10 min. In comparison to Figure 4a, the α-Al particles in Figure 5a are larger and present fewer colors, meaning that fewer individual particles of varying crystal orientation formed within this process. This is due to the fact that the higher pouring temperature resulted in fewer nucleation sites and more dendritic structure. This subject was explained in another publication [14].

Some of the particles in Figure 5 belong to one primary dendrite as they show the same color contrast. This is regardless of changing the misorientation angle criteria. An example is particles 1, 2, 3, and 4. By increasing the grain misorientation criterion from 5 to 10 or even 15-degrees, particles 1-4 still have the same color contrast. However, by increasing the misorientation to 10 degrees, particles E and E1 showed similar color to particles 1–4 (Figure 5d,e). This indicates similar origin and disintegration mechanism for all of the adjacent particles of similar color contrast. The same rationale is applicable for particles C1–C3 meaning that they are originated from the same dendrite. C1–C3 appear as three distinct particles when the misorientation is five degrees. This may indicate these sub grain particles are separated from each other by five degree or less. When the misorientation criterion increases to 10 or 15 degrees, they show the same color which means they were originated from the same dendrite (considering that the sample was electromagnetically stirred and dendrites cramped and/or broken during this process, however they all have their origin in one dendrite). In other words, what it is being said here is the fact that although the particles are detected as individual particles with clear grain boundaries, they may be connected from underneath the plane of polish or broken down by stirring action during semi-solid processing. Nonetheless, the most important finding here is the similarity of polarized light microscopy of anodized specimen with those of EBSD characterization and grain mapping, particularly when a critical misorientation angle of 10 degrees is implemented. Table 3 shows additional EBSD information on some specific particles and whether they are detectable as individual grains in color metallography.

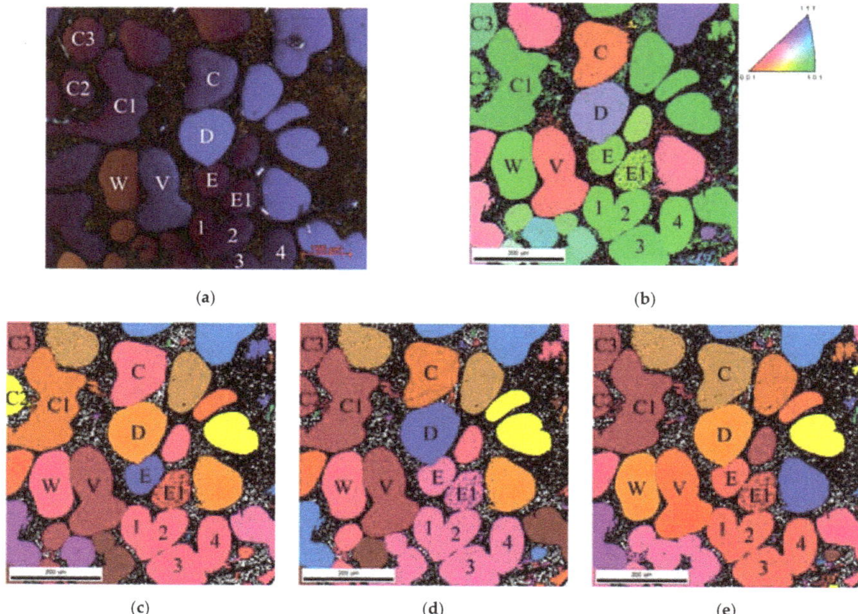

Figure 5. Thixocast sample, copper mold, poured at 690 °C; (**a**) polarized light image with sensitive tint plate, polished and anodized with Barker's reagent, (**b**) IPF map, (**c**) 5° grain map, (**d**) 10° grain map, (**e**) 15° grain map.

Table 3. EBSD data for thixocast sample, copper mold, poured at 690 °C (Y, Yes; N, No).

Grain Boundary	Misorientation	EBSD 5° Misorientation	EBSD 10° Misorientation	EBSD 15° Misorientation	Color Metallography
CD	58	Y	Y	Y	Y
DE	51.6	Y	Y	Y	Y
VW	38.4	Y	Y	Y	Y
1 2	2.4	N	N	N	N
3 4	4.2	N	N	N	N

Representatives of as-cast EMS sand billets are shown in Figures 6 and 7. The application of electromagnetic stirring results in forced convection of the bulk liquid which encourages dendrites fragmentation as well as dendrite arm root re-melting due to the thermal and solutal convection [1]. It is also important to consider that in contrast to the conventional casting, electromagnetic stirring of the melts with higher superheat results in more re-melting of the nuclei [18].

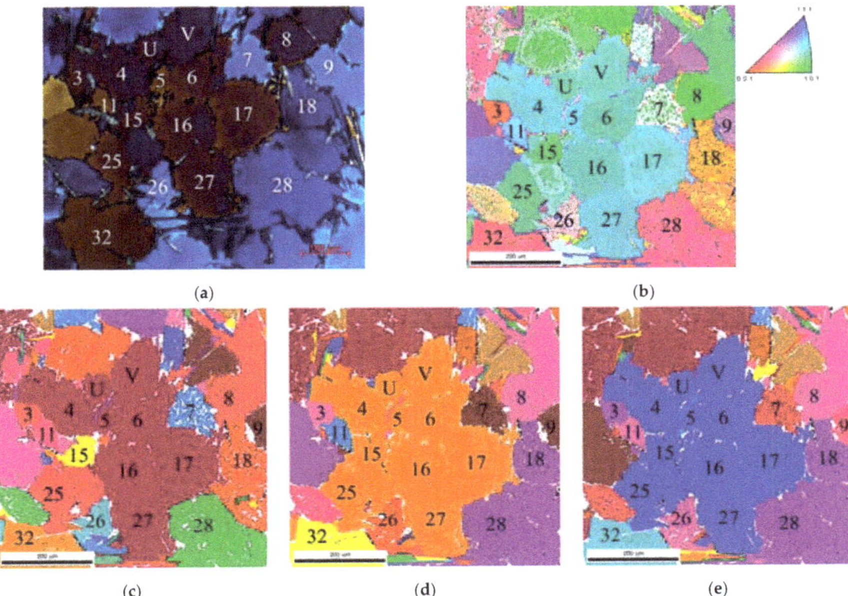

Figure 6. As-cast EMS billets, sand mold, poured at 630 °C; (**a**) polarized light image with sensitive tint plate, polished and anodized with Barker's reagent, (**b**) IPF map, (**c**) 5° grain map, (**d**) 10° grain map, (**e**) 15° grain map.

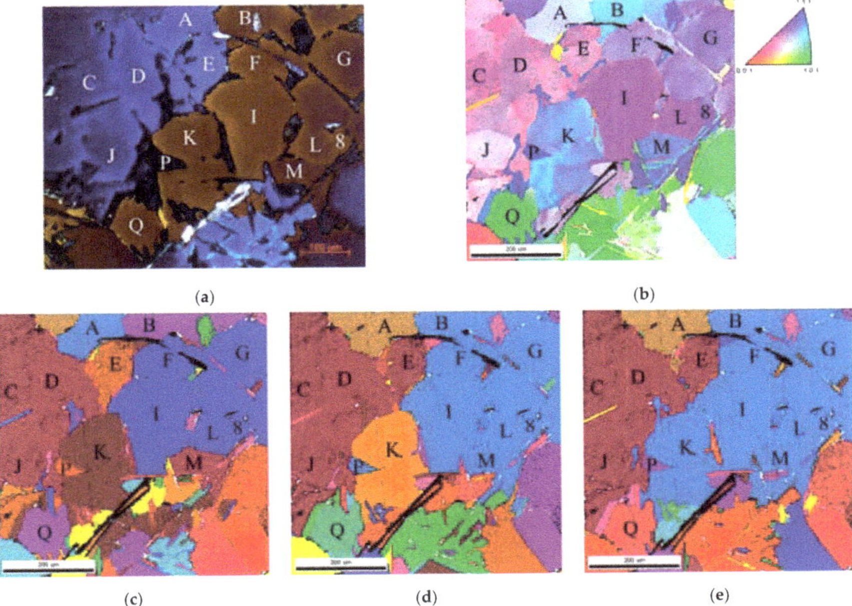

Figure 7. As-cast EMS billets, sand mold, poured at 690 °C; (**a**) polarized light image with sensitive tint plate, polished and anodized with Barker's reagent, (**b**) IPF map, (**c**) 5° grain map, (**d**) 10° grain map, (**e**) 15° grain map.

Quality of the colored images depends not only to the anodizing technique (e.g., solution, voltage, time, temperature which all were constant for these samples), but also to the alloying elements and manufacturing process. By isothermally holding the samples, the image quality and color differentiation improves (compare Figures 4–7).

As a result, detection of the individual grains in color metallography is a challenging task. For instance, in Figure 6, particles 3 and 4 are two different grains according to EBSD grain maps; however, polarized light microscopy is unable to detect the differences between the two grains (Figure 6a). By increasing the critical misorientation angle to 10 degrees, some particles such as 4–15, 15–16, 18–28, and 16–25 present similar color in EBSD grain mapping. Table 4 provides additional information on the area presented in Figure 6.

Table 4. EBSD data for as-cast EMS billets, sand mold, poured at 630 °C (Y, Yes; N, No).

Grain Boundary	Misorientation	EBSD 5° Misorientation	EBSD 10° Misorientation	EBSD 15° Misorientation	Color Metallography
UV	2.9	N	N	N	N
U4	6.8	N	N	N	N
U5	8.4	N	N	N	N
V5	5.9	N	N	N	N
V6	3.8	N	N	N	N
3 11	58.5	Y	Y	Y	Y
3 4	55.8	Y	Y	Y	N
4 5	10.2	N	N	N	N
4 15	10.3	Y	N	N	N
4 11	35.6	Y	Y	Y	Y
5 6	5	N	N	N	N
5 15	12.3	Y	N	N	N
5 16	3.4	N	N	N	N
6 7	55.3	Y	Y	Y	Y
6 16	2.3	N	N	N	N
6 17	8.3	N	N	N	N
7 8	50.9	Y	Y	Y	Y
7 17	55.8	Y	Y	Y	Y
8 9	53	Y	Y	Y	Y
8 18	43.9	Y	Y	Y	Y
9 18	60	Y	Y	Y	Y
11 15	28.2	Y	Y	Y	Y
15 16	8.4	Y	N	N	N
15 25	4	Y	N	N	N
16 17	6.5	N	N	N	N
16 27	3.7	N	N	N	N
16 26	41.3	Y	Y	Y	Y
17 18	44.8	Y	Y	Y	Y
17 28	51	Y	Y	Y	Y
18 28	14.7	Y	N	N	N
25 26	40.9	Y	Y	Y	Y
26 27	41.8	Y	Y	Y	Y
27 28	53.2	Y	Y	Y	Y

The same is applicable in Figure 7 for as-cast EMS billets, sand mold, poured at 690 °C. Particles A and E, A and D, K and Q are different grains according to EBSD; however, polarized light microscopy is unable to detect the grains. By increasing the critical grain misorientation to 10 degrees, particles B–F, D–E, I–M, and M–L exhibit similar color (see Table 5), once more confirming that a 10-degree misorientation is an acceptable grain detection limit for EBSD technique (see Figure 6).

Table 5. EBSD data for as-cast magnetic stirring machine (EMS) billets, sand mold, poured at 690 °C (Y, Yes; N, No).

Grain Boundary	Misorientation	EBSD 5° Misorientation	EBSD 10° Misorientation	EBSD 15° Misorientation	Color Metallography
AB	34.1	Y	Y	Y	Y
AE	53.4	Y	Y	Y	N
AD	53.1	Y	Y	Y	N
BF	21.7	Y	N	N	N
BE	39.6	Y	Y	Y	Y
CD	5.3	N	N	N	N
DE	8.2	Y	N	N	N
DJ	4.8	N	N	N	N
DK	52.2	Y	Y	Y	Y
DI	43.1	Y	Y	Y	Y
EI	43.4	Y	Y	Y	Y
EF	49.4	Y	Y	Y	Y
FG	8.8	N	N	N	N
GL	6.4	N	N	N	N
G8	10.3	N	N	N	N
IL	3.9	N	N	N	N
IK	15.2	Y	Y	N	N
IM	14	Y	N	N	N
KP	19.6	Y	Y	Y	Y
KQ	36.2	Y	Y	Y	N
LM	11.4	Y	N	N	N
L8	6.8	N	N	N	N

Various areas were analyzed in these four samples and Figures 8 and 9 presents the comparison graphs. In these graphs, misorientations were categorized in different bins; each bin spans over 5-degree misorientation. To develop the graphs in Figures 8a and 9a, the total number of boundaries identified through each technique were counted and separated into their respective misorientation intervals. Figures 8b and 9b represent the number fraction of the boundaries in each interval and was determined by dividing the number of boundaries in each interval found in Figures 8a and 9a by the total number of boundaries identified. For the thixocast samples, the number and fraction of grain boundaries using both color metallography and EBSD follow each other closely and have the same trend. In general, the EBSD technique detects more boundaries than color metallography. This is evident in the graph for the number of identified boundaries (Figures 8a and 9a). However, the fraction graphs support the idea that we could definitely rely on the results of the color metallography as the trend is similar.

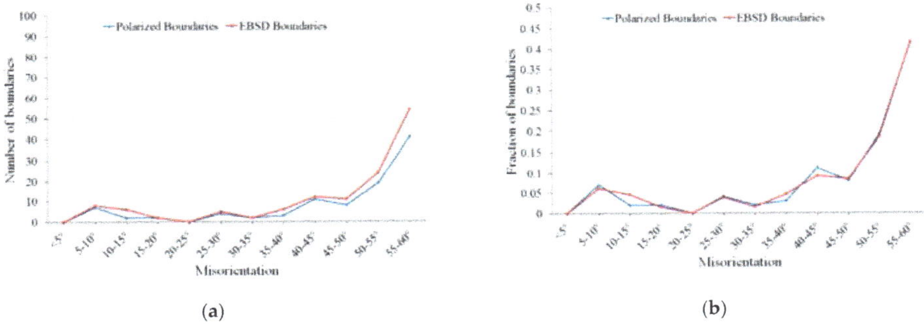

(a) (b)

Figure 8. Number of identified boundaries (**a**) and fraction of boundaries (**b**) based on the misorientation angles, thixocast sample.

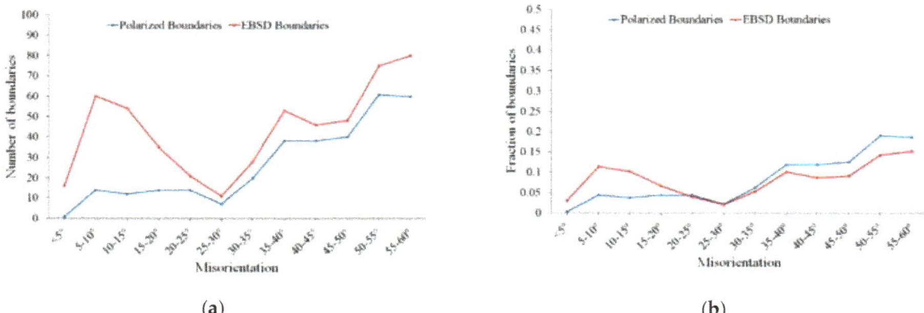

Figure 9. Number of identified boundaries (**a**) and fraction of boundaries (**b**) based on the misorientation angles, as-cast EMS billets, sand mold billets.

In the case of as-cast EMS sand billets, the number and fraction of grain boundaries in both color metallography and EBSD follow each other closely and have the same trend. However, comparing Figures 8 and 9, the difference between color metallography and EBSD results in as-cast EMS samples are more than the thixocast samples. This is due to the color distinction in these samples. EBSD is more accurate for samples having complications during color metallography and EBSD detects far more boundaries than were detected using polarized light and sensitive tint after anodizing. Below five degrees, color microscopy was unable to detect the difference between any boundaries as the difference in color was not significant enough to detect. Whilst EBSD did not detect any boundaries below five degrees either, this was due to the critical misorientation angle being set to five degrees and not an inherent limitation of EBSD.

When reviewing the results of grain boundary identification through color metallography and EBSD, whilst the end results were comparable, the time spent acquiring the data was significantly different. As explored in Table 1, the total time to analyze a single region with color metallography was 100 min faster than through EBSD analysis. The higher testing time associated with EBSD can be linked to the longer sample preparation time required to develop high quality scans as well as the upwards of 75 min required to produce a scan. Samples can be anodized and then analyzed through optical microscopy in as little as 20 min hence, for the time it takes to scan and analyze a single EBSD sample, four samples could be anodized and analyzed through color metallography. When comparing the cost of carrying out each technique, polishing costs are higher for EBSD than for color metallography due to the longer polishing times. Whilst this higher polishing cost is somewhat offset by the requirement of an anodizing solution, the capital expenditure and booking fees are significantly higher for an SEM than an optical microscope.

4. Conclusions

In order to study the capability of color metallography as a reliable microstructural characterization technique in distinguishing the primary particles, anodizing-polarized light microscopy was compared with the EBSD technique for Al–Si samples produced with different casting procedures;

1. Anodizing-polarized light microscopy combination is a reliable characterization method for microstructural analysis of as-cast alloys.
2. Color metallography is a faster and more cost-effective method for microstructural analysis of as-cast alloys, although more data is acquired through EBSD analysis.
3. It was found that a 10 degree misorientation criterion is an acceptable misorientation level for EBSD grain mapping.
4. The number and fraction of grain boundaries detected and resolved for both techniques are basically the same and have similar trend when plotted against the degree of grain misorientation.

Author Contributions: Conceptualization, S.N.; methodology, S.N., validation, S.N. and G.V.V.; formal analysis, S.N. and A.R.; investigation, A.R.; resources, S.N. and R.G.; data curation, A.R.; writing—original draft preparation, S.N.; writing—review and editing, S.N., A.R., R.G. and G.V.V.; visualization, S.N. and A.R.; supervision, S.N.; project administration, S.N.; funding acquisition, R.G.

Funding: This research received no external funding.

Acknowledgments: The authors acknowledge the facilities, and the scientific and technical assistance, of the Australian Microscopy and Microanalysis Research Facility at Flinders University. This work was performed in part at the OptoFab node of the Australian National Fabrication Facility utilizing Commonwealth and SA State Government funding.

Conflicts of Interest: The authors declare no conflict of interest.

References

1. Nafisi, S.; Ghomashchi, R. *Semi-Solid Processing of Aluminum Alloys*; Springer: Berlin, Germany, 2016; ISBN 978-3-319-40333-5.
2. Niroumand, B.; Xia, K. 3D Study of the Structure of Primary Crystals in a Rheocast Al-Cu Alloy. *Mater. Sci. Eng. A* **2000**, *283*, 70–75. [CrossRef]
3. Ito, Y.; Flemings, M.C.; Cornie, J.A. Rheological Behavior and Microstructure of Al-6.5 wt % Si Alloy. In *Nature and Properties of Semi-Solid Materials*; Sekhar, J.A., Dantzig, J., Eds.; TMS: Warrendale, PA, USA, 1991; pp. 3–17.
4. Suery, M. *Mise en forme des alliages métalliques à l'état semi-solide*; Lavoisier Publication: Paris, France, 2002.
5. Limodin, N.; Salvo, L.; Suery, M.; DiMichiel, M. In situ and real-time 3-D microtomography investigation of dendritic solidification in an Al–10 wt.% Cu alloy. *Acta Mater.* **2009**, *57*, 2300–2310. [CrossRef]
6. Xu, W.; Ferry, M.; Mateescu, N.; Cairney, J.M.; Humphreys, F.J. Techniques for generating 3-D EBSD microstructures by FIB tomography. *Mater. Charact.* **2007**, *58*, 961–967. [CrossRef]
7. Zaafarani, Z.; Raabe, D.; Singh, R.N.; Roters, F.; Zaefferer, S. Three-dimensional investigation of the texture and microstructure below a nanoindent in a Cu single crystal using 3D EBSD and crystal plasticity finite element simulations. *Acta Mater.* **2006**, *54*, 1863–1876. [CrossRef]
8. Voort, G.V. *Metallography, Principles and Practice*; ASM International: New York, NY, USA, 1999.
9. Voort, G.V. *ASM Handbook, Volume 09: Metallography and Microstructures*; ASM International: Materials Park, OH, USA, 2004.
10. Oxford Instruments. Available online: www.ebsd.com (accessed on 25 April 2019).
11. OIM Analysis Tutorials, Ametek Inc. Available online: www.edax.com (accessed on 25 April 2019).
12. Vander Voort, G.F. Metallographic specimen preparation for electron backscattered diffraction. Part I. *Pract. Metallogr.* **2011**, *48*, 454–473. [CrossRef]
13. Vander Voort, G.F. Metallographic specimen preparation for electron backscattered diffraction. Part II. *Pract. Metallogr.* **2011**, *48*, 527–543.
14. Nafisi, S.; Ghomashchi, R. Microstructural Evolution of Electromagnetically Stirred Feedstock SSM Billets during Reheating Process. *Metall. Microstruct. Anal.* **2013**, *2*, 96–106. [CrossRef]
15. Flemings, M.C. Behavior of metal alloys in the semi-solid state. *Metal. Trans. A* **1991**, *22*, 952–981. [CrossRef]
16. Kirkwood, D.H. Semisolid metal processing. *Int. Mater. Rev.* **1994**, *39*, 173–189. [CrossRef]
17. Fan, Z. Semisolid metal processing. *Int. Mater. Rev.* **2002**, *47*, 49–85. [CrossRef]
18. Nafisi, S.; Emadi, D.; Shehata, M.T.; Ghomashchi, R. Effects of Electromagnetic Stirring and Superheat on the Microstructural Characteristics of Al-Si-Fe Alloy. *Mater. Sci. Eng. A* **2006**, *432*, 71–83. [CrossRef]

Article

Microstructure, Mechanical Properties and Wear Behavior of the Rheoformed 2024 Aluminum Matrix Composite Component Reinforced by Al$_2$O$_3$ Nanoparticles

Jufu Jiang [1,*], Guanfei Xiao [1], Changjie Che [1] and Ying Wang [2]

[1] School of Materials Science and Engineering, Harbin Institute of Technology, Harbin 150001, China; guanfeixiao@163.com (G.X.); 18092233726@163.com (C.C.)
[2] School of Mechatronics Engineering, Harbin Institute of Technology, Harbin 150001, China; wangying1002@hit.edu.cn
* Correspondence: jiangjufu@hit.edu.cn; Tel.: +86-187-4601-3176

Received: 3 May 2018; Accepted: 14 June 2018; Published: 15 June 2018

Abstract: The 2024 nanocomposite reinforced with Al$_2$O$_3$ nanoparticles was fabricated by the ultrasonic assisted semisolid stirring (UASS) method and rheoformed into a cylinder component. Microstructure, mechanical properties, and wear behavior of the rheoformed composite components were investigated. The results showed that the composite components with complete filling status and a good surface were rheoformed successfully. The deformation of semisolid slurries was mainly dominated by flow of liquid incorporating solid grains (FLS), sliding between solid grains (SSG), and plastic deformation of solid grains (PDS). Mechanical properties of the rheoformed composite components were influenced by stirring temperature, stirring time, and volume fraction of Al$_2$O$_3$ nanoparticles. The optimal ultimate tensile strength (UTS) of 358 MPa and YS of 245 MPa were obtained at the bottom of the rheoformed composite components after a 25-min stirring of composite semisolid slurry with 5% Al$_2$O$_3$ nanoparticles at 620 °C. Enhancement of mechanical properties was attributed to high density dislocations and dislocation tangles and uniform dispersed Al$_2$O$_3$ nanoparticles in the aluminum matrix. Natural ageing led to the occurrence of needle-like Al$_2$CuMg phase and short-rod-like Al$_2$Cu phase. UTS of 417 MPa and YS of 328 MPa of the rheoformed composite components were achieved after T6 heat treatment. Improvement of mechanical properties is due to the more precipitated needle-like Al$_2$CuMg phase and short-rod-like Al$_2$Cu phase. Wear resistance of the rheoformed composite components was higher than that of the rheoformed matrix component. Wear resistance of the rheoformed composite component increased with an increase in Al$_2$O$_3$ nanoparticles from 1% to 7%. A slight decrease in wear rate resulted from 10% Al$_2$O$_3$ nanoparticles due to greater agglomeration of Al$_2$O$_3$ nanoparticles. A combination mechanism of adhesion and delamination was determined according to worn surface morphology.

Keywords: 2024 aluminum matrix composites; rheoformed; Al$_2$O$_3$ nanoparticles; microstructure; mechanical properties

1. Introduction

Metal matrix composites (MMC) have exhibited some obvious advantages such as high specific strength, high specific stiffness, and good wear resistance [1–3]. Fabrication technology of MMC involves stirring casting [4,5], powder metallurgy [6–8], squeeze casting [9–11], and semisolid stirring [12–14]. In addition, selective laser melting (SLM) was employed to fabricate high-performance alloys and MMC [15–17]. As a typical MMC, aluminum matrix composite reinforced by ceramic particles (AMCCP) also exhibited some important applications in the automotive and aerospace industries [18,19]. The reinforced ceramic particles are composed of micro-sized ceramic particles and nano-sized ceramic particles. In recent years, aluminum matrix composites reinforced with nano-sized ceramic particles (AMCNCP) have attracted researchers' attention because of higher strength, increased dimensional stability, high thermal stability, high modulus, and good wear resistance as compared to conventional materials [20]. Raju et al. [21] evaluated fatigue of nano-sized Al_2O_{3p}/2024 composite and found that it was slightly increased as compared to matrix material. Raturi et al. [22] reported mechanical, tribological, and micro structural behavior of the Al 7075 matrix reinforced with nano Al_2O_3 particles and concluded that tensile, impact, and flexural strength of the composite were enhanced as compared with the matrix alloy. Sajjadi et al. [23] investigated the fabrication and mechanical properties of A356 composite reinforced with micro and nano-sized Al_2O_3 particles by a developed compocasting method. The results showed that the hardness of the composites increased with increasing particle weight fraction and decreasing particle size.

However, it is very difficult to disperse uniformly nano-sized ceramic particles in the matrix alloy due to higher surface energy and specific surface area as compared to micro-sized ceramic particles. Therefore, some novel methods were developed to realize uniform dispersion of nano-sized ceramic particles such as Al_2O_3 and SiC. For example, uniform dispersion of Al_2O_3 nanoparticles in matrix alloy was achieved successfully via incorporating milled powders of Al_2O_3 nanoparticles and aluminum or copper into A356 alloy melt [24,25]. Acoustic streaming and cavitation created by ultrasonic wave led to a uniform dispersion of nano-sized SiC particles in molten A356 aluminum alloy [26,27]. Semisolid stirring and ultrasonic wave were joined together to obtain a uniform dispersion of nano-sized SiC particles in 7075 aluminum matrix. It was attributed to the controllable viscosity of semisolid slurries and acoustic streaming and cavitation created by ultrasonic wave [28].

Matrix materials of AMCNCP mainly have been focused on A356 [24–27], 7075 [28], A357 [29], and 6061 [8] aluminum alloys, but 2024 matrix material has not been studied in detail. The present investigation will deal with microstructure and mechanical properties of 2024 aluminum matrix composite reinforced with Al_2O_3 nanoparticles.

2. Materials and Methods

2.1. Fabrication of 2024 Aluminum Matrix Composite Reinforced with Al_2O_3 Nanoparticles

Commercial 2024 aluminum alloy was used as matrix material. Its chemical composition was determined via an Axios pw4400 X-ray fluorescence spectrometer and contained 4.52 wt % Cu, 1.51 wt % Mg, 0.56 wt % Mn, 0.18 Si wt %, 0.12 wt % Fe, 0.02 wt % Zn, and a balance of Al. α-Al_2O_3 nanoparticles with an average size of 60 nm were used as reinforcement of the composite. Solidus temperature of 529 °C and liquidus temperature of 650 °C were achieved from a differential scanning calorimetry (DSC) test. Figure 1 gives a schematic diagram of fabrication and rheoforming of 2024 matrix composite semisolid slurry. As shown in Figure 1, there were three main procedures in the fabrication and rheoforming of composite semisolid slurry. In the first procedure, α-Al_2O_3 nanoparticles with an average size of 60 nm parceled by pure aluminum foil were added into the melt after 2024 aluminum alloy was melted at 670 °C and held for 20 min. The XRD pattern of as-received Al_2O_3 nanoparticles shows the presence of α-Al_2O_3 peaks (Figure 2). The melt with α-Al_2O_3 nanoparticles was treated for 10 min via an ultrasonic device. In the second procedure, melt with α-Al_2O_3 nanoparticles was stirred and cooled to the predefined semisolid temperature, and then isothermally stirred for the required time, as shown in Table 1.

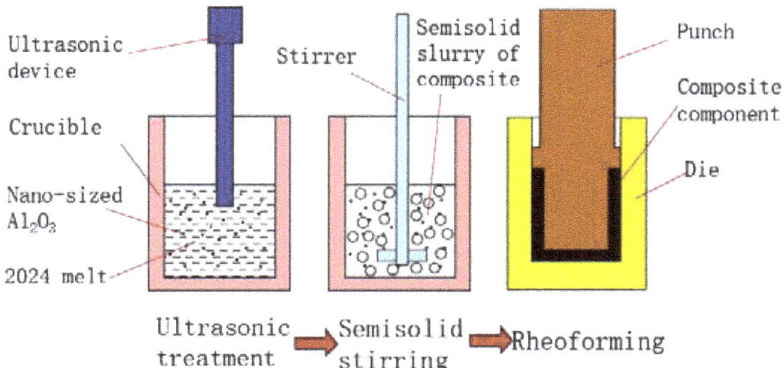

Figure 1. Schematic diagram of fabricating and rheoforming the semisolid slurry of composite.

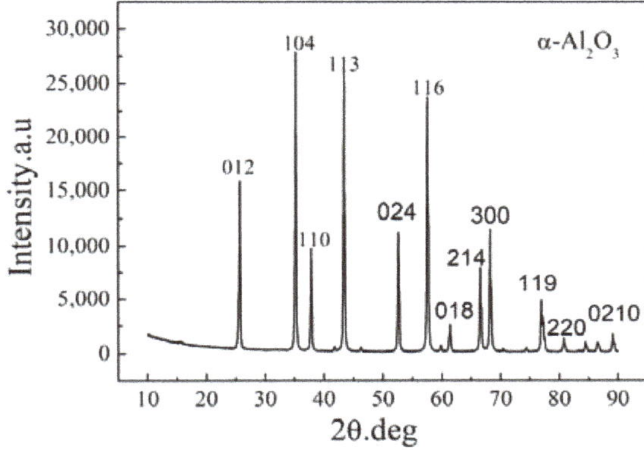

Figure 2. XRD pattern of as-received Al_2O_3 nanoparticles.

Table 1. Experimental scheme of the rheoformed 2024 aluminum matrix composite reinforced by Al_2O_3 nanoparticles and original 2024 components.

Serial Number	Stirring Time (Min)	Stirring Temperature (°C)	Al₂O₃ Volume Fraction (%)	Ultrasonic Treatment Time (Min)	Force (kN)	Dwell Time (s)	Preheated Temperature of Die (°C)
1	5	620	5	10	2000	20	400
2	10	620	5	10	2000	20	400
3	15	620	5	10	2000	20	400
5	20	620	5	10	2000	20	400
6	25	620	5	10	2000	20	400
7	30	620	5	10	2000	20	400
8	25	610	5	10	2000	20	400
9	25	615	5	10	2000	20	400
10	25	625	5	10	2000	20	400
11	25	630	5	10	2000	20	400
12	25	620	0	10	2000	20	400
13	25	620	1	10	2000	20	400
14	25	620	3	10	2000	20	400
15	25	620	7	10	2000	20	400
16	25	620	10	10	2000	20	400
17	25	610	0	10	2000	20	400
18	25	615	0	10	2000	20	400
19	25	625	0	10	2000	20	400
20	25	630	0	10	2000	20	400

In the third procedure, the fabricated semisolid slurries of the composite were carried directly into the die cavity with a preheated temperature of 400 °C and rheoformed (i.e., semisolid slurry was directly formed into the final part under some pressure) under a force of 2000 kN. The detailed experimental scheme was shown in Table 1. Two composite components under the same process parameters were rheoformed in order to improve the accuracy of tensile test. Therefore, thirty composite components were rheoformed successfully. Ten original 2024 components without Al_2O_3 nanoparticles were also rheoformed in order to compare microstructure, mechanical properties, and wear behavior with composite components.

2.2. Microstructure Observation and Measurement of Mechanical and Wear Properties

The microstructural specimens cut from composite components were firstly ground with 200, 400, 600, 800, 1200, and 2000 grit papers and then polished with 0.1 μm diamond paste. The specimens were etched for about 10 s by Keller's reagent (4 mL HF, 6 mL HCL, 8 mL HNO3 and 82 mL water) and observed by using Olympus GX71 optical microscope (OM, Olympus Coporation, Toyko, Japan), Quanta 200 FEG scanning electron microscope (SEM, FEI, Hillsboro, OR, USA), and talos f200x transmission electron microscopy (TEM, FEI, Hillsboro, OR, USA) equipped with an energy dispersive X-ray spectrometer (EDX). Transmission electron microscopy specimens were fabricated via cutting 1 mm slices from the rheoformed composite component with a wire cutting machine, and then mechanically ground to a thickness of 100 μm. Then 3 mm diameter disks were cut from the thin slices by punching. Ion milling was carried out on these 3 mm diameter disks.

Tensile specimens cut from composite components were machined into standard tensile specimens according to ASTM Standard Test Methods for Tension Testing of Metallic Materials, E8M [30]. The sampled location and drawing of tensile specimens were indicated in Figure 3. Eight specimens were obtained from side wall of the two rheoformed composite components under the same process parameters and four specimens were obtained from the bottom. Four side-wall tensile specimens and two bottom specimens were directly carried out on tensile test at room temperature. The other four side-wall specimens and two bottom specimens were firstly treated via T6 heat treatment involving the solution treatment for 2 h at 490 °C and ageing for 10 h at 190 °C and then used as a tensile test at room temperature. The tensile strength of the side wall reported in this paper was obtained from the average value of data of four side-wall specimens. The tensile strength of bottom was achieved from the average value of data of two bottom specimens. The dry sliding wear tests were carried out on a pin-on-disc wear-testing apparatus. The disc was made from 5Cr15 steel. After the rheoformed matrix and composite components were formed, they were machined into the samples with dimensions of φ6 × 15 mm for the dry sliding wear tests. The process parameters of wear test involved a distance of 1000 m, a speed of 0.8 m/s, and a load of 30 N.

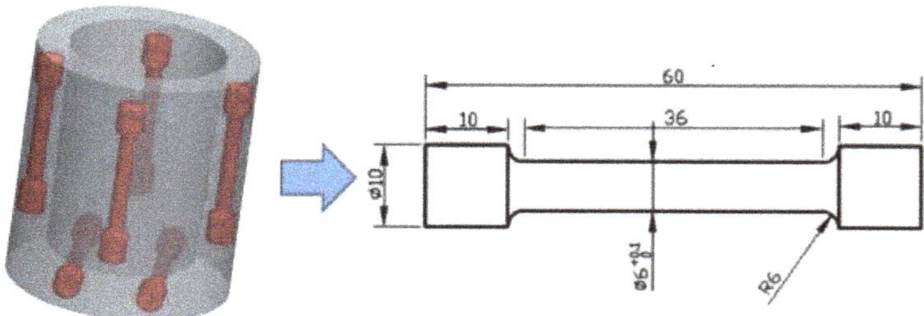

Figure 3. Sampled location and drawing of tensile specimens.

3. Results and Discussion

3.1. Macrograph and Microstructure of the Rheoformed Composite Component

Figure 4 presented the whole and half-sectional macrographs of the rheoformed composite component reinforced by 5 vol % Al_2O_3 nanoparticles at 620 °C and for 25 min stirring time.

As shown in Figure 4a, complete filling status and good surface quality were obtained from the rheoformed composite component. No obvious porosity and incomplete filling status were found in the half-sectional macrograph of the rheoformed composite component (Figure 4b). It illustrates that densified microstructure was obtained in the rheoformed composite components. The densified microstructure is beneficial to improve the mechancial properties of the rheoformed composite components. In order to characterize the microstructure and mechanical properties in different locations of the rheformed composite components, the microstructural specimens were achieved from the locations A to D (Figure 4b), and tensile specimens were obtained from the side wall and bottom (Figure 3).

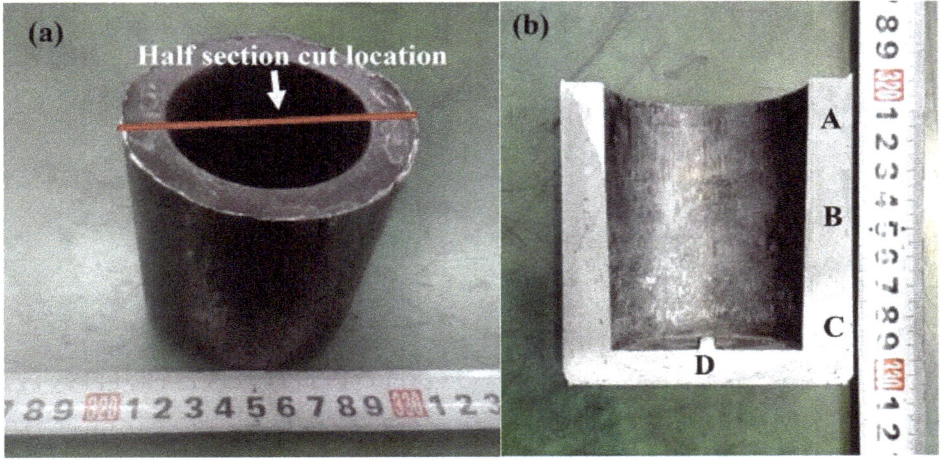

Figure 4. Whole and half-sectional macrographs of rheoformed 2024 aluminum matrix composite component reinforced by 5 vol % Al_2O_3 nanoparticles at 620 °C and for 25 min stirring time (**a**) whole macrograph; (**b**) half-sectional macrograph.

Microstructure in different locations of the rheoformed composite components is shown in Figure 5. As indicated in Figure 5a,b, microstructure in locations A and B consisted of near spheroidal grains and liquid phase. It illustrates that no obvious plastic deformation occurred in the solid grains in locations A and B during the rheoforming process. The microstructure in locations C and D consisted of elongated solid grains and liquid phase (Figure 5c,d). It indicates that obvious plastic deformation along flowing direction of semisolid slurries occurred in the solid grains in locations C and D. There are four deformation mechanisms in the semisolid processing, liquid flow (LF), flow of liquid incorporating solid grains (FLS), sliding between solid grains (SSG), and plastic deformation of solid grains (PDS) [31]. When the semisolid slurry was rheoformed in the die cavity, it showed a backward extrusion mode. The flow front of the semisolid slurry in locations A and B is a free surface [27], indicating the lowest resistance to flow. The flow velocity of liquid phase is higher than that of solid phase. It led to more liquid phase existed in locations A and B. As a consequence, deformation compatibility of liquid phase is higher than that of semisolid slurry in locations C and D. Therefore, deformation in locations A and B depends on flow of liquid incorporating solid grains (FLS) and sliding between solid grains (SSG).

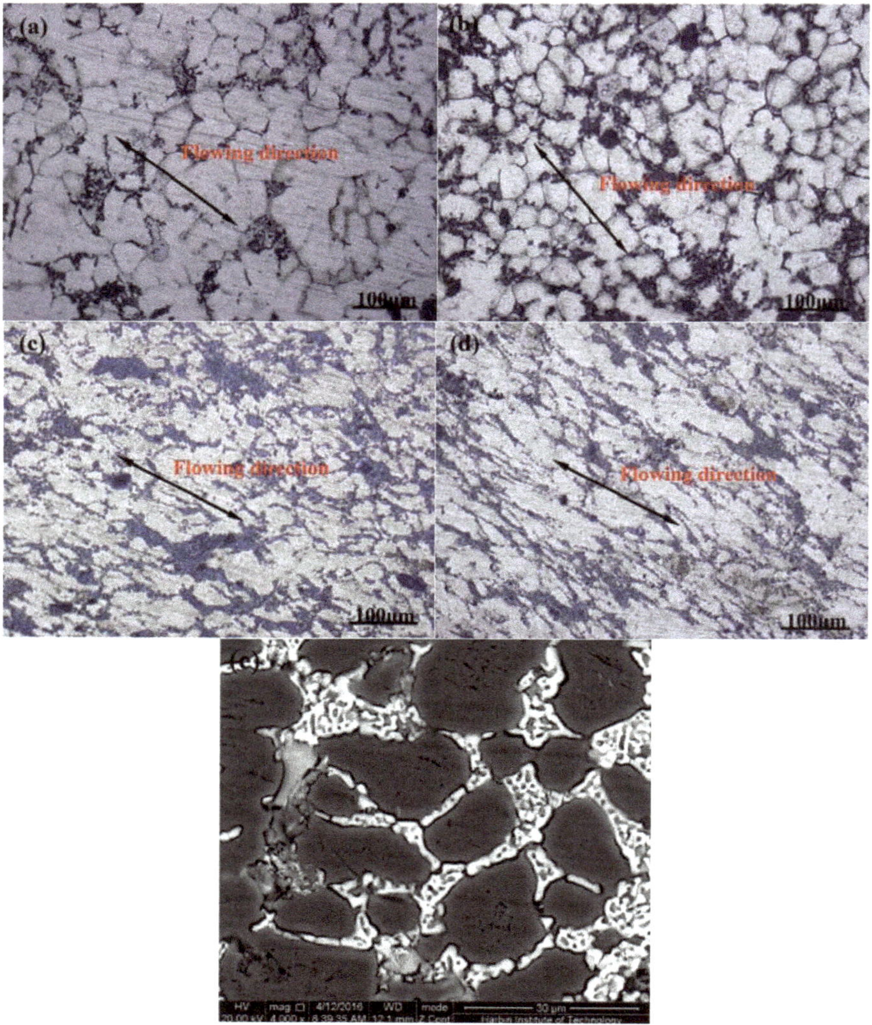

Figure 5. Microstructure of 2024 aluminum matrix composite component reinforced by 5 vol % Al$_2$O$_3$ nanoparticles rheoformed at 620 °C and for 25 min stirring time. (**a**) Optical microscope (OM) image in location A of the composite component; (**b**) OM image in location B of the composite component; (**c**) OM image in location C of the composite component; (**d**) OM image in location D of the composite component; (**e**) SEM image in location B of the composite component.

It led to near spheroidal solid grains with no obvious plastic deformation. However, the deformation in locations C and D was mainly dominated by plastic deformation of solid grains (PDS). Consequently, the elongated solid grains were created along the flowing direction. In addition, it can be noted that the deformation degree in location D is higher than that in location C due to low fraction liquid. Figure 5e shows the SEM image of the rheoformed composite component. It illustrates further that the microstructure in location B consisted of near spheroidal grains due to dependence of the deformation on FLS and SSG.

The microstructure in various locations of the rheoformed original 2024 is similar to that of the composite component (Figure 6). As a consequence, the deformation mechanisms in locations A and B rely on FLS and SSG. The deformation mechanisms in locations C and D depend on PDS. The SEM image in location B of 2024 matrix component is presented in Figure 6e. As shown in Figure 6e, similar to the composite, the microstructure of the rheoformed original 2024 consisted of near spheroidal grains and liquid phase. It illustrates that the deformation mechanism depends on the FLS and SSG. However, the size of the solid grains of the rheoformed composite component is obviously smaller than that of the rheoformed original 2024. It was attributed to the action of nano-sized Al_2O_3 particles as heterogeneous nuclei during the solidification of aluminum alloy [24].

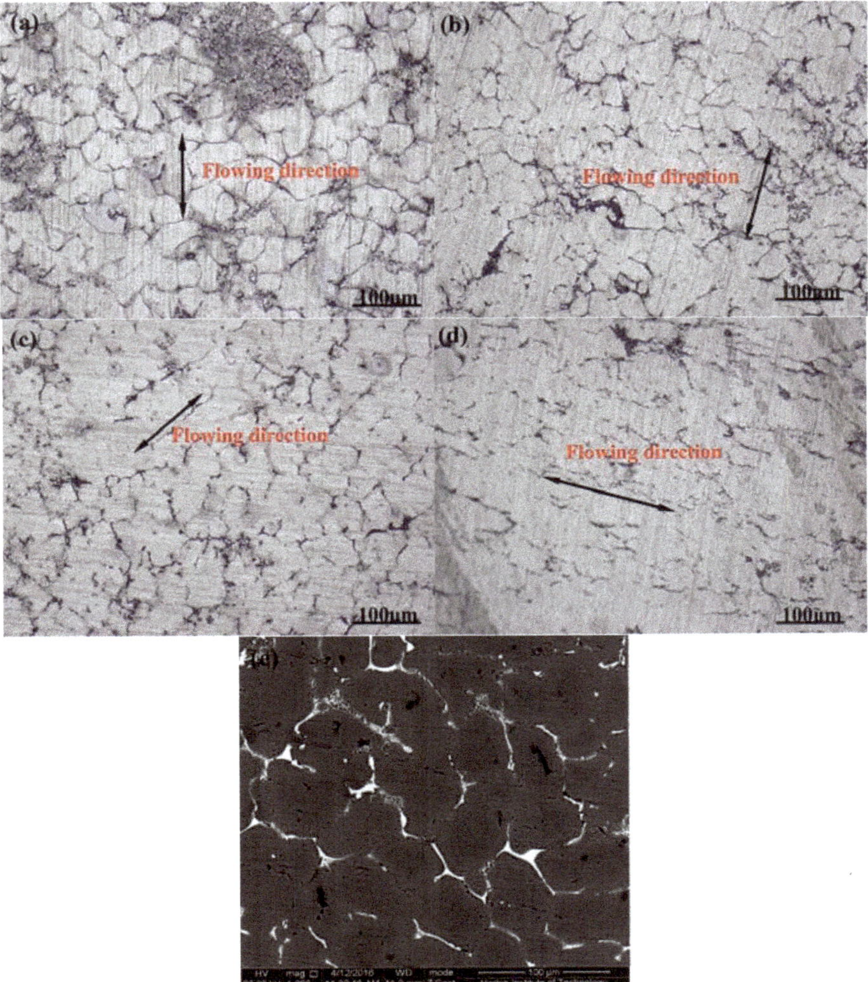

Figure 6. Microstructure of the original 2024 rheoformed at 620 °C and for 25 min stirring time. (**a**) OM image in location A of the matrix component; (**b**) OM image in location B of the matrix component; (**c**) OM image in location C of the matrix component; (**d**) OM image in location D of the matrix component; (**e**) SEM image in location B of 2024 matrix component.

3.2. Influence of Stirring Temperature and Stirring Time on Mechanical Properties

Figure 7 shows mechanical properties of the rheoformed composite components reinforced by 5 vol % Al_2O_3 nanoparticles and matrix components for 25 min stirring time at different stirring temperatures. As indicated in Figure 7, ultimate tensile strength (UTS), yield strength (YS), and elongation all increase and then decrease with an elevated stirring temperature. The highest UTS of 315 MPa in the side wall and the highest UTS of 358 MPa at the bottom were obtained at 620 °C. The highest YS of 238 MPa in the side wall and the highest YS of 245 MPa at the bottom were obtained at 620 °C. The highest elongation of 5.3% in the side wall and the highest elongation of 5.6% at the bottom were all obtained at 620 °C. Similar to the composite components, the optimal mechanical properties were also obtained at 620 °C. It illustrated that 620 °C was the optimal stirring temperature to obtain the highest mechanical properties. In addition, UTS and elongation at the bottom were higher than the side wall. It is due to the fact that severe plastic deformation occurred in the bottom location of the rheoformed composite components (Figure 4d). The UTS and YS of the matrix components are lower than those of the composite components. The YS in the side wall of the composite and matrix components was close to that at the bottom.

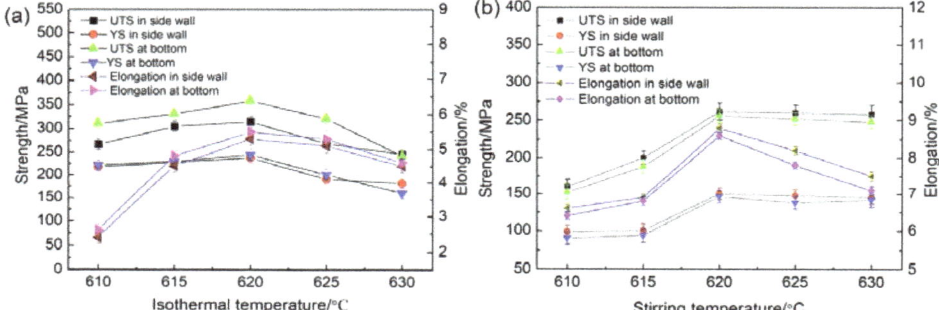

Figure 7. Influence of different stirring temperature on mechanical properties of rheoformed composite components reinforced by 5 vol % Al_2O_3 nanoparticles for 25 min stirring time and matrix components (**a**) composite components; (**b**) matrix components.

Microstructure evolution with elevating stirring temperature was given in Figure 8. The microstructure specimens were cut from location B as shown in Figure 4b. The microstructure consisted of near spheroidal solid grains and liquid phase due to dependence of deformation mechanism on FLS and SSG. Low stirring temperature leads to coarse spheroidal grains. Even some obvious dendrites were found in the microstructure of the rheoformed composite parts. It has an adverse influence on the mechanical properties of the rheoformed composite parts. With the increase in stirring temperature, the grain size of spheroidal grains decreased. The average grain sizes obtained from image analysis are 57 µm, 76 µm, 47 µm, 55 µm, and 54 µm respectively when the stirring temperatures are 610 °C, 615 °C, 620 °C, 625 °C, and 630 °C.

Therefore, the average size of solid grains in the microstructure at 620 °C is obviously smaller than those at other stirring temperatures. According to the Hall-Petch effect [31], fine grains can lead to an increase in YS and UTS. Furthermore, stirring temperatures higher than 625 °C lead to aggregation of liquid phase, which is detrimental to mechanical properties. When stirring temperature is higher than 625 °C, more aggregation of liquid phase also reduced controllable viscosity of semisolid slurries due to the lack of solid grains. The dispersion effect of Al_2O_3 nanoparticles was reduced due to decreased viscosity of semisolid slurries. As a result, greater agglomeration of Al_2O_3 nanoparticles occurred in the microstructure, leading to a decrease in mechanical properties of the rheoformed composite components.

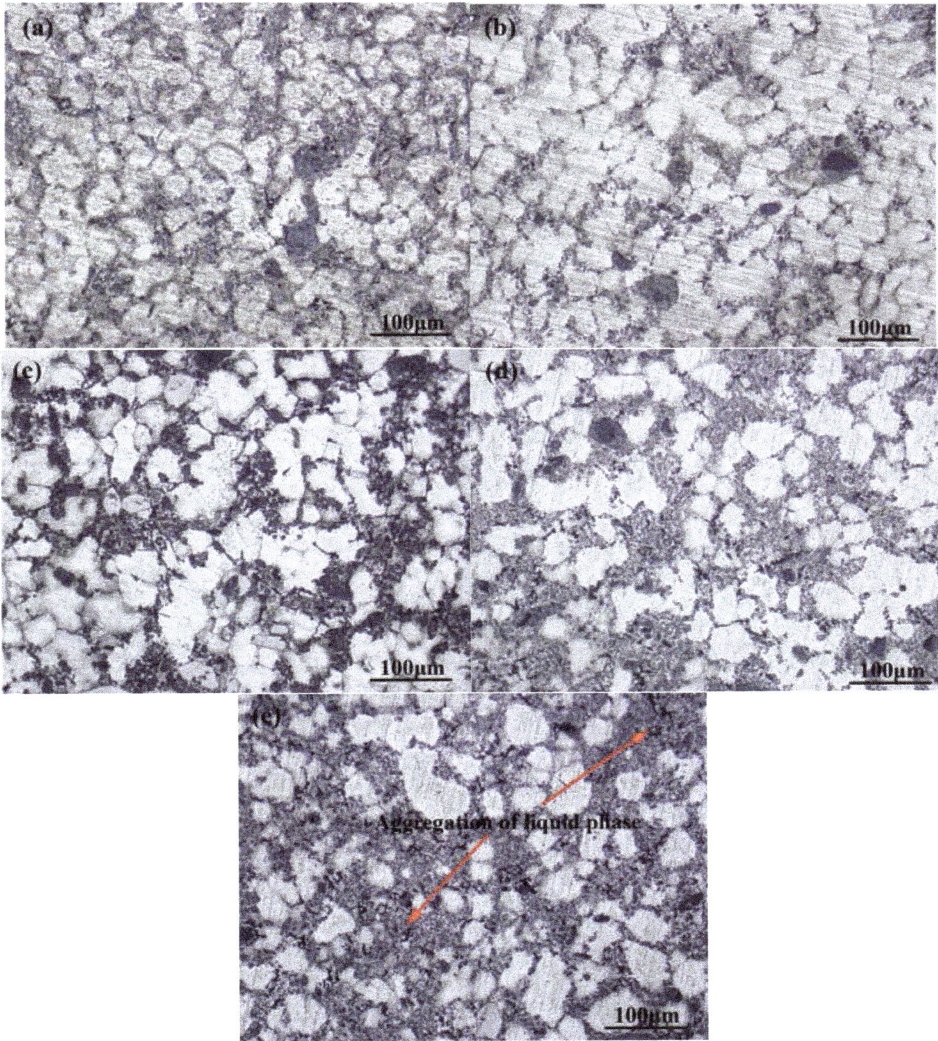

Figure 8. OM microstructure in location B of rheoformed 2024 aluminum matrix composite component reinforced by 5 vol % Al$_2$O$_3$ nanoparticles for 25 min stirring time at various stirring temperatures (**a**) 610 °C; (**b**) 615 °C; (**c**) 620 °C; (**d**) 625 °C; (**e**) 630 °C.

Figure 9 shows mechanical properties of the rheoformed composite component reinforced by 5 vol % Al$_2$O$_3$ nanoparticles at 620 °C for different stirring times. As shown in Figure 7, mechanical properties of rheoformed composite components increased significantly when stirring time increased from 5 min to 25 min. Mechanical properties of the rheoformed composite component changed slightly when stirring time increased from 25 min to 30 min. Similar results were found in the rheoformed cylindrical part of the 7075 aluminum matrix composite reinforced with nano-sized SiC particles [28]. The highest mechanical properties including UTS of 358 MPa, YS of 245 MPa, and elongation of 5.6%

were achieved at bottom of the rheoformed cylindrical part of the composite reinforced with Al_2O_3 nanoparticles for 25 min stirring time at 620 °C.

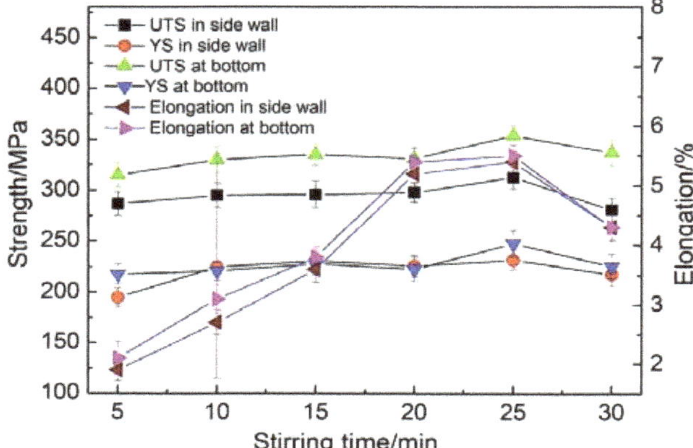

Figure 9. Influence of different stirring time on mechanical properties of rheoformed 2024 aluminum matrix composite component reinforced by 5 vol % Al_2O_3 nanoparticles at 620 °C.

Increasing stirring time resulted in a fine-grained microstructure of composite semisolid slurries (Figure 10). As indicated in Figure 10a, coarse solid grains of more than 200 μm were found in the microstructure when stirring time was 5 min. With an increase in stirring time, solid grains were refined significantly (Figure 10b–e). When stirring time was 25 min, the average size of solid grains was about 46 μm. Fine-grained microstructure can improve mechanical properties of the rheoformed composite components due to the Hall-Petch effect [32]. However, the grain size of solid grains changed slightly when stirring time increased from 25 min to 30 min. As a result, the mechanical properties of the rheoformed composite components also changed slightly, or even showed a slight decrease.

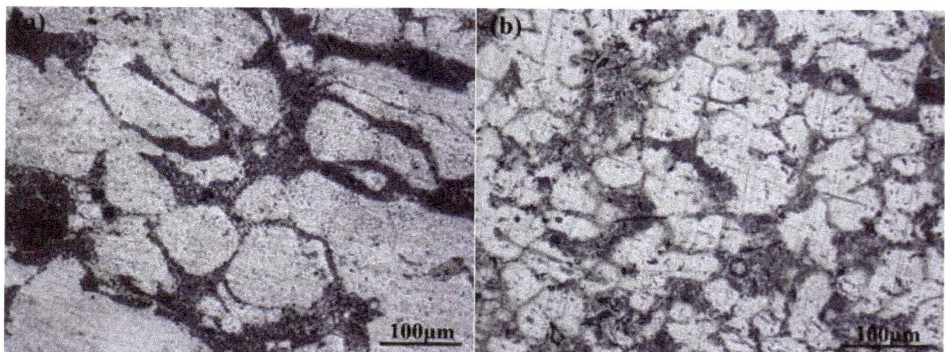

Figure 10. *Cont.*

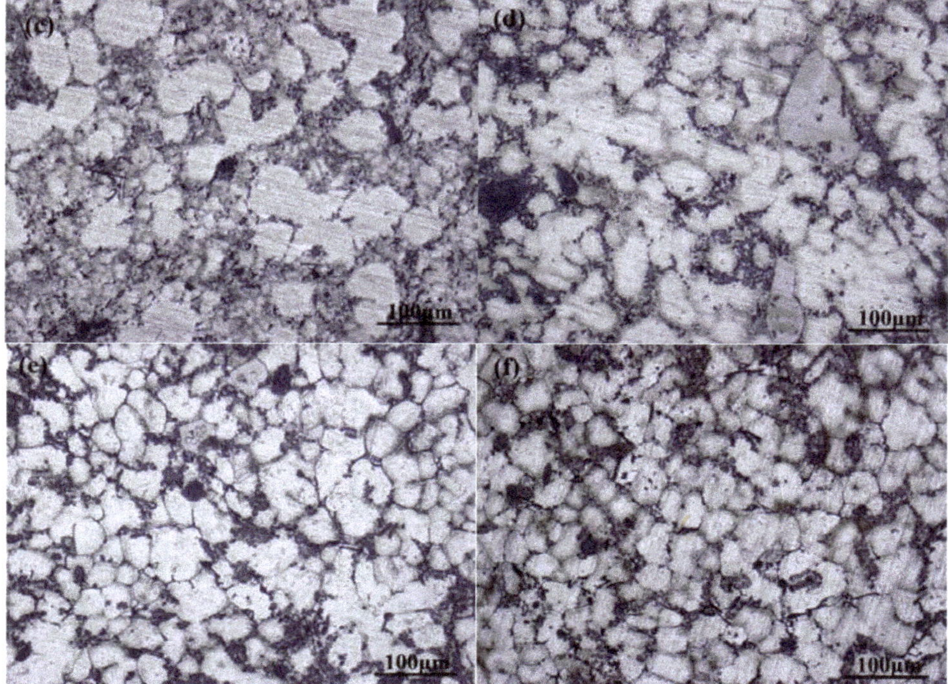

Figure 10. Influence of different stirring time on OM microstructure in location B of the rheoformed 2024 aluminum matrix composite component reinforced by 5 vol % Al_2O_3 nanoparticles at 620 °C (**a**) 5 min; (**b**) 10 min; (**c**) 15 min; (**d**) 20 min; (**e**) 25 min; (**f**) 30 min.

3.3. Influence of Volume Fraction of Al_2O_3 Nanoparticles on Mechanical Properties

Figure 11 shows the influence of volume fraction of Al_2O_3 nanoparticles on mechanical properties of the rheoformed composite components. As indicated in Figure 11, UTS and YS of the rheoformed composite components increased when the volume fraction of Al_2O_3 nanoparticles increased from 0 to 5%. UTS values in the side wall and at the bottom of the matrix components are 268 MPa and 272 MPa respectively. When the volume fraction of Al_2O_3 nanoparticles increased to 5%, they reached 315 MPa and 358 MPa respectively. The increasing degrees of UTS and YS are 17.5% and 31.7% respectively. Al_2O_3 nanoparticles act as barriers of dislocations mobility, leading to an improvement of UTS and YS [33]. In addition, mismatch of coefficient of thermal expansion (CTE) between the matrix and reinforcement phase, load transfer from matrix to reinforcement phase and Orowan strengthening mechanism also play an important role in improving the mechanical properties [34–36]. Al_2O_3 nanoparticles also acted as heterogeneous nuclei for the aluminum alloy matrix, leading to a grain-refined strengthening effect [24,37,38]. The improvement of mechanical properties may be related to the residual stress and fracture toughness at the interface between the Al_2O_3 and the Al matrix. However, it is very difficult to measure the residual stress and fracture toughness at the interface between the Al_2O_3 and the Al matrix, and it could be helpful to employ nanoindentation and pillar splitting techniques as reported by Matteo Ghidelli et al. [39,40].

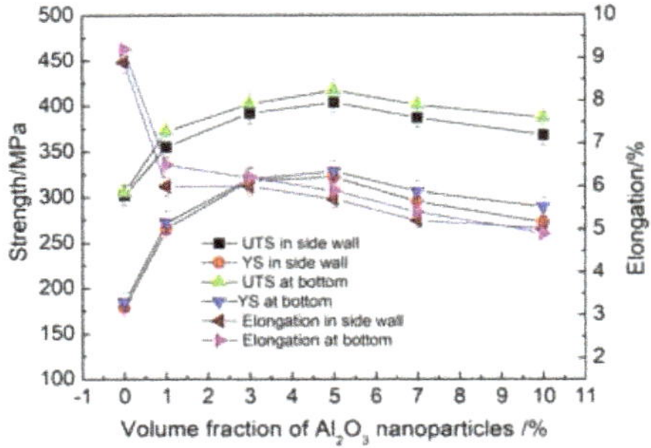

Figure 11. Influence of volume fraction of nano-sized Al_2O_3 on mechanical properties of rheoformed 2024 aluminum matrix composite component reinforced by 5 vol % Al_2O_3 nanoparticles at 620 °C for 25 min stirring time.

In addition, it can be noted that UTS and YS of the rheoformed composite components decreased slightly when the volume fraction of Al_2O_3 nanoparticles was more than 5%. It was attributed to greater agglomeration of Al_2O_3 nanoparticles that occurred in the composite due to a large volume fraction of Al_2O_3 nanoparticles [37]. A similar phenomenon was found in the research of Mazahery et al. [41] and Su et al. [37]. Mechanical properties in the present study were higher than those of Su et al. [37]. It may be due to the effect of a different fabrication method and different component shape on mechanical properties. Elongation of the rheoformed composite components is lower than that of the matrix component. In addition, elongation of the rheoformed composite components decreased with increasing volume fraction of Al_2O_3 nanoparticles. UTS at the bottom of the rheoformed composite components was higher as compared to that of the side wall. Figure 12 depicts the microstructure of rheoformed composite reinforced by nano-sized Al_2O_3 particles and matrix components at 620 °C and for 25 min stirring time.

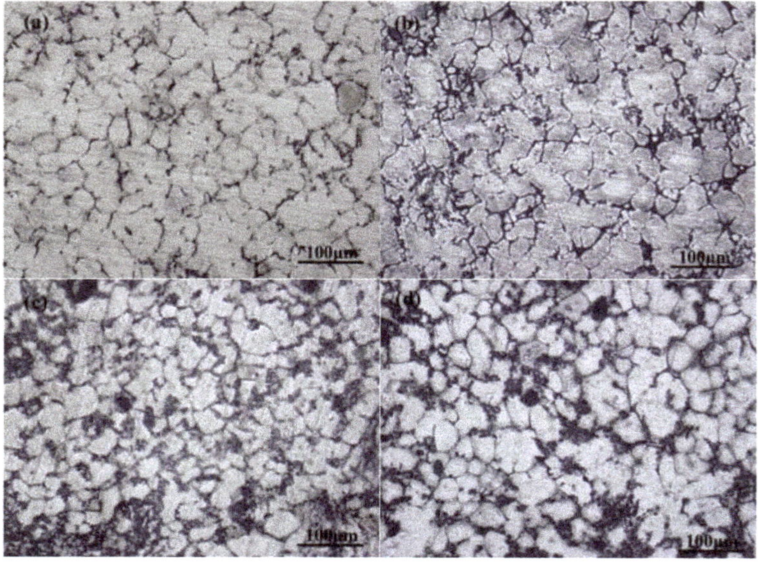

Figure 12. *Cont.*

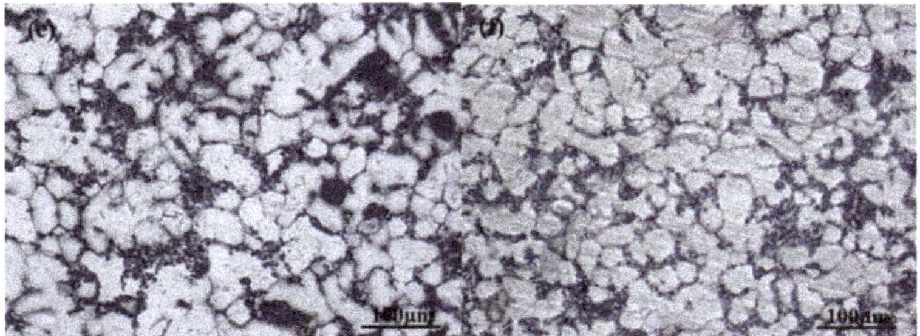

Figure 12. OM microstructure of rheoformed 2024 aluminum matrix composite component reinforced by different volume fraction of Al_2O_3 nanoparticles at 620 °C and for 25 min stirring time (**a**) 2024 matrix; (**b**) 1%; (**c**) 3%; (**d**) 5%; (**e**) 7%; (**f**) 10%.

Average grain size of the rheoformed composite components was smaller than that of the matrix component. It is due to the increasing heterogeneous nucleation for the aluminum matrix created by Al_2O_3 nanoparticles [37,38]. It is helpful to improve the UTS and YS of the rheoformed composite components due to a grain-refined strengthening effect [24,37,38].

3.4. Microstructure Characterization of the Rheoformed Composite Components

Figure 13 shows TEM micrographs at bottom of the rheoformed composite component reinforced by 5 vol % Al_2O_3 nanoparticles. As shown in Figure 13a–e, Al_2O_3 nanoparticles distributed uniformly in the matrix alloy. It is due to the fact that cavitation and acoustic streaming created via ultrasonic wave dispersed Al_2O_3 nanoparticles uniformly [26,42]. In addition, further dispersion of Al_2O_3 nanoparticles was obtained via controllable viscosity of semisolid slurries [27–29]. However, few Al_2O_3 nanoparticles were found in the TEM microstructure at a stirring temperature of 630 °C (Figure 13f). It may be due to the fact that greater agglomeration of Al_2O_3 nanoparticles leads to nonuniform dispersion of Al_2O_3 nanoparticles. This result also provided good evidence for the decreased mechanical properties of the rheoformed composite components at 630 °C (Figure 7a).

High density dislocations and dislocation tangles were found in the TEM images. Deformation of composite semisolid slurries depends on flow of liquid incorporating solid grains (FLS), sliding between solid grains (SSG), and plastic deformation of solid grains (PDS). PDS dominated the deformation of the semisolid slurry at the bottom. Therefore, plastic deformation at the bottom mainly relied on dislocation mobility. As a result, some dislocations were created due to plastic deformation, as shown in Figure 13. Twin crystal was noticed in the TEM microstructure (Figure 13b). It indicated that twinning also occurred in the plastic deformation of solid grains of semisolid slurries. Furthermore, some sub-grains were found in the twin crystal. It is due to the fact that dynamic recovery occurred in rheoformed composite components during the cooling course. Furthermore, Al_2O_3 nanoparticles were surrounded by these high density dislocations (Figure 13a,c,d).

Al_2O_3 nanoparticles acted as barriers of dislocations, leading to an enhancement of the mechanical properties of the rheoformed composite components [33]. It demonstrates that two strengthening mechanisms including dislocation strengthening caused by PDS and interaction between Al_2O_3 nanoparticles and dislocations play an important role in improving the mechanical properties of the rheoformed composite components together.

EDX analysis of Al_2O_3 nanoparticles on microstructure of rheoformed composite components reinforced was shown in Figure 14. Al_2O_3 nanoparticles, Al_2Cu phase, and $MgAl_2O_4$ phase were determined via mapping of Al, Cu, Mg, and O elements.

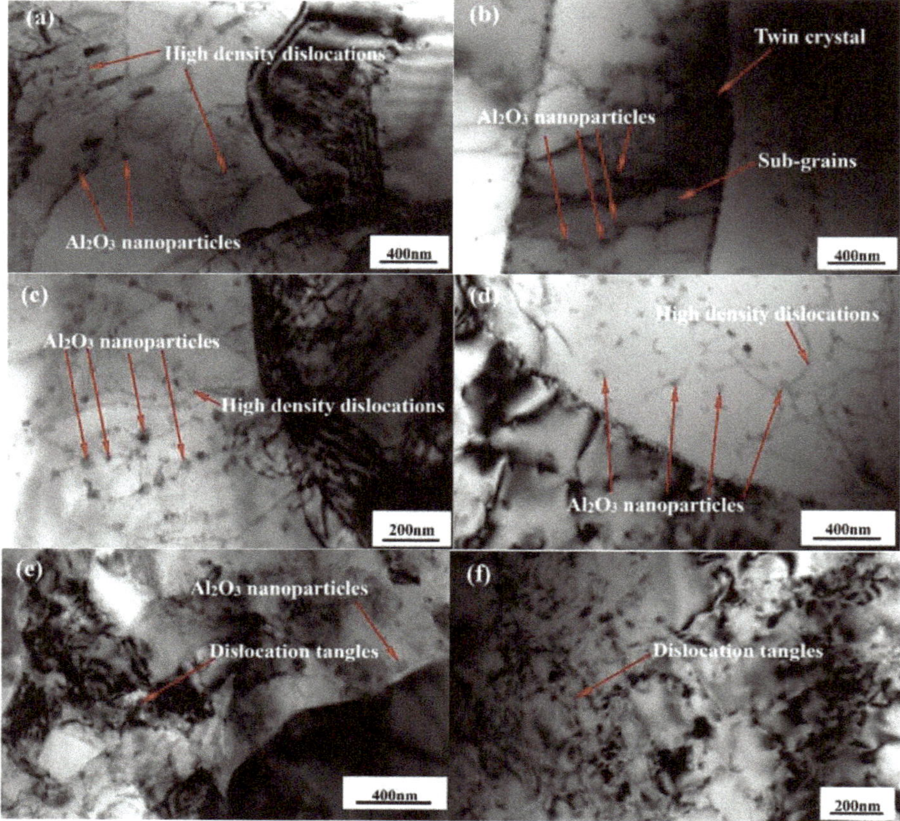

Figure 13. TEM micrographs of rheoformed composite component reinforced by 5 vol % Al$_2$O$_3$ nanoparticles at (**a**,**b**) 610 °C; (**c**,**d**) 620 °C and (**e**,**f**) 630 °C.

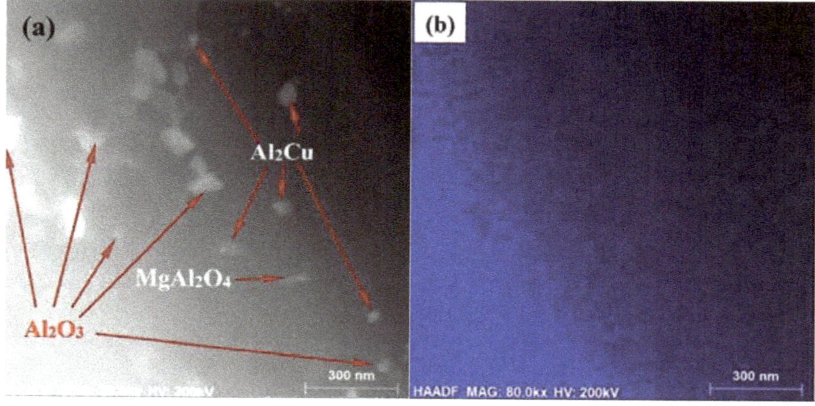

Figure 14. *Cont.*

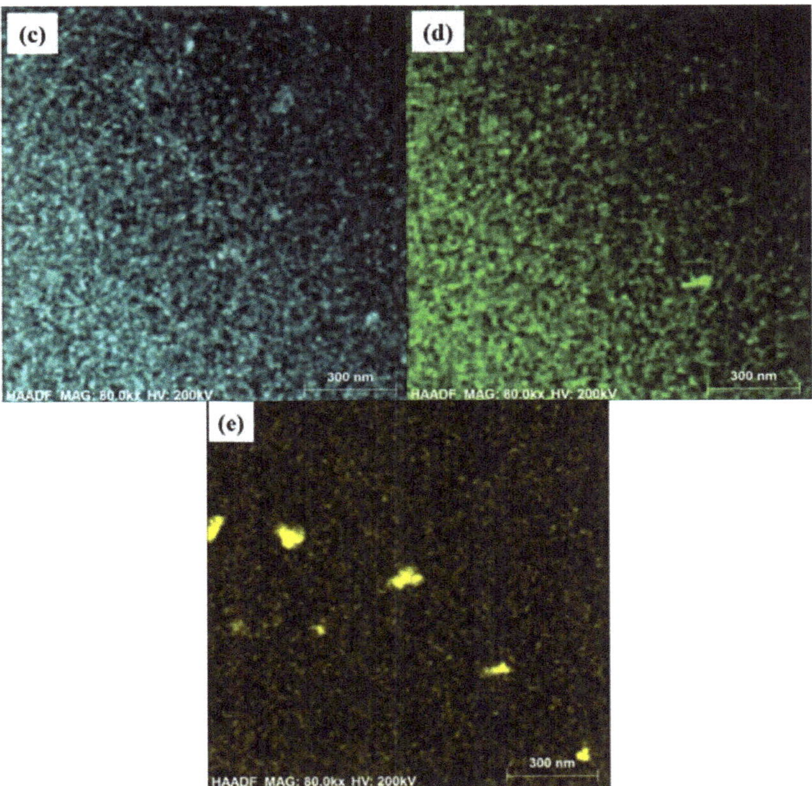

Figure 14. Energy dispersive X-ray spectrometer (EDX) analysis of nano-sized Al_2O_3 on microstructure of rheoformed 2024 aluminum matrix composite component reinforced by 5 vol % Al_2O_3 nanoparticles at 615 °C for for 25 min stirring time (**a**) TEM micrograph; (**b**) Al; (**c**) Cu; (**d**) Mg; (**e**) O.

As indicated in Figure 14, uniform dispersion of Al_2O_3 nanoparticles was noted in the TEM image due to double dispersion of cavitation and acoustic streaming created by ultrasonic wave and controllable viscosity of semisolid slurries. Existence of Al_2Cu phase was due to natural ageing that occurred at room temperature. $MgAl_2O_4$ phase illustrates that an interface reaction occurred in the interface of Al and Al_2O_3 nanoparticles due to existence of Mg element. $MgAl_2O_4$ phase has some advantages such as low density, high melting point, good wear resistance, good heat stability, and high mechanical properties [43,44]. Furthermore, relatively good wetting was obtained at the interface Al, $MgAl_2O_4$, and Al_2O_3 [45]. Therefore, the existence of $MgAl_2O_4$ phase has no effect on mechanical properties of the rheoformed composite components.

Figure 15 shows TEM microstructure and SAED of rheoformed composite components reinforced by 5 vol % Al_2O_3 nanoparticles for 25 min stirring time at 615 °C and 625 °C. As indicated in Figure 15, short-rod-like Al_2Cu phase and needle-like Al_2CuMg phase were determined via selected area electron diffraction (SAED). The second phases such as Al_2Cu phase and Al_2CuMg act as the role of strengthening the 2024 aluminum alloy. After the rheoformed composite components were cooling down to room temperature, these second phases precipitated in the 2024 matrix and strengthened it due to natural ageing that occurred in the components. These second phases also hindered the dislocation movement while the semisolid slurries were deformed. As a result, the mechanical properties of the rheoformed composite component were improved.

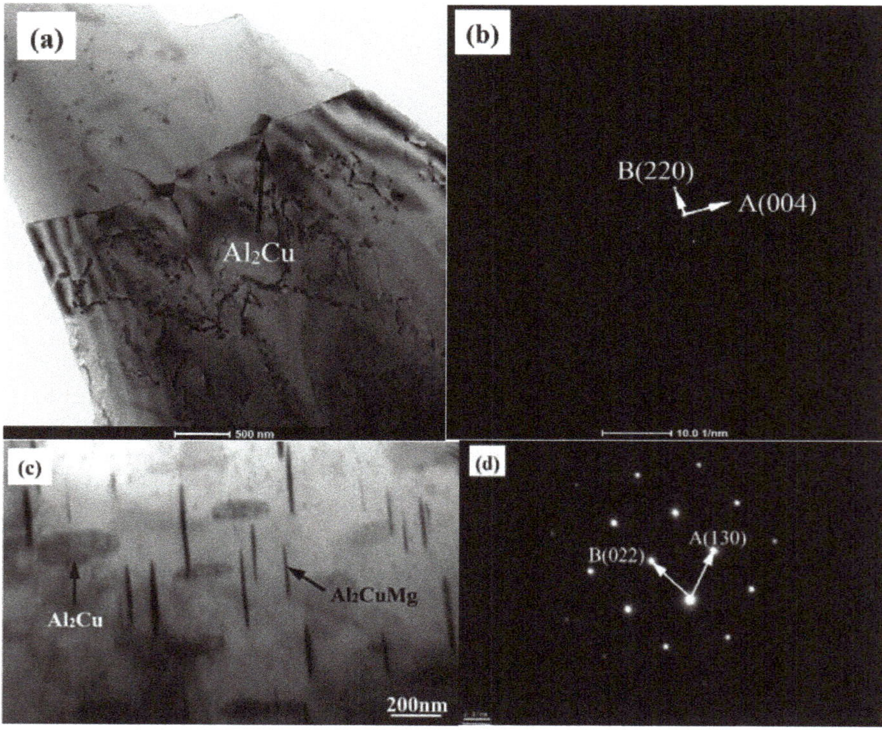

Figure 15. TEM microstructure and selected area electron diffraction (SAED) of rheoformed 2024 aluminum matrix composite component reinforced by 5 vol % Al$_2$O$_3$ nanoparticles for 25 min stirring time at 615 °C and 625 °C (**a**) TEM image at 615 °C (**b**) SAED at 615 °C (**c**) TEM image at 625 °C (**d**) SAED at 625 °C.

SEM image and EDX analysis in location B of the rheoformed composite component are presented in Figure 16. As shown in Figure 16a, the microstructure consisted of spheroidal solid grains (grey black color) and liquid phase (white color). It illustrates that the deformation mechanism of semisolid slurry in location B belongs to flow of liquid incorporating solid grains (FLS), sliding between solid grains (SSG). The atom ratio of aluminum and copper at grain boundary is close to 2:1 according to the Figure 16b. It illustrated that the liquid phase at the grain boundary was mainly composed of θ phase (Al$_2$Cu). The spheroidal solid grains contained a large amount of Al elements and a small amount of Cu and Mg elements, as show in Figure 16c. It illustrates that the spheroidal solid phase is α-Al phase. The α-Al phase is a solid solution of copper and magnesium in aluminum. Existence of O element was attributed to oxidation occurred in the grinding course of the specimens.

3.5. Influence of T6 Heat Treatment on Mechanical Properties

Mechanical properties of the rheoformed composite components reinforced by Al$_2$O$_3$ nanoparticles at 620 °C and for 25 min stirring time after T6 heat treatment are indicated in Figure 17. As indicated in Figure 17, mechanical properties of the rheoformed composite components were improved significantly after T6 heat treatment. UTS of 417MPa and YS of 328 MPa were achieved at the bottom of the rheoformed composite components at 620 °C. UTS and YS of the composite components with T6 heat treatment were 16.5% and 20.6% respectively, higher than those of the composite component without T6 heat treatment. Elongation of the composite components with T6 heat treatment was increased by 5.6% as compared to the composite component without T6 heat treatment. UTS and YS of the composite components with T6 heat treatment were increased by 36.7% and 49.1% respectively as compared to the matrix component with T6 heat treatment.

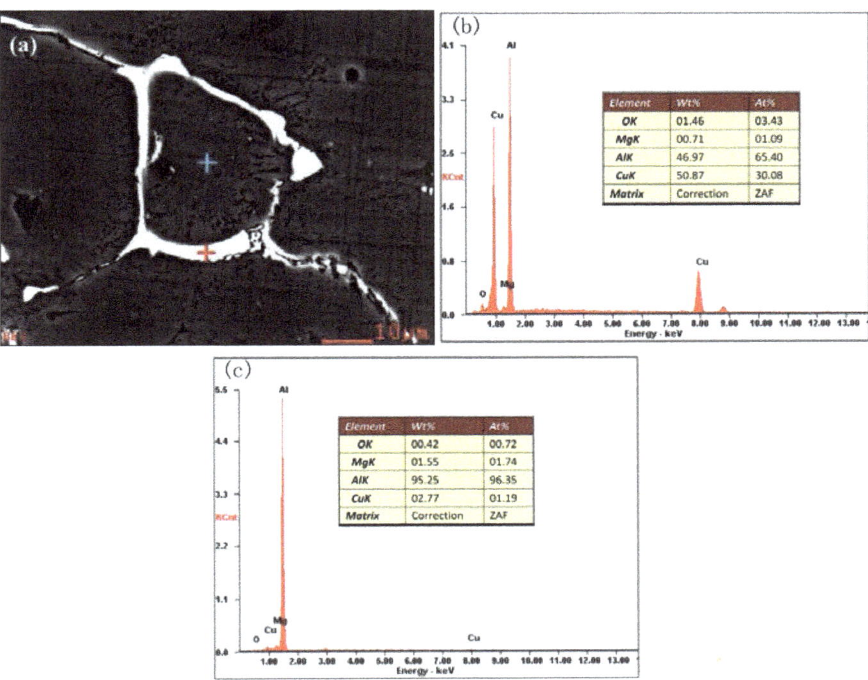

Figure 16. SEM image and EDX analysis of the rheoformed matrix components (**a**) SEM image; (**b**) EDX in location marked with red cross; (**c**) EDX in location marked with blue cross.

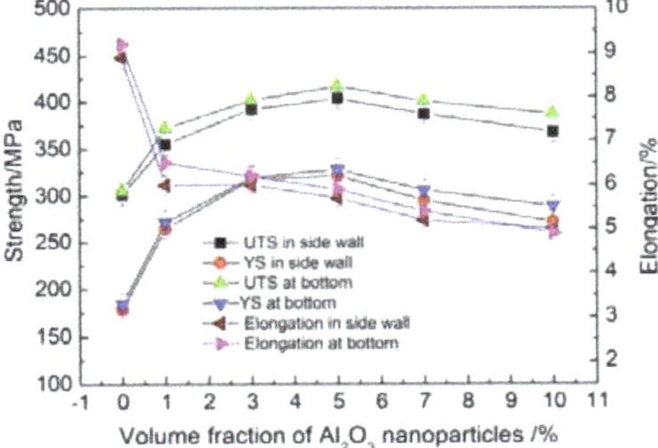

Figure 17. Mechanical properties of rheoformed 2024 aluminum matrix composite component reinforced by Al_2O_3 nanoparticles at 620 °C and for 25 min stirring time after T6 heat treatment.

TEM and EDX of the rheoformed composite components reinforced by 5 vol % Al_2O_3 nanoparticles at 620 °C after T6 heat treatment are shown in Figure 18. Short-rod-like Al_2Cu phase and needle-like Al_2CuMg phase were found in the TEM image of the rheoformed composite components.

The length and width of the short-rod-like Al_2Cu phase are about 100 nm and 65 nm respectively. As for needle-like Al_2CuMg phase, its length varies from 150 nm to 200 nm. Its width is in a range from 20 nm to 50 nm. The mechanical properties of the rheoformed composite component after T6 treatment were improved because of the needle-like Al_2CuMg phase and the short-rod-like Al_2Cu phase.

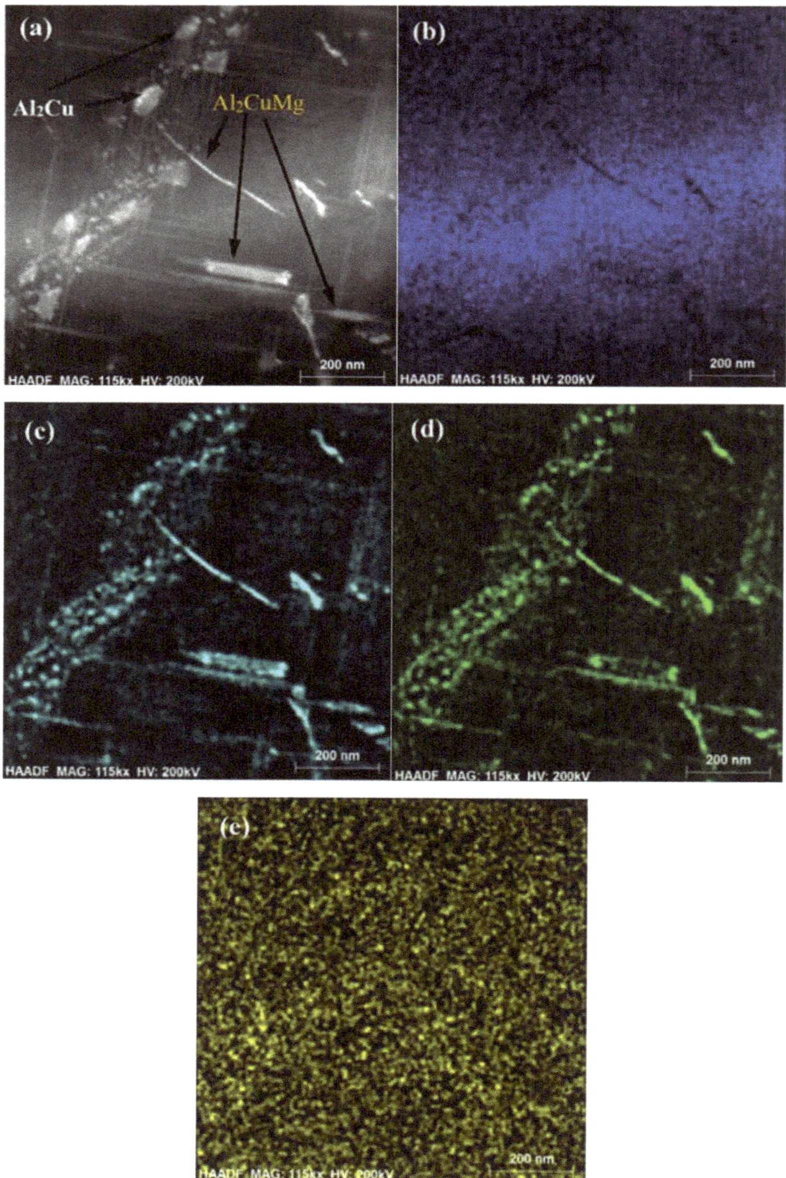

Figure 18. TEM and EDX of rheoformed 2024 aluminum matrix composite component reinforced by 5 vol % Al_2O_3 nanoparticles at 620 °C after T6 heat treatment (**a**) TEM image; (**b**) Al; (**c**) Cu; (**d**) Mg; (**e**) O.

Figure 19 shows the XRD patterns of the rheoformed composite components with T6 and without T6 treatment. The XRD pattern of the rheoformed composite components without T6 showed the presence of the Al peaks, Al_2Cu peaks, and Al_2O_3 peaks.

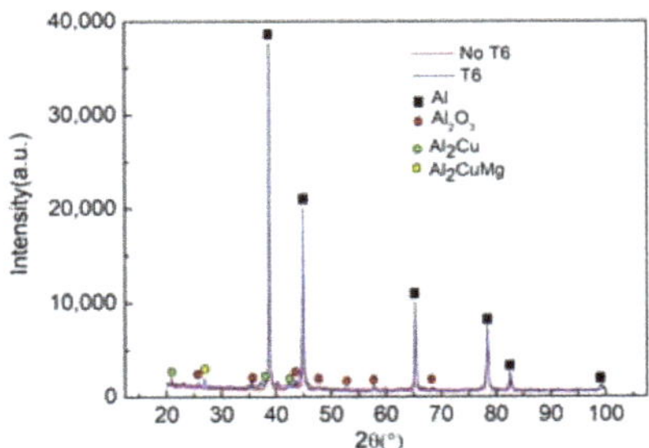

Figure 19. XRD analysis of rheoformed 2024 aluminum matrix composite component reinforced by 5 vol % Al_2O_3 nanoparticles at 620 °C for 25 min stirring time before and after T6 heat treatment.

Al_2CuMg phase besides Al phase, Al_2Cu phase, and Al_2O_3 nanoparticles was also detected in the XRD pattern of the rheoformed composite components with T6. It illustrates that more needle-like Al_2CuMg phase besides short-rod-like Al_2Cu phase precipitated in the rheoformed composite components with T6. As a result, mechanical properties of the rheoformed composite components with T6 were improved significantly as compared to the rheoformed composite components without T6.

3.6. Wear Behavior of the Rheoformed Composite Components

Figure 20 shows wear rate of the rheoformed composite components reinforced by different volume fraction Al_2O_3 nanoparticles at 620 °C for 25 min stirring time. As shown in Figure 20, wear resistance of the rheoformed composite components increased significantly as compared to that of the rheoformed matrix components. Furthermore, wear resistance of the rheoformed composite components increased when volume fraction of Al_2O_3 nanoparticles increased from 1% to 7%. The research of Alhawari et al. [46] also showed that wear resistance of the composite part formed via semisolid processing was higher than that of the composite part via conventional stirring casting. When volume fraction of Al_2O_3 nanoparticles reached 10%, wear rate of the rheoformed composite components decreased slightly as compared to that of the rheoformed composite components with 5% Al_2O_3 nanoparticles. Greater agglomeration of Al_2O_3 nanoparticles leads to difficult uniform dispersion of Al_2O_3 nanoparticles. It led to a decrease in effective dispersion of Al_2O_3 nanoparticles in the 2024 matrix, reducing wear resistance.

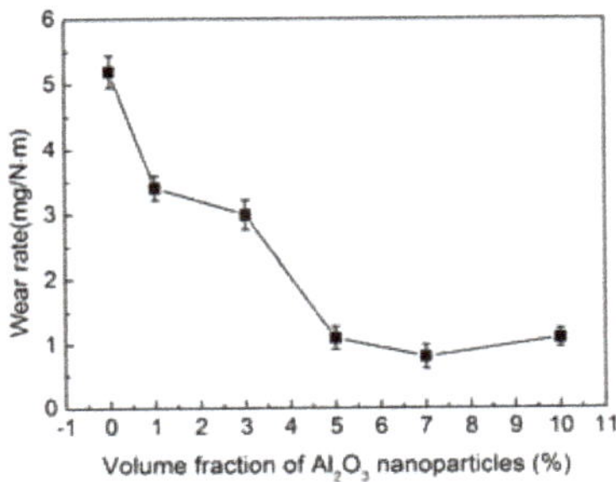

Figure 20. Wear rate of rheoformed 2024 aluminum matrix composite component reinforced by different volume fraction Al_2O_3 nanoparticles at 620 °C for 25 min stirring time.

The specimens subjected to wear test have been examined via SEM (Figure 21). The surface exhibited some clear longitudinal abrasive grooves due to the ploughing effects of harder steel asperities. With an increase of volume fraction of Al_2O_3 nanoparticles, the depth of ploughing grooves became shallow. It indicates that the composite's resistance to wear increases. It is due to the fact that the increase in volume fraction of Al_2O_3 nanoparticles led to an increase in the hardness of the composite. Increase of the hardness of the material is beneficial to improve the resistance to wear [47]. In addition, a delamination was found in the microstructure of the worn surface. It illustrates that the dominant wear mechanism was a combination of adhesion and delamination mechanisms, similar to the findings of Alhawari [46].

Figure 21. Wear morphology of rheoformed 2024 aluminum matrix composite component reinforced by Al_2O_3 nanoparticles with different volume fraction at 620 °C for 25 min stirring time (**a**) 0; (**b**) 1%; (**c**) 3%; (**d**) 5%; (**e**) 7%; (**f**) 10%.

4. Conclusions

(1) 2024 aluminum matrix composite components reinforced by Al_2O_3 nanoparticles were rheoformed successfully. Complete filling status and good surface quality were achieved in the rheoformed composite components. Microstructure at the top and middle side wall consisted of near spheroidal grains and liquid phase, indicating dependence of deformation on liquid incorporating solid grains (FLS) and sliding between solid grains (SSG). However, obvious elongated grains were noted in the low side wall and at bottom of the rheoformed composite components. It illustrated that deformation in these locations was dominated by plastic deformation of solid grains (PDS).

(2) Mechanical properties of the rheoformed composite components were influenced by stirring temperature, stirring time, and volume fraction of Al_2O_3 nanoparticles of composite semisolid slurries. The optimal UTS of 358 MPa and YS of 245 MPa were obtained at the bottom of the rheoformed composite components with 5% Al_2O_3 nanoparticles at 620 °C for 25 min stirring time. The increasing degrees of UTS are 17.5% and 31.7% as compared to the matrix component. Uniform dispersed Al_2O_3 nanoparticles and high density dislocations and dislocation tangles caused by PDS led to an improvement of mechanical properties. Needle-like Al_2CuMg phase and short-rod-like Al_2Cu phase were found in the microstructure of the rheoformed composite components due to natural ageing. $MgAl_2O_4$ phase has no effect on mechanical properties due to good wetting and high properties.

(3) T6 heat treatment led to an improvement of mechanical properties of the rheoformed composite components. UTS of 417MPa and YS of 328 MPa were achieved at bottom of the rheoformed composite components with 5% Al_2O_3 nanoparticles at 620 °C for 25 min stirring time. UTS, YS, and elongation of the composite components with T6 heat treatment were increased by 16.5%, 20.6%, and 5.6% respectively as compared to the composite component without T6 heat treatment. UTS and YS of the composite components with T6 heat treatment were increased by 36.7% and 49.1% respectively as compared to the matrix components with T6 heat treatment. Improvement of mechanical properties of the rheoformed composite components with T6 was attributed to a large amount of precipitated needle-like Al_2CuMg phase and short-rod-like Al_2Cu phase.

(4) Wear resistance of the rheoformed composite components increased as compared to that of the matrix component. Furthermore, wear resistance of the rheoformed composite components increased with an increase of Al_2O_3 nanoparticles from 1% to 7%. A slight decrease in wear rate of the rheoformed composite components resulted from 10% Al_2O_3 nanoparticles due to a decrease in effective dispersion of Al_2O_3 nanoparticles caused by greater agglomeration. The delamination and shallow ploughing grooves illustrate that the dominant wear mechanism was a combination mechanism of adhesion and delamination. To sum up, the optimal process parameters to obtain best comprehensive mechanical properties and resistance to wear are a stirring temperature of 620 °C, a stirring time of 25 min, and a volume fraction of 5% nano-sized Al_2O_3 nanoparticles.

Author Contributions: J.J. designed most experiments, analyzed the results, and wrote this manuscript. G.X. and Y.L. performed most experiments. Y.W. helped analyze the experimental data and gave some constructive suggestions on this manuscript.

Funding: This research was funded by the Natural Science Foundation of China (NSFC) under Grant No. 51375112.

Conflicts of Interest: The authors declare no conflicts of interest.

References

1. Mandal, D.; Viswanathan, S. Effect of heat treatment on microstructure and interface of SiC particle reinforced 2124 Al matrix composite. *Mater. Charact.* **2013**, *85*, 73–81. [CrossRef]

2. Mindivan, H.; Kayali, E.S.; Cimenoglu, H. Tribological behavior of squeeze cast aluminum matrix composites. *Wear* **2008**, *265*, 645–654. [CrossRef]

3. Bharath, V.; Nagaral, M.; Auradi, V.; Kori, S.A. Preparation of 6061Al-Al$_2$O$_3$ MMC's by stir casting and evaluation of mechanical and wear properties. *Procedia Mater. Sci.* **2014**, *6*, 1658–1667. [CrossRef]

4. Natarajan, N.; Vijayarangan, S.; Rajendran, I. Wear behaviour of A356/25SiC$_p$ aluminium matrix composites sliding against automobile friction material. *Wear* **2006**, *261*, 812–822. [CrossRef]

5. Prabu, S.B.; Karunamoorthy, L.; Kathiresan, S.; Mohan, B. Influence of stirring speed and stirring time on distribution of particles in cast metal matrix composite. *J. Mater. Process. Technol.* **2006**, *171*, 268–273. [CrossRef]

6. Tjong, S.C.; Ma, Z.Y. High-temperature creep behaviour of powder-metallurgy aluminium composites reinforced with SiC particles of various sizes. *Compos. Sci. Technol.* **1999**, *59*, 1117–1125. [CrossRef]

7. Bajpai, G.; Purohit, R.; Rana, R.S.; Rajpurohit, S.S.; Rana, A. Investigation and testing of mechanical properties of Al-nano SiC composites through cold isostatic compaction process. *Mater. Today Proc.* **2017**, *4*, 2723–2732. [CrossRef]

8. Tatar, C.; Özdemir, N. Investigation of thermal conductivity and microstructure of the α-Al$_2$O$_3$ particulate reinforced aluminum composites (Al/Al$_2$O$_3$-MMC) by powder metallurgy method. *Phys. B Condensed Matter* **2010**, *405*, 896–899. [CrossRef]

9. Zhang, Q.; Wu, G.H.; Chen, G.Q.; Jiang, L.T.; Luan, B.F. The thermal expansion and mechanical properties of high reinforcement content SiCp/Al composites fabricated by squeeze casting technology. *Compos. Part A-Appl. S.* **2003**, *34*, 1023–1027. [CrossRef]

10. Chou, S.N.; Huang, J.L.; Lii, D.F.; Lu, H.H. The mechanical properties of Al$_2$O$_3$/aluminum alloy A356 composite manufactured by squeeze casting. *J. Alloy. Compd.* **2006**, *419*, 98–102. [CrossRef]

11. Chen, W.P.; Liu, Y.X.; Yang, C.; Zhu, D.Z.; Li, Y.Y. (SiCp + Ti)/7075Al hybrid composites with high strength and large plasticity fabricated by squeeze casting. *Mater. Sci. Eng. A* **2014**, *609*, 250–254. [CrossRef]

12. Wang, L.; Qiu, F.; Zou, Q.; Yang, D.L.; Tang, J.; Gao, Y.Y.; Li, Q.; Han, X.; Shu, S.L.; Chang, F.; et al. Microstructures and tensile properties of nano-sized SiC$_p$/Al-Cu composites fabricated by semisolid stirring assisted with hot extrusion. *Mater. Charact.* **2017**, *131*, 195–200. [CrossRef]

13. Guan, L.N.; Geng, L.; Zhang, H.W.; Huang, L.J. Effects of stirring parameters on microstructure and tensile properties of (ABO$_w$ + SiC$_p$)/6061Al composites fabricated by semi-solid stirring technique. *Trans. Nonferrous Met. Soc. China* **2011**, *21*, s274–s279. [CrossRef]

14. Sameer, K.D.; Sumanm, K.N.S.; Tara, S.C.; Ravindra, K.; Palash, P.; Venkata, S.S.B. Microstructure, mechanical response and fractography of AZ91E/Al$_2$O$_3$ (p) nano composite fabricated by semi solid stir casting method. *J. Magn. Alloys* **2017**, *5*, 48–55.

15. Prashanth, K.G.; Scudino, S.; Chaubey, A.K.; Löber, L.; Wang, P.; Attar, H.; Schimansky, F.P.; Pyczak, F.; Eckert, J. Processing of Al–12Si–TNM composites by selective laser melting and evaluation of compressive and wear properties. *J. Mater. Res.* **2016**, *31*, 55–65. [CrossRef]

16. Prashantha, K.G.; Shahabia, H.S.; Attara, H.; Srivastavac, V.C.; Ellendtd, N.; Uhlenwinkeld, V.; Eckerta, J.; Scudino, S. Production of high strength Al$_{85}$Nd$_8$Ni$_5$Co$_2$alloy by selective laser melting. *Addit. Manuf.* **2015**, *6*, 1–5. [CrossRef]

17. Attar, H.; Bönisch, M.; Calin, M.; Zhang, L.C.; Scudino, S.; Eckert, J. Selective laser melting of in situ titanium—Titanium boride composites: Processing, microstructure and mechanical properties. *Acta Mater.* **2014**, *76*, 13–22. [CrossRef]

18. Cui, Y.N.; Wang, L.F.; Ren, J.Y. Multi-functional SiC/Al Composites for aerospace applications. *Chin. J. Aeronaut.* **2008**, *21*, 578–584.

19. Lee, S.S.; Yeo, J.S.; Hong, S.H.; Yoon, D.J.; Na, K.H. The fabrication process and mechanical properties of SiCp/Al-Si metal matrix composites for automobile air-conditioner compressor piston. *J. Mater. Process. Technol.* **2001**, *113*, 202–208. [CrossRef]

20. Koli, D.K.; Agnihotri, G.; Purohit, R. A review on properties, behaviour and processing methods for Al-nano Al$_2$O$_3$ composites. *Procedia Mater. Sci.* **2014**, *6*, 567–589. [CrossRef]

21. Raju, P.R.M.; Siriyala, R.; Raju, K.S.; Raju, R.V.R. Evaluation of fatigue life of Al2024/Al$_2$O$_3$ particulate nano composite fabricated using stir casting technique. *Mater. Today Proc.* **2017**, *4*, 3188–3196. [CrossRef]

22. Raturi, A.; Mer, K.K.S.; Pant, P.K. Synthesis and characterization of mechanical, tribological and micro structural behaviour of Al 7075 matrix reinforced with nano Al$_2$O$_3$ particles. *Mater. Today Proc.* **2017**, *4*, 2645–2658. [CrossRef]

23. Sajjadi, S.A.; Parizi, M.T.; Ezatpour, H.R.; Sedghi, A. Fabrication of A356 composite reinforced with micro and nano Al_2O_3 particles by a developed compocasting method and study of its properties. *J. Alloy. Compd.* **2012**, *511*, 226–231. [CrossRef]
24. Akbari, M.K.; Baharvandi, H.R.; Mirzaee, O. Fabrication of nano-sized Al_2O_3 reinforced casting aluminum composite focusing on preparation process of reinforcement powders and evaluation of its properties. *Compos. Part B-Eng.* **2013**, *55*, 426–432. [CrossRef]
25. Akbari, M.K.; Baharvandi, H.R.; Mirzaee, O. Nano-sized aluminum oxide reinforced commercial casting A356 alloy matrix: Evaluation of hardness, wear resistance and compressive strength focusing on particle distribution in aluminum matrix. *Compos. Part B-Eng.* **2013**, *52*, 262–268. [CrossRef]
26. Yang, Y.; Lan, J.; Li, X.C. Study on bulk aluminum matrix nano-composite fabricated by ultrasonic dispersion of nano-sized SiC particles in molten aluminum alloy. *Mater. Sci. Eng. A* **2004**, *380*, 378–383. [CrossRef]
27. Kandemir, S.; Atkinson, H.V.; Weston, D.P.; Hainsworth, S.V. Thixoforming of A356/SiC and A356/TiB_2 nanocomposites fabricated by a combination of green compact nanoparticle incorporation and ultrasonic treatment of the melted compact. *Metall. Mater. Trans. A* **2014**, *45*, 5782–5798. [CrossRef]
28. Jiang, J.F.; Wang, Y. Microstructure and mechanical properties of the rheoformed cylindrical part of 7075 aluminum matrix composite reinforced with nano-sized SiC particles. *Mater. Des.* **2015**, *79*, 32–41. [CrossRef]
29. Kandemir, S. Microstructure and mechonicol properties of A357/SiC nonocomposites fobricated by ultrosonic covitotion-based dispersion of ball-milled nonoparticles. *J. Compo. Mater.* **2017**, *51*, 395–404. [CrossRef]
30. ASTM Standard E8M. *Standard Test Methods for Tension Testing of Metallic Materials [Metric]*; ASTM International: West Conshohocken, PA, USA, 2008.
31. Chen, C.P.; Tsao, C.-Y.A. Semi-solid deformation of non-dendrtic structure-I. Phemonological behavior. *Acta Mater.* **1997**, *45*, 1955–1968. [CrossRef]
32. Kubotak, K.; Mabuchi, M.; Higashi, K. Review processing and mechanical properties of fine-grained Mg alloys. *J. Mater. Sci.* **1999**, *34*, 2255–2262. [CrossRef]
33. Sajjadi, S.A.; Ezatpour, H.R.; Beygi, H. Microstructure and mechanical properties of Al-Al_2O_3 micro and nanocomposites fabricated by stir casting. *Mater. Sci. Eng. A* **2011**, *528*, 8765–8771. [CrossRef]
34. Yar, A.A.; Montazerian, M.; Abdizadeh, H.; Baharvandi, H.R. Microstructure and mechanical properties of aluminum alloy matrix composite reinforced with nano-particle MgO. *J. Alloy. Compd.* **2009**, *484*, 400–404. [CrossRef]
35. Nguyen, Q.B.; Gupta, M. Enhancing compressive response of AZ31B using nano-Al_2O_3 and copper additions. *J. Alloy. Compd.* **2010**, *490*, 382–387. [CrossRef]
36. Sastry, S.; Krishnab, M.; Uchil, J. A study on damping behaviour of aluminite particulate reinforced ZA-27 alloy metal matrix composites. *J. Alloy. Compd.* **2001**, *314*, 268–274. [CrossRef]
37. Su, H.; Gao, W.L.; Feng, Z.H.; Lu, Z. Processing, microstructure and tensile properties of nano-sized Al_2O_3 particle reinforced aluminum matrix composites. *Mater. Des.* **2012**, *36*, 590–596. [CrossRef]
38. Zhong, X.L.; Wong, W.L.E.; Gupta, M. Enhancing strength and ductility of magnesium by integrating it with aluminum nanoparticles. *Acta Mater.* **2007**, *55*, 6338–6344. [CrossRef]
39. Ghidelli, M.; Sebastiani, M.; Collet, C.; Guillemet, R. Determination of the elasticmoduli and residual stresses of freestanding Au-TiW bilayer thin films by nanoindentation. *Mater. Des.* **2016**, *106*, 436–445. [CrossRef]
40. Ghidelli, M.; Sebastiani, M.; Johanns, K.E.; Pharr, G.M. Effects of indenter angle on micro-scale fracture toughness measurement by pillar splitting. *J. Am. Ceram. Soc.* **2017**, *100*, 5731–5738. [CrossRef]
41. Mazahery, A.; Abdizadeh, H.; Baharvandi, H.R. Development of high-performance A356/nano-Al_2O_3 composites. *Mater. Sci. Eng. A* **2009**, *518*, 61–64. [CrossRef]
42. Lan, J.; Yang, Y.; Li, X.C. Microstructure and microhardness of SiC nanoparticles reinforced magnesium composites fabricated by ultrasonic method. *Mater. Sci. Eng. A* **2004**, *386*, 284–290. [CrossRef]
43. Korgul, P.; Wilson, D.R.; Lee, W.E. Microstructural analysis of corroded alumina-spinel castable refractories. *J. Eur. Ceram. Soc.* **1997**, *17*, 77–84. [CrossRef]
44. Ghosha, A.; Sarkar, R.; Mukerjee, B.; Das, S.K. Effect of spinel content on the properties of magnesia-spinel composite refractory. *J. Eur. Ceram. Soc.* **2004**, *24*, 2079–2085. [CrossRef]
45. Zang, J. Wetting and Adhesion at Al/MgAl2O4 Interface and the Effect of Substrate Crystallographic Orientation. Ph.D. Thesis, Jilin University, Changchun, China, 2014.

46. Alhawari, K.S.; Omar, M.Z.; Ghazali, M.J.; Salleh, M.S.; Mohammed, M.N. Wear properties of A356/Al_2O_3 metal matrix composites produced by semisolid processing. *Procedia Eng.* **2013**, *68*, 186–192. [CrossRef]
47. Ehtemam-Haghighi, S.; Prashanthb, K.G.; Attar, H.; Chaubey, A.K.; Caod, G.H.; Zhang, L.C. Evaluation of mechanical and wear properties of Ti-xNb-7Fe alloys designed for biomedical applications. *Mater. Des.* **2016**, *111*, 592–599. [CrossRef]

Article

Thixoforming of an Fe-Rich Al-Si-Cu Alloy—Thermodynamic Characterization, Microstructural Evolution, and Rheological Behavior

Gabriela Lujan Brollo, Cecília Tereza Weishaupt Proni and Eugênio José Zoqui *

Materials and Manufacturing Department, Faculty of Mechanical Engineering, University of Campinas—UNICAMP, Campinas, SP 13083-860, Brazil; gbrollo@fem.unicamp.br (G.L.B.); wyliah@gmail.com (C.T.W.P.)
* Correspondence: zoqui@fem.unicamp.br; Tel.: +55-19-35213296; Fax: +55-19-35213722

Received: 28 February 2018; Accepted: 20 April 2018; Published: 9 May 2018

Abstract: Thixoforming depends on three factors: (a) the thermodynamic stability of the solid-to-liquid transformation in the presence of temperature fluctuations; (b) the size and morphology of the solid particles in the liquid in the semisolid state; and (c) the rheology of the semisolid slurry during formation. In this study, these parameters were characterized for an Al-Si-Cu alloy with a high Fe content (B319+Fe alloy). Fe is usually found in raw metal produced by recycling, and its removal increases processing costs. This study is an attempt to use this lower-cost alloy for the thixoforming route. Thermodynamic analysis was performed by numerical simulation (under Scheil conditions) and the application of the differentiation method (DM) to differential scanning calorimetry (DSC) curves recorded during heating cycles up to 700 °C at 5, 10, 15, 20, and 25 °C/min. The processing window was evaluated by comparing the results of the DM and those of the analysis of open-die thixoforged samples after isothermal heat treatment at 575, 582, 591, and 595 °C for 0, 30, 60, and 90 s. The microstructural and rheological behavior of the semisolid slurry was analyzed at 591 and 595 °C for all four soak times. Isothermal heat treatment caused the refinement and spheroidization of the solid phase. Good agreement between the predicted thermodynamic behavior and the microstructural behavior of the thixoformed B319+Fe alloy samples was observed. Although the alloy exhibited a coarse microstructure, it was microstructurally and rheologically stable at all temperatures and for all the soak times studied, indicating that B319+Fe is a promising raw material for thixoforming.

Keywords: semisolid processing; thixoformability; Fe-rich Al-Si-Cu alloy

1. Introduction

Semisolid materials (SSM) processing is the formation of metallic alloys in the semisolid state and can be an advantageous alternative to conventional metal forming and casting operations [1,2]. SSM processing can be performed by cooling liquid metal to the semisolid state in a range of operations called rheocasting [3] or by heating solid metal to the semisolid condition in a range of operations called thixoforming [4].

Thixoforming is carried out at approximately 0.3–0.6 liquid fractions and consists essentially of three main consecutive steps: (a) controlled heating from the solid to the semisolid condition at a specific heating rate until the desired thixoforming temperature is reached; (b) isothermal treatment at the chosen processing temperature for a specific soak time; and (c) formation of the semisolid slurry into a die with the pre-form (near-net-shape) or form (net-shape) of the final product [5].

It is crucial in thixoforming that (a) the solid particles in the liquid have a refined, spheroidal microstructure and (b) the solid-liquid mixture is stable. The former requirement ensures that the semisolid slurry has a thixotropic rheology, which is responsible for the advantageous smooth, laminar die-filling characteristic of SSM processing [6–9]. The raw-metal microstructure, heating rate, processing temperature, soak time, and shear rate during formation play a crucial role in the microstructure and, consequently, the rheology of the semisolid slurry. A stable solid-liquid mixture, i.e., a controllable microstructure, rheology, and liquid fraction in the presence of temperature variations during formation, ensures that SSM processing is reproducible. Achieving this stable mixture requires certain thermodynamic characteristics, which in turn depend on the chemical composition of the alloy and the kinetics of the thixoforming process. An alloy possessing these characteristics is said to be thixoformable [10–12].

Although SSM technology has been in use for approximately 40 years, the range of alloys used in thixoforming processes is still limited mainly to the aluminum alloys A356 and A357 [13]. The Semisolid Processing Group in the Faculty of Mechanical Engineering at the State University of Campinas (UNICAMP) has undertaken several research efforts to increase the number of potential alloys that can be used in SSM processing. These include studies of systems such as Al-Si, Al-Si-Cu, Al-Si-Mg, and Fe-C-Si using many pre-processing routes, including electromagnetic and ultrasound stirring and equal channel angular pressing (ECAP) [14–20]. The present study represents a new attempt to expand the range of raw materials suitable for SSM processing and focuses on Al alloys with high levels of Fe, an element usually found in alloys produced using recycled metal and whose removal increases processing costs [21,22]. The study analyses the suitability for thixoforming of an Al-Si-Cu alloy with a high level of Fe. The differentiation method (DM) [23,24] is used, and the microstructure and rheology of the semisolid slurry are analyzed. As the focus of the study is the processing of this alloy; the mechanical properties and microstructure of the final thixoformed product are not discussed here but will form the subject of a future study.

2. Experimental Procedure

The chemical composition of the Al-Si-Cu-Fe alloy studied here is shown in Table 1 and was determined with an Anacon BILL optical emission spectrometer (OES) (Anacon, São Paulo, Brazil). The alloy has the same chemical composition as commercial B319 alloy except for the Fe content, which is greater than the maximum permitted for this grade (1.2 wt.%) [25]. It is therefore referred to here as "B319+Fe". The alloy was melted and poured in a semi-continuous casting system consisting of a refrigerated copper mold (30 mm D × 180 mm L). Grain refinement was achieved by adding an Al-5.0 wt.%Ti-1.0 wt.%B master alloy until the liquid alloy had a Ti content of 0.2 wt.%, and then applying magnetic stirring to the solidifying ingots with an 8 kW 14 G induction coil around the casting die.

Table 1. Chemical composition of the B319+Fe alloy studied here as measured by OES and the expected range (min-max) for a conventional B319 alloy in the literature [25].

Element	Content (wt.%)		
	Literature (min.)	Measured	Literature (max.)
Si	5.5	6.48 ± 0.140	6.5
Cu	3.0	3.53 ± 0.090	4.0
Fe	0.0	1.30 ± 0.021	1.2
Mg	0.1	0.21 ± 0.003	0.5
Zn	0.0	0.91 ± 0.016	1.0
Mn	0.0	0.35 ± 0.016	0.8
Sn	-	0.04 ± 0.017	-
Ni	0.0	0.03 ± 0.003	0.5
Cr	-	0.02 ± 0.001	-
Res *	0.0	0.12 ± 0.003	0.5
Al (Bal.)	84.8	87.0 ± 0.146	91.4

* The sum of residual elements.

Thermal analysis to determine the experimental f_l vs. T curve for the alloy was performed with a NETZSCH STA 409C DSC system (Netzsch, Bayern, Germany). The STA 409C has a resolution of 720,000 digits, a data acquisition rate of 20 data s^{-1}, and a sensitivity of 710 mV, giving a measurement accuracy of more than 0.5 °C. The cylindrical DSC sample (max. 5 mm D × 3 mm H) weighs approximately 20 mg. The sample was heated to 700 °C at 5, 10, 15, 20, and 25 °C/min. The heat flow rate (here called *HFR*) and temperature were monitored with thermocouples so that *HFR* vs. T curves could be generated. All DSC tests were carried out using the same specimen to prevent differences in the chemical composition along the billet from influencing the results. Prior to data acquisition, the DSC sample was heated to 700 °C and then cooled to room temperature at the lowest rate (5 °C/min), to also guarantee chemical homogeneity along the specimen. The differentiation method (DM) [23] was applied to the DSC data to determine the temperatures corresponding to the solidus, liquidus, main eutectic knee, and limits of the thixoforming working window. To this end, the *HFR* data were differentiated with respect to temperature, giving the variation in *HFR* with temperature during the heating process (d*HFR*/dT vs. T curves). The f_l vs. T relationship was determined with NETZSCH Protheus® thermal analysis software (V4.8.3, Netzsch, Bayern, Germany) by applying the corrected Flynn method [24] of integration of partial areas under the DSC curves. The use of the corrected Flynn method is justified by the presence of tertiary phases on the DSC curves. Origin® software (V8.0951, OriginLab Corporation, Northampton, MA, USA) was used for direct differentiation of the f_l vs. T curve to obtain the df_l/dT (sensitivity) vs. T curve.

Thermodynamic simulation was performed with Thermo-Calc® (V5.01.61, Foundation for Computacional Thermodynamics, Stockolm, Sweden) to generate the f_l vs. T curve for the alloy under non-equilibrium Scheil conditions [26]. The results of this simulation were used as a reference for the experimental curves. The percentages of Al, Cu, Mg, Fe, and Ti were considered in the simulation. The TTAL5 database was used.

Thermal treatment to analyze the microstructural evolution of the B319+Fe alloy in the semisolid state was performed using cylindrical samples (20 mm D × 30 mm H) taken from the as-cast ingots. The samples were heated to approximately 550 °C at an initial heating rate of 100 °C/min and then further heated to 575, 582, 591, or 595 °C at a rate of approximately 18–22 °C/min. After being kept at the holding temperature for 0, 30, 60, or 90 s, the samples were immediately quenched in water. Temperature was monitored with a thermocouple placed within a hole drilled at the upper flat surface of the cylinder. The thermocouple tip matched the geometrical center of the sample (10 mm D × 15 mm H). A Norax 8 kHz 25 kW induction furnace (NoraxCanada, Montreal, QC, Canada) with a temperature accuracy within ±2 °C was used.

Microstructural homogeneity along the transversal section of the ingots was achieved in approximately 80% of the correspondent length. Approximately 10% of the length presented coarser grains, due to the lower heat flow in the center of the ingot, and 10% (borders) presented smaller grains due to contact with the refrigerated surface during casting. Therefore, micrographs used in this work were taken from this intermediate homogeneous region between the center and borders.

The microstructures of the as-cast and heat-treated samples were characterized by conventional black and white (B&W) and polarized-light color microscopy using a Leica DM ILM optical microscope (Leica Microsystems, Wetzlar, Germany). For color microscopy, the samples were etched electrolytically in 1.8% HBF_4 at 0.6 A and 30 V for 180 s under stirring. Polarizing filters were used to obtain color images of the grains so that grains with the same crystal orientation had similar colorings. Observations were performed in the longitudinal direction of the ingots. Each primary phase observed separately in a conventional B&W micrograph is referred to here as a "globule". It is assumed that adjacent globules with the same coloring in a polarized optical micrograph possess the same interconnected "skeleton" structure in three dimensions (3D). This structure is in turn represented by the grain observed in the two-dimensional (2D) images [27]. A single grain is outlined in white in Figure 1a and the corresponding six globules are outlined in red in Figure 1b. A complete explanation of the relationship between grain and globule can be found in an earlier work [28].

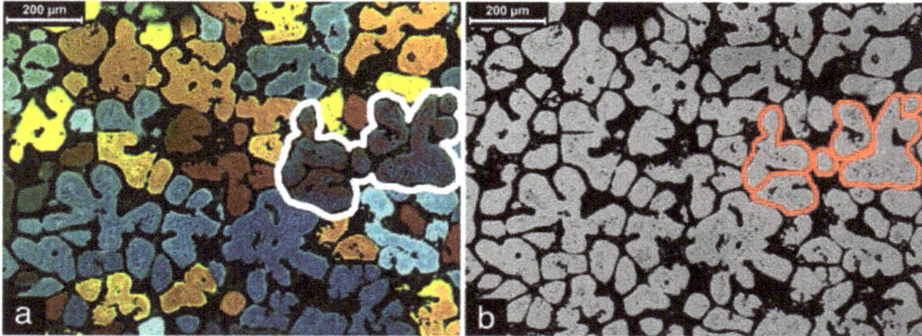

Figure 1. Micrograph of the B319+Fe alloy. (**a**) Polarized light optical micrograph highlighting grains (a single grain is outlined in white); (**b**) conventional B&W optical micrograph highlighting globules (six globules of the same grain are outlined in red).

Grain size (G_S) and globule size (GL_S) were determined using the Heyn intercept method [29]. The shape factor (circularity, C) was calculated using Image-J$^®$ 1.40 g software (V1.40, National Institutes of Health, Bethesda, MD, USA) and the equation $C = (4\pi A)/P^2$, where A is the area and P the perimeter ($C = 1$: sphere; $C \to 0$: needle). Particles smaller than 10 μm were not considered in the calculation. Statistical distributions for the microstructural parameters were determined because of the large standard errors. The statistical populations (n) analyzed were approximately $n = 80$ for G_s and GL_s and $n = 300$ for circularity. The standard error was obtained by dividing the standard deviation by $\sqrt{n}$ for each parameter.

The phases in the alloy were identified using secondary electron (SE), backscattered electron (BSE), and energy-dispersive spectroscopy (EDS) scanning electron microscopy (SEM). The size of polyhedral α-Fe particles (P_s) present in the microstructure was considered to be the maximum Feret diameter (measured using Image-J$^®$ 1.40 g software). This is the normal distance between two parallel tangents touching the particle outline and is also known as the maximum caliper diameter. The fraction of the polyhedral α-Fe phase (f_P) was also estimated as the ratio of the area occupied by these particles to the total area of the respective optical microscopy images (using Image-J$^®$ 1.40 g software). The statistical populations analyzed for the polyhedral α-Fe phase were approximately $n = 150$ for P_s and $n = 4$ for f_P. The "frozen" liquid fraction after quenching was estimated as the ratio of the area occupied by the microstructure surrounding primary-phase globules to the total area of the respective optical microscopy images (using Image-J$^®$ 1.40 g software). The statistical population for these measurements was $n = 5$.

Compression tests were performed on the alloy in the semisolid state using an instrumented mechanical press, which also works as a compression rheometer [30]. The thixoforming samples (30 mm D × 30 mm H) were taken from the as-cast billets and heated to the semisolid state in the same manner and under the same conditions as in the thermal treatment experiments. After the target temperature had been reached and the holding time had elapsed, an engineering strain of 0.8 was imposed at a strain rate of approximately 4.2 s^{-1}. The output data from the mechanical press were used to calculate the engineering stress (σ in MPa) vs. engineering strain (e) curve as well as the apparent viscosity (μ in Pa.s) vs. shear rate ($\dot{\gamma}$ in s^{-1}) curve based on the following equations [1,31]:

$$e = 1 - \frac{h}{h_0} \tag{1}$$

$$\sigma = \frac{Fh_0}{V}(1 - e) = \frac{Fh}{V} \tag{2}$$

$$\mu = -\frac{2\pi F h^5}{3 V^2 \left(\frac{dh}{dt}\right)} \tag{3}$$

$$\dot{\gamma} = -\left(\sqrt{\frac{V}{\pi}}\right) \frac{\left(\frac{dh}{dt}\right)}{2 h^{2.5}} \tag{4}$$

where F is the load, h and h_0 are the instantaneous and initial heights of the sample, respectively, V is the constant volume of the sample, and t is the time.

3. Results and Discussion

3.1. Microstructural Characterization

The microstructure of the as-cast B319+Fe alloy is shown in Figure 2 as a reference for the heat-treated conditions. Figure 2a shows grains, and Figure 2b globules. The microstructure is essentially formed by a dendritic-like Al-rich primary phase and an intergranular Si-rich main eutectic. The latter is shown in detail in Figure 2d (arrow 4). Several tertiary eutectic compounds are found in the interdendritic space, such as the acicular β-Al_5FeSi phase (Figure 2c, arrow 1), platelets of the θ-$Al_2Cu(Si)$ phase (Figure 2c, arrow 2), preferentially nucleated around β-phase, and clusters of the Cu-rich Al-Al_2Cu-Si phase (Figure 2e, arrow 5) [21,32–35]. Also, the complex $Al_{15}(FeMnCr)_3Si_2$ compound (known as α-Fe phase) is present in the form of both Chinese script (Figure 3d, arrow 3) and polyhedral (Figure 3e, arrow 6) morphologies [22,36–38]. The latter morphology is found in abundance in the microstructure of the B319+Fe alloy. These coarse polyhedral particles are indicated by blue arrows in Figure 2b.

Chemical identification of these phases is shown in Table 2, based on the points shown in Figure 3. The distribution of elements along the microstructure is shown in Figure 4. The region of reference (Figure 4a) (produced by backscattered electron (BSE) SEM) is analyzed in terms of Al (Figure 4b), Si (Figure 4c), Cu (Figure 4d), Fe (Figure 4e), Zn (Figure 4f), Mn (Figure 4g), Mg (Figure 4h), and Cr (Figure 4i) (SEM with energy-dispersive spectroscopy (EDS)).

Punctual analysis of the Al-rich primary phase (points A in Figure 3), approx. 98 at. % Al, is consistent with the expected composition for this solid solution. Al-Si main eutectic (points B in Figure 3) also presents an Al-to-Si ratio of ~0.7, which is compatible with the eutectic region presenting narrow Si particles distributed in the Al-rich matrix.

The Al-to-Cu ratio of ~2.2 for point D (Figure 3) is coherent with the stoichiometric relation expected for the θ-$Al_2Cu(Si)$ phase, i.e., Al-to-Cu ratio of 2.0. Likewise, the Al-to-Si ratio of ~8, Al-to-Fe ratio of ~6, and Fe-to-Si ratio of ~1.2 found for the polygonal particles (point C in Figure 3) are consistent with the stoichiometric relation expected for the α-Fe phase, i.e., the Al-to-Si ratio of 7.5, Al-to-Fe ratio of 5.0, and Fe-to-Si ratio of 1.5. Manganese, eventually present in the α-Fe, was also found in the polygonal particles, but at a smaller fraction (Al-to-Mn ratio of ~12.7) than theoretically expected (Al-to-Mn ratio of 5.0). Chromium is also found in this phase, but at an extremely small fraction. It can be seen (Figure 4e,g) that practically all of the Fe and Mn content in the alloy is concentrated in the α-Fe polygonal particles.

Punctual analysis of the Chinese script component (point E) also indicates that it may exist in the α-Fe phase, due to the presence of Mn and Cr. However, the Al-to-Si,Fe ratios in this phase are higher than expected for both α-Fe and β phases, probably due to the limited dimension of this structure compared with the electronic beam reach, leading to Al-rich primary phase scanning in the same measurement. For the same reason, an analysis of the narrow acicular β-Al_5FeSi phase was not performed.

The means $\pm$ standard errors of the microstructural parameters for the as-cast condition were: $G_S = 146 \pm 4$ μm, $GL_S = 39 \pm 1$ μm, $G_S/GL_S = 3.8 \pm 0.1$, $C = 0.39 \pm 0.01$, $P_S = 17.0 \pm 0.7$ μm (α-Fe polyhedrons size), and $f_P = 0.019 \pm 0.008$ (α-Fe polyhedrons fraction).

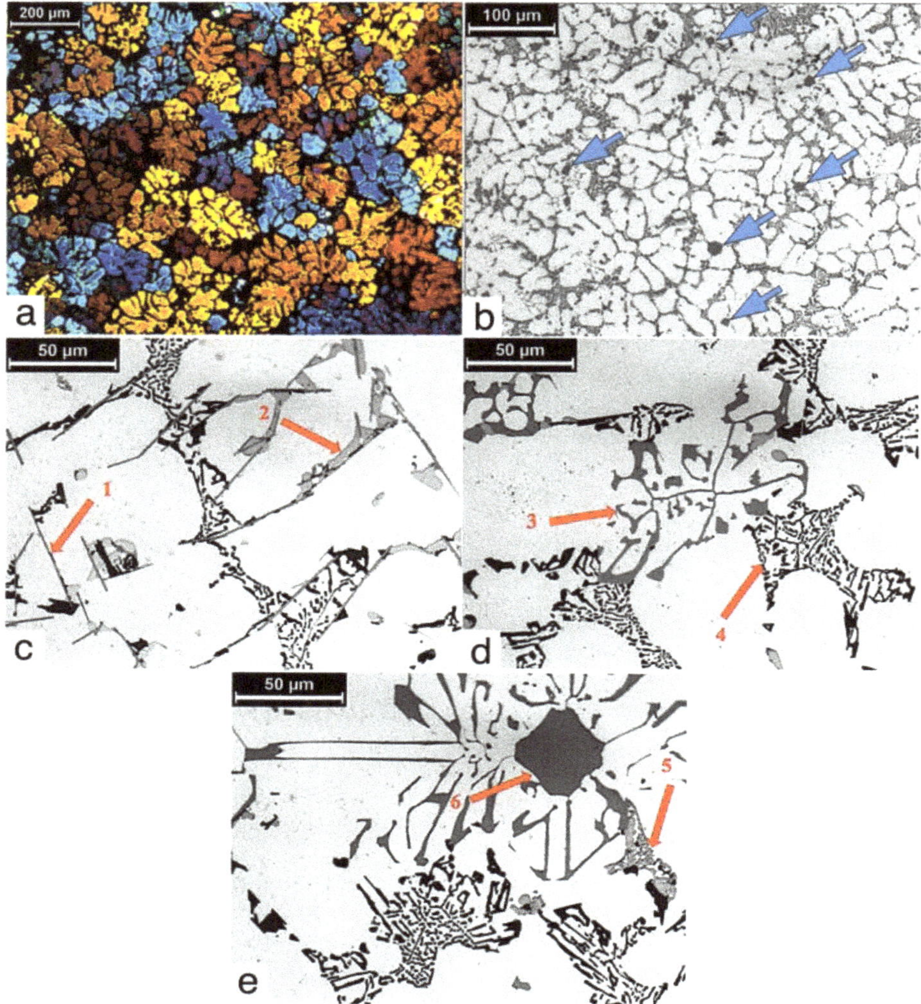

Figure 2. Micrographs of the as-cast B319+Fe alloy. (**a**) Polarized light optical micrograph highlighting grains; conventional B&W optical micrographs highlighting (**b**) globules and α-Fe polyhedral particles (blue arrows), (**c**) acicular β-Al_5FeSi phase (arrow 1) and plates of the θ-Al_2Cu phase (arrow 2) nucleated around β phase, (**d**) Chinese script α-Fe ($Al_{15}(MnFeCr)_3Si_2$) phase (arrow 3) and Al-Si eutectic (arrow 4), (**e**) clusters of the Al-Al_2Cu-Si phase (arrow 5) and polyhedrons of the α-Fe ($Al_{15}(MnFeCr)_3Si_2$ / $Al_{15}(MnFe)_3Si_2$) phase (arrow 6).

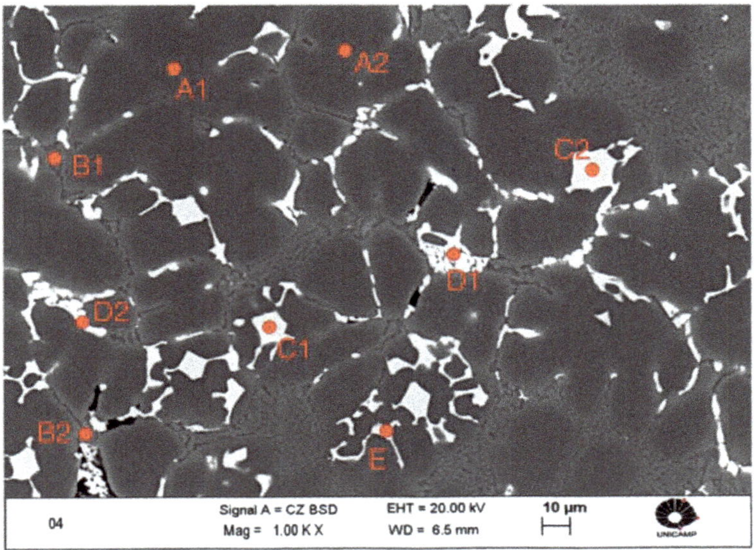

Figure 3. SE-SEM micrographs of the as-cast B319+Fe alloy showing points used for punctual chemical analysis: α-Al (points A), Al-Si eutectic (points B), α-Al$_{15}$(MnFeCr)$_3$Si$_2$ particles (points C), θ-Al$_2$Cu/Al-Al$_2$Cu-Si clusters (points D), Chinese script α-Al$_{15}$(MnFeCr)$_3$Si$_2$ (point E).

Table 2. MEV-EDS chemical composition of the points presented in Figure 3.

Phase	Point	Element (at.%) *							
		Mg	Al	Si	Cr	Mn	Fe	Cu	Zn
α-Al	A1	-	98.33	0.82	-	-	-	0.35	0.28
	A2	-	98.15	0.95	-	-	-	0.41	0.29
Eutectic (Al-Si)	B1	-	40.98	58.04	-	-	-	0.70	0.28
	B2	-	36.09	63.23	-	-	-	0.68	-
Polyhedrons α-Al$_{15}$(MnFeCr)$_3$Si$_2$	C1	-	71.63	9.72	0.72	5.57	11.66	0.70	-
	C2	-	71.37	9.47	0.74	5.65	11.81	0.73	-
θ-Al$_2$Cu/ Al-Al$_2$Cu-Si	D1	0.92	66.99	2.04	-	-	-	29.79	-
	D2	1.47	66.30	2.66	-	-	0.36	28.90	-
Chinese Script α-Al$_{15}$(MnFeCr)$_3$Si$_2$	E	-	82.46	7.39	0.23	2.58	6.65	0.69	-

* Residual elements also present in some structures are Ni (max. 0.29 at.%), Ti (max. 0.22 at.%), and V (max. 0.24 at.%).

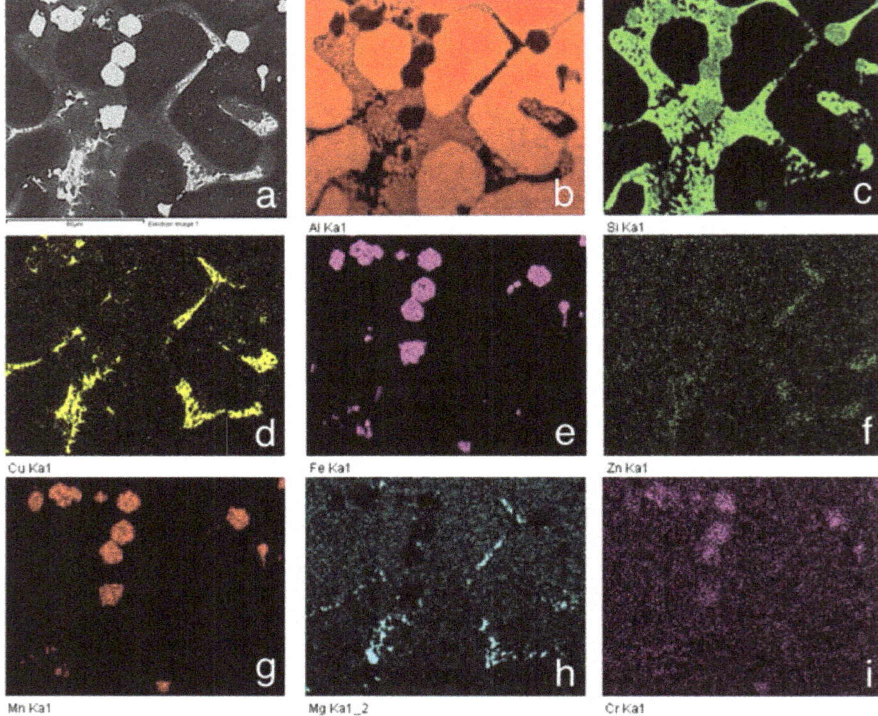

Figure 4. SEM micrographs of the as-cast B319+Fe alloy. (**a**) Backscattered electron (BSE) scanning electron micrograph showing zoomed-in α-Fe polyhedral particles (white polygons); energy-dispersive spectroscopy (EDS) scanning electron micrograph of the same region shown in (**a**), revealing the distribution of Al (**b**), Si (**c**), Cu (**d**), Fe (**e**), Zn (**f**), Mn (**g**), Mg (**h**), and Cr (**i**).

3.2. Thermodynamic Characterization using Calculation of Phase Diagrams (CALPHAD) Simulation

Figure 5 presents the DSC heat flow rate (*HFR*) vs. *T* curves (Figure 5a) and the corresponding derivative of *HFR* with relation to temperature (d*HFR*/d*T*) vs. *T* curves (Figure 5b), liquid fraction (f_l) vs. *T* curves (Figure 5c), and sensitivity (df_l/d*T*) vs. *T* curves (Figure 5d) for the B319+Fe alloy for all DSC heating conditions analyzed. The f_l vs. *T* and df_l/d*T* vs. *T* curves for the Scheil condition are shown for reference (grey curves in Figure 5c,d, respectively). The phase transformations during the melting of the B319+Fe alloy predicted by the Scheil curve are as follows: starting at 492 °C (the solidus), the melting of the tertiary eutectic phases formed by (1) $Al_5Cu_2Mg_8Si_6$, (2) Al_2Cu, and (3) Al_7Cu_2Fe, which melt completely by 520 °C, occur when $f_l = 0.10$. This is followed by melting of the (4) main Si-rich eutectic, which is responsible for the knee-like shape in the f_l *vs.* *T* curve (564 °C and $f_l = 0.53$). Subsequently, the melting of the (5) Fe-rich AlFeSi β-phase particles (601 °C and $f_l = 0.97$) and, finally, the melting of the (6) Al-rich α phase indicate the total melting of the alloy at 603 °C (the liquidus).

The DSC curves (Figure 5a) clearly show two endothermic peaks, one probably related to the melting of both the Al-rich α phase and the β-Al-Fe-Si phase (the intense peak at higher temperature ranges) and another probably related to the melting of both the Si-rich eutectic phase and the tertiary phase with a higher melting range, Al_7Cu_2Fe (the intense peak at lower temperature ranges). A third discrete peak is also identifiable at the extreme left of the curve related to one of the (or the superposition of several) remaining tertiary phases. These peaks are identified by arrows in Figure 5a.

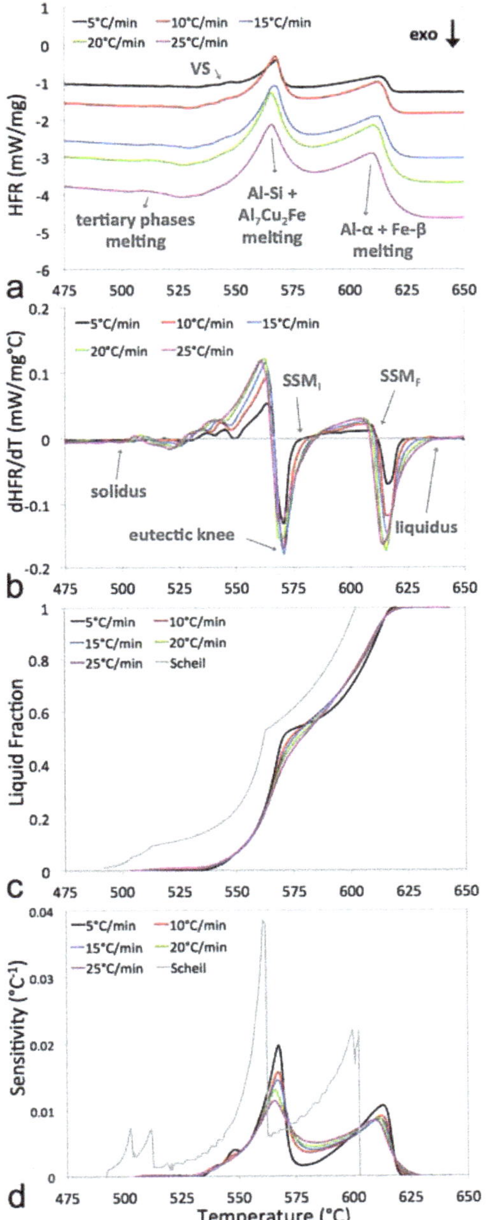

Figure 5. *HFR* (**a**), d*HFR*/d*T* (**b**), f_l (**c**), and df_l/d*T* (**d**) vs. *T* curves for the B319+Fe alloy at several heating rates. The values predicted by numerical simulation under the Scheil conditions are shown for reference (grey curves). The arrows indicate phase transformations and temperatures of interest identified by the differentiation method (DM). The VS arrow indicates a "visible superposition" of phase transformations in the DSC curves.

The hypothesis of multiple phases melting at the first and third DSC peaks is justified by the proximity of the temperature ranges predicted by the Scheil conditions for the phase transformations considered here (α-Al + β-Al-Fe-Si and $Al_5Cu_2Mg_8Si_6$ + Al_2Cu, respectively). The superposition in the second peak (Al-Si eutectic + Al_7Cu_2Fe) is not only supported by the Scheil prediction but also clearly visible in the DSC curves in the form of waves on the left of this peak indicated by an arrow and VS (visible superposition) in Figure 5a.

The DSC derivative curves (Figure 5b) are used in the DM to identify the temperatures of interest in the thixoformability analysis of the B319 alloy: the solidus, liquidus, working window (T_{SSMI} and T_{SSMF}), and eutectic knee. These temperatures are indicated by arrows in Figure 5b. The superposition of phases in the second peak seen in the DSC curves (Figure 5a) is also visible in the differentiated curves (Figure 5b). This is to be expected as these patterns are inherited from the original curve.

The faster the heating rate, the softer the knee corresponding to the eutectic transformation in the f_l vs. T curves (Figure 5c). This smoothing is a result of the delay in the phase transformations and the increase in the temperature range in which these occur as the distance from the equilibrium conditions increases [23]. The sensitivity curves (Figure 5d) also tend to become less pronounced as the heating rate increases. The peaks in the curves decrease and extend over a wider temperature range as the distance from equilibrium increases. This graphical behavior is inherited from the original f_l vs. T curves.

Using the temperatures of the eutectic knee and thixoforming interval obtained by the DM (Figure 5b), the respective liquid fractions ($f_{l\ Knee}$, $f_{l\ SSMI}$ and $f_{l\ SSMF}$) were identified on the f_l vs. T curves (Figure 5c). To establish an adequate working window, the maximum sensitivity (df_l/dT_{max}) in the proposed interval (between T_{SSMI} and T_{SSMF}) was determined from the sensitivity graphs (Figure 5d). This information was then used to assess the thermodynamic stability and determine whether the $0.03\ °C^{-1}$ criterion was met [10]. The resulting data are shown in Table 3 and were used to plot the semisolid range, working window, and knee point for the B319+Fe alloy in terms of temperature (Figure 6a) and liquid fraction (Figure 6b) for all of the heating rates analyzed. The following remarks can be made in relation to Figures 5 and 6 and Table 3:

(a) Semisolid interval. There is an increase of 24.3 °C in the semisolid interval as the heating rate increases (Figure 6a). At 5 °C/min, $\Delta T_{s\text{-}l}$ = 118.2 °C and at 25 °C/min $\Delta T_{s\text{-}l}$ = 142.5 °C. This trend is a result of the delaying and extending of the melting reaction as kinetic effects become stronger. Furthermore, higher heating rates result in a larger driving force (superheating) for melting, making the reaction start sooner, i.e., faster curves have lower solidus points than slower ones, resulting in larger semisolid intervals [23]. The main factor responsible for the enlargement of the semisolid interval is the increase in the liquidus temperature ($\Delta T_{liquidus}$ = 17.5 °C between heating rates of 5 and 25 °C/min), followed by the smaller decrease in the solidus temperature ($\Delta T_{solidus}$ = -6.8 °C between heating rates of 5 and 25 °C/min). This result shows that the delaying effect of the melting reaction (resulting in a higher liquidus) plays a greater role in enlarging the semisolid range than the larger driving force (lower solidus) observed with the rise in the heating rate.

The Scheil condition is nearer equilibrium than any of the experimental heating rates. Then, the simulated Scheil curve lies on the left of the experimental curves and presents the smallest semisolid interval ($\Delta T_{s\text{-}l}$ = 111.0 °C), respecting the above-discussed trend of progressive delaying in the temperature range due to kinetic effects.

(b) Eutectic knee. Despite the changes in the shape of the curves caused by kinetic effects, the temperature corresponding to the knee remains almost unaltered (Figure 6a) and has a mean value ($\pm$ standard error) of T_{knee} = 570.8 $\pm$ 0.13 °C. However, because of the delayed/extended melting reaction and smoothing of the knee, the f_l vs. T curves around the knee (Figure 6c) move progressively toward higher temperatures (right) as the heating rate increases. This effect causes the knee liquid fraction to be reduced by 0.11 as the heating rate increases from 5 °C/min ($f_{l\ Knee}$ = 0.51) to 25 °C/min ($f_{l\ Knee}$ = 0.40). The Scheil curve, which is nearer equilibrium than any of the experimental curves, is shown on the extreme left of the graph near the curves corresponding to the lowest heating rates.

(c) Thixoforming working window. For all of the conditions analyzed, the sensitivity values in Table 3 are below the 0.03 $°C^{-1}$ limit, so the original SSM working window suggested by the DM was maintained. The thixoforming interval for the B319+Fe alloy becomes slightly smaller as the heating rate increases, decreasing by 7.0 °C (32.0 °C to 25.0 °C) between the lowest and highest heating rates. This result contrasts with the trend for the total semisolid interval to increase with the increasing heating rate (item (a) above). A possible explanation for this is that the melting range increases at the beginning of the reaction (decrease in solidus) and the end of the reaction (increase in liquidus) whereas the phase transformations between these points occur closer together. The Scheil curve (condition nearer equilibrium than any of the experimental curves) also follows the above-discussed trend, presenting the larger thixoforming interval (36.0 °C). This result is also due to the abrupt drop of sensitivity above the knee for this condition, which in turn is a consequence of the sharp knee shape on the numerical f_l curve, leading the T_{SSMI} (point of minimum sensitivity) to be remarkably closer to the knee and contributing to the working window enlargement.

Table 3. Key thixoforming parameters obtained by the DM for B319+Fe alloy at different heating rates. The values predicted by numerical simulation (Scheil conditions) are shown for reference.

Rate (°C/min)	$T_{solidus}$ (°C)	T_{Knee} (°C)	T_{SSMI} (°C)	T_{SSMF} (°C)	$T_{liquidus}$ (°C)	$f_{l\ Knee}$	$f_{l\ SSMI}$	$f_{l\ SSMF}$	df_l/dT_{max} (°C^{-1})
Scheil	492.0	564.0	564.0	600.0	603.0	0.53	0.53	0.95	0.0023
5	507.8	571.0	581.0	613.0	626.0	0.51	0.55	0.92	0.0012
10	507.8	570.3	582.8	612.9	630.3	0.45	0.57	0.92	0.00085
15	505.9	570.9	583.3	610.9	633.4	0.43	0.57	0.89	0.00085
20	503.3	570.8	583.3	610.8	641.0	0.40	0.54	0.90	0.00083
25	501.0	571.0	586.0	611.0	643.5	0.40	0.57	0.90	0.00083

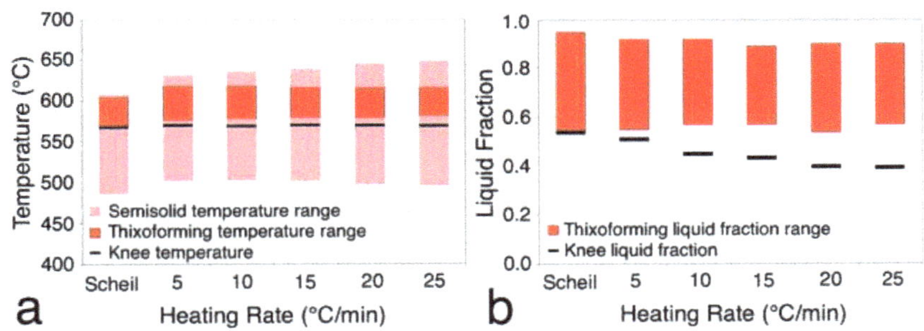

Figure 6. Key thixoforming parameters for B319+Fe alloy at different heating rates: temperature (**a**) and liquid fraction (**b**). The values predicted by numerical simulation are shown for reference (leftmost column in each chart).

A remark should be made concerning the equivalence between the DSC and sensitivity curves: the liquid fraction (dimensionless) is obtained by the integration of the DSC curve with relation to the temperature. Since the DSC data represents the heat flow rate, *HFR* (dimension = heat × time^{-1}), a direct integration of *HFR* would result in a first-order integral $\int HFR\ dT$ with dimension = heat × time^{-1} × temperature. To transform this integral into the liquid fraction, a base line is set under/above the endothermic/exothermic region of the *HFR* curve where the melting/solidification occur and is subtracted from the curve to disregard the noisy *HFR*. Furthermore, normalization from 0 (solidus) to 1 (liquidus) precedes this step to enable the analysis of the liquid as a fraction.

Then, the *DSC* vs. *T* curve represents the derivative of the f_l vs. *T* curve, after the reverse algebraic operations of normalization and subtraction of a defined base line. In turn, the sensitivity (dimension = temperature^{-1}) is obtained via direct derivation of the liquid fraction with relation to the temperature, i.e., df_l/dT. Therefore, both *HFR* and sensitivity represent derivatives of the f_l with relation to T, with different values in the y-axis due to the algebraic operations used to obtain the f_l from *HFR* data. As a consequence, both *HFR* and sensitivity curves present the same graphical behavior with respect to the x-axis and differ in the y-axis only by a certain constant value.

3.3. Evaluation of the Thixoforming Processing Window

Based on the results discussed in Section 3.2, four temperatures were tested to evaluate the thixoformability of the B319+Fe alloy: 575, 582, 591, and 595 °C. These temperatures are highlighted in the magnified f_l vs. T curves of the B319+Fe alloy in Figure 7. Only heating rates of 15, 20, and 25 °C/min are shown because these are close to the final heating rate used in the thermal treatments (18–22 °C/min). The eutectic knee (purple) and lower limit of the SSM working window (green) are also highlighted on the curves. Thixoformed (compressed) disks after isothermal heat treatment at each of these four temperatures for 0, 30, 60, and 90 s are shown to the right of the graph.

The lowest temperature (575 °C) is below the suggested SSM working window and is also extremely close to the knee for the experimental conditions analyzed. In contrast, the second-lowest temperature, 582 °C, is in the intermediate region between the knee and the lower limit for SSM processing (T_{SSMi}) and is thus still in the zone that is theoretically unstable for SSM processing. Finally, the two remaining temperatures (591 and 595°C) are within the interval suggested by the DM (T_{SSMi} and T_{SSMF}) for all cycles, where the sensitivity values are lowest ($S < 0.01$ °C^{-1}), allowing easy control of SSM processing.

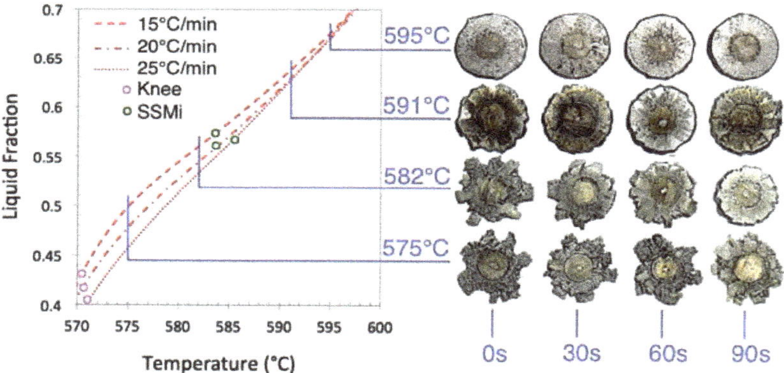

Figure 7. f_l vs. T curves (in red) derived from DSC data for B319+Fe alloy at heating rates of 15, 20, and 25 °C/min. The four temperatures chosen for the thixoforming tests (575, 582, 591, and 595 °C) are indicated by horizontal blue lines. Thixoformed disks after isothermal heat treatment at these temperatures for 0, 30, 60, and 90 s are shown to the right of the graph. The eutectic knee (purple circles) and lower limit of the SSM working window (green circles) are also highlighted on the curves.

The liquid fraction for the four temperatures tested is presented in Table 4, taken directly from the Scheil curve and DSC curves for 15 to 25 °C/min and represented by the area surrounding primary phase globules in an optical micrograph of the heat-treated and subsequently quenched samples. According to previous work [7], secondary particle growth during quenching should be ignored for the microstructural measurements of G_s and GL_s shown here as secondary particle growth has been reported to be negligible (<4 µm) in relation to the magnitude of the particles analyzed here for the liquid fraction prior to quenching and the estimated cooling rate during quenching in this study (<0.7 and ~130 °C/s, respectively).

However, the liquid fraction measured via %area in the micrographs varies at most 0.04 from 575 °C to 595 °C, in contrast to the variation of 0.23 predicted for the Scheil condition, 0.21–0.18 for DSC at 15–25 °C/min. Figure 8 highlights the similarity in f_l measured via micrographs showing the areas corresponding to primary phase globules (black) and assumed "frozen" liquid (red) for two extreme conditions of heat treatment, 575 °C-0 s (Figure 8a) and 595 °C-90 s (Figure 8b).

Table 4. Liquid fraction of B319+Fe alloy samples heat-treated for 0–90 s at four temperatures and subsequently water quenched.

Temperature (°C)	Liquid Fraction					
	Scheil (0 s) [1]	DSC [2] (0 s) [1]	Micrograph Area Fraction [3]			
			0 s	30 s	60 s	90 s
575	0.61	0.46–0.50	0.40	0.41	0.42	0.35
582	0.68	0.53–0.56	0.43	0.35	0.40	0.41
591	0.78	0.63–0.64	0.42	0.40	0.43	0.35
595	0.85	0.67–0.68	0.44	0.42	0.38	0.37

[1] Scheil and DSC curves represent continuous heating (0 s of hold time); [2] Liquid fraction for the heating rate range of 25 °C/min (lower f_l) to 15 °C/min (higher f_l); [3] Standard error for micrograph area fraction was max. ±0.01 for all measurements.

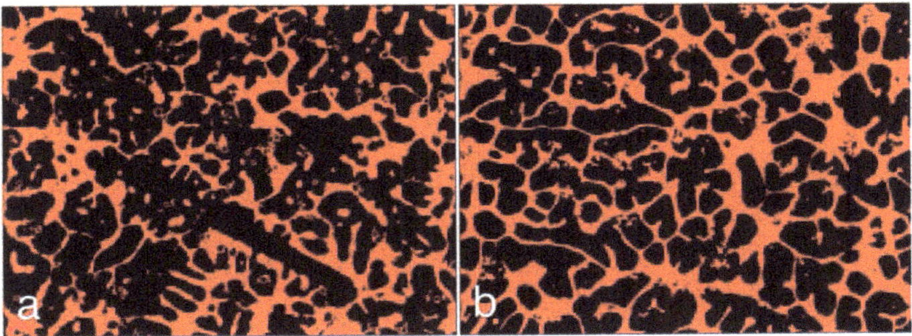

Figure 8. Micrograph highlighting globules (black area) and "frozen" liquid (red area) for B319+Fe alloy after heat treatments using two extreme combinations of temperature and soak time: 575 °C-0 s (**a**) and 595 °C-90 s (**b**).

Since the predicted liquid fraction (Scheil and DSC) and the appearance of the thixoformed disks are evidently different amongst the tested temperatures (from brittle and asymmetric at lower temperatures to more homogeneous, compact, and symmetric at higher temperatures), the liquid fraction for these temperatures during forming cannot be similar, which indicates that a relevant secondary growth occurred during quenching and that the analysis of the liquid fraction directly from the microstructures does not represent the accurate "frozen" liquid fraction at the four temperatures of partial melting.

Figure 9 shows the microstructure (grains—color micrographs; globules—B&W micrographs) of samples heat-treated isothermally for 60 s at each of these temperatures (575 °C—Figure 9a,b; 582 °C—Figure 9d,e; 591 °C—Figure 9g,h; 595 °C—Figure 9j,k) and the corresponding thixoformed disks (Figure 9c,f,i,l). A soak time of 60 s was chosen because this is usually sufficient to allow diffusion to take place and a reasonable overview of the microstructural evolution in the semisolid state to be obtained [20]. Table 4 shows the mean values of the microstructural parameters for these conditions.

In general, the microstructures are similar for all of the heat treatments when a 60-s soak time is used (Figure 9 and Table 5). The differences in the appearance of the disks are probably caused by the different liquid fractions present at each temperature: the low liquid fraction at the two lowest temperatures (575 °C and 582 °C) leads to inadequate flow of the semisolid slurry during compression, resulting in disks with a brittle appearance after thixoforming (even after 90 s of isothermal heat treatment). At 575 °C the effect of the low liquid fraction is compounded by the possible presence of sharp particles in the eutectic phase, resulting in completely asymmetric disks. The two highest temperatures (591 °C and 595 °C) yielded more homogeneous, symmetric, compact thixoformed disks, probably because the semisolid slurry flowed more smoothly between the die. These two soak temperatures were therefore chosen for the analysis. Interestingly, the results reveal good agreement between the thermodynamic behavior predicted by the DM (based on the DSC curves) and the data obtained from semisolid forging of the B319+Fe alloy, i.e., only the two highest temperatures (591 °C and 595 °C) are considered good options for SSM processing by both approaches.

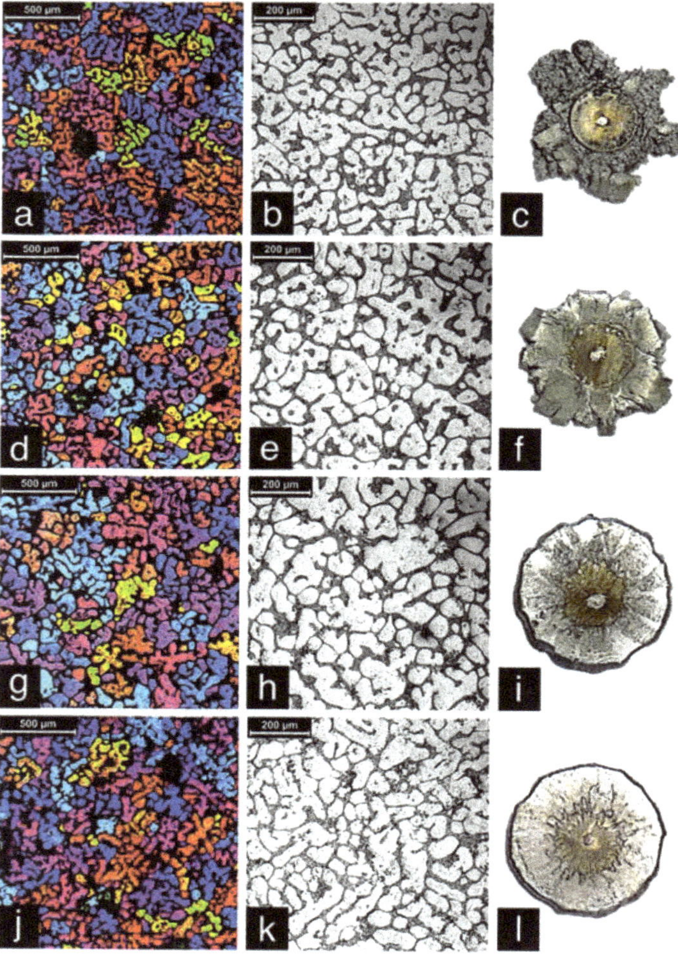

Figure 9. Micrographs showing the microstructure of semisolid B319+Fe alloy (grains—color; globules—B&W) heat-treated isothermally for 60 s at 575 °C (**a**,**b**), 582 °C (**d**,**e**), 591 °C (**g**,**h**), and 595 °C (**j**,**k**), and the respective thixoformed disks (**c**,**f**,**i**,**l**).

Table 5. Microstructural parameters of B319+Fe alloy samples heat-treated for 60 s at four temperatures.

Temperature (°C)	Microstructure			
	G_s (μm)	GL_s (μm)	G_s/GL_s	Circularity
575	153 ± 32	86 ± 15	1.8 ± 0.5	0.39 ± 0.18
582	146 ± 23	95 ± 17	1.5 ± 0.3	0.45 ± 0.19
591	186 ± 49	108 ± 23	1.7 ± 0.6	0.41 ± 0.20
595	183 ± 52	69 ± 10	2.7 ± 0.8	0.41 ± 0.19

3.4. Microstructural Evolution During Isothermal Heat Treatment

Figures 10 and 11 show grains and globules, respectively, in the B319+Fe alloy after heating to 591 (Figure 10a–d) and 595 °C (Figure 10e–h) (i.e., the semisolid state) and soaking for 0, 30, 60, and 90 s (respectively) followed by water quenching. The polyhedral α-Fe particles in the alloy are highlighted in red in Figure 11. The mean values of the microstructural parameters are given in Table 5 for all of the analyzed conditions.

The colored micrographs (Figure 10) show a significant amount of α-Al phase particles with similar colors close to each other, indicating a high level of three-dimensional connection amongst the globules. When no soaking is applied (Figure 10a,e), a coarse dendritic-like morphology inherited from the original cast microstructure (Figure 2) is evident. As the soak time increases (30–90 s) (Figure 10b–d,f–h), progressively greater grain refinement is observed, resulting in a visible reduction in G_S (Table 6). Grain refinement occurs because of the breakdown of the dendritic arms following the dissolution of thinner particles caused by Ostwald ripening during partial melting [30]. An exception to this trend is the 595 °C-90 s condition (Figure 10h), in which the long soak time at a high temperature allowed extensive coarsening (either by coalescence or Ostwald ripening), leading to exaggerated grain growth [39–42].

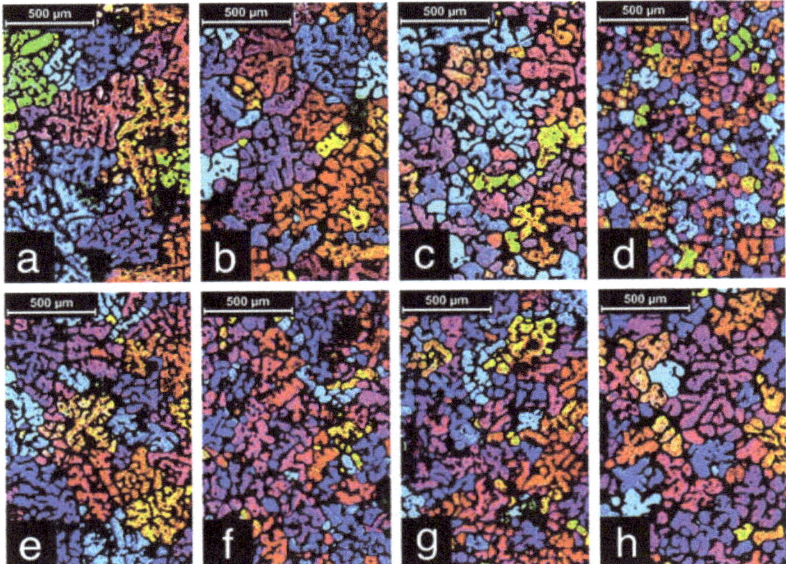

Figure 10. Colored micrographs showing grains in semisolid B319+Fe alloy after isothermal heat treatment at 591 °C (**a–d**) and 595 °C (**e–h**) for 0 (**a,e**), 30 (**b,f**), 60 (**c,g**), and 90 s (**d,h**).

As observed for the grain analysis, the conventional B&W micrographs (Figure 11) show progressive, discrete refinement and spheroidization of the α-Al phase globules with increasing soak time. Spheroidization can also be observed from the circularity measurements in Table 6 and is due to the tendency for the surface-area-to-volume ratio of the particles to decrease as a result of atomic diffusion during coarsening. Particle refinement is a consequence of the same phenomenon discussed in connection with the results for grains. Again, the 595 °C-90 s condition is an exception, with larger globules caused by excessive heating (Figure 11h).

Neighboring globules become increasingly interconnected as the soak time increases. This is expected since the distance between neighboring particles decreases progressively and in some cases even becomes zero as a direct consequence of coarsening. Interconnected particles are particularly visible for the 90-s soak time (Figure 11d,h). Interconnectivity between the solid particles is undesirable because it contributes to the formation of a cohesive network that hampers the flow of the semisolid metal [20].

Another feature observed in the micrographs is the increase in the thickness of the liquid film between the solid particles as the soak time and temperature increase. A higher liquid fraction is a direct effect of increasing temperature and leads to a thicker liquid film throughout the entire microstructure. Furthermore, the formation of large agglomerates along with smaller particles in the microstructure as a result of Ostwald ripening (which is enhanced with the increase in soak time and temperature) promotes not only thickening of the liquid film but also segregation of the liquid, leading to regions with liquid "pockets" [43], as clearly seen for the 595 °C-90 s condition (Figure 11h). The ratio G_s/GL_s (Table 6) gives an indication of the degree of connection between α-phase globules, i.e., the complexity of the three-dimensional dendritic "skeleton". The decrease in this parameter with increasing soak time for both temperatures is a consequence of grain refinement, which makes the dimensions of the grains increasingly similar to those of the globules ($G_T/GL_T \rightarrow 1$).

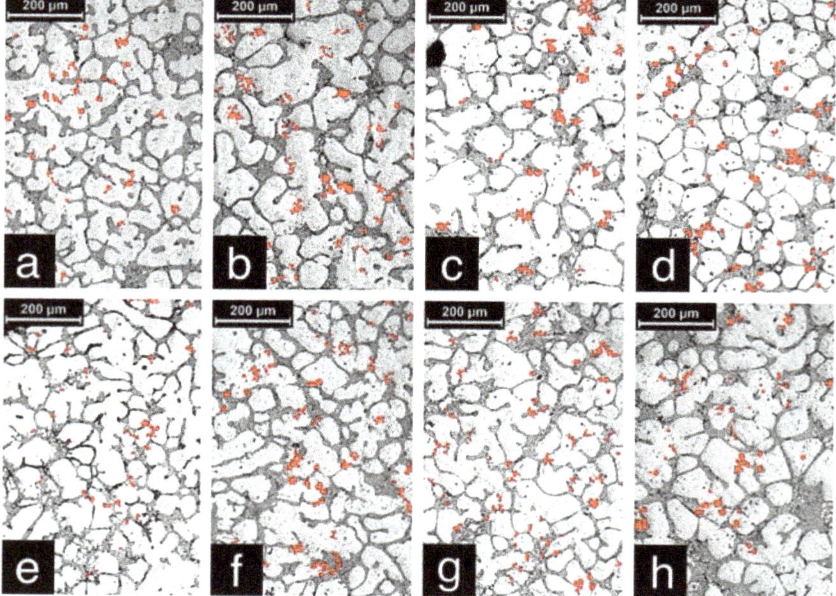

Figure 11. B&W micrographs showing globules in semisolid B319+Fe alloy after isothermal heat treatment at 591 °C (**a–d**) and 595 °C (**e–h**) for 0 s (**a,e**), 30 s (**b,f**), 60 s (**c,g**), and 90 s (**d,h**). Polyhedral α-Fe particles are highlighted in red.

Table 6. Microstructural parameters of B319+Fe alloy after isothermal heat treatment in the semisolid state using different temperatures and soak times.

Condition		G_s (µm)	GL_s (µm)	G_s/GL_s	Circularity	P_s (µm)	f_P
	0 s	285 ± 64	112 ± 36	2.6 ± 1.0	0.45 ± 0.009	21.5 ± 1.0	0.020 ± 0.008
591 °C	30 s	228 ± 44	115 ± 25	2.0 ± 0.6	0.45 ± 0.010	21.5 ± 0.9	0.027 ± 0.003
	60 s	186 ± 49	109 ± 23	1.7 ± 0.6	0.44 ± 0.010	23.0 ± 1.1	0.025 ± 0.003
	90 s	128 ± 26	75 ± 12	1.7 ± 0.4	0.51 ± 0.007	18.0 ± 0.5	0.029 ± 0.004
	0 s	263 ± 68	89 ± 19	3.0 ± 1.0	0.41 ± 0.007	14.7 ± 0.6	0.023 ± 0.003
595 °C	30 s	186 ± 40	68 ± 12	2.7 ± 0.8	0.46 ± 0.007	18.2 ± 0.7	0.026 ± 0.003
	60 s	183 ± 52	69 ± 10	2.7 ± 0.8	0.44 ± 0.007	17.1 ± 0.6	0.022 ± 0.003
	90 s	192 ± 49	103 ± 23	1.9 ± 0.6	0.49 ± 0.010	19.1 ± 1.0	0.022 ± 0.003

Up to this point, the microstructural evolution of the B319+Fe alloy has been discussed in relation to soak time (from 0 to 90 s). Analysis in terms of temperature shows that the higher of the two heat treatment temperatures studied (591 °C and 595 °C) leads to slightly more refined grains and globules, with the exception of the 595 °C-90 s condition, which produces the exaggerated coarsening already discussed. Circularity values are similar for both temperatures. G_s/GL_s is higher at 595 °C than at 591 °C, indicating that globule refinement increases more quickly with increasing temperature than grain refinement does, i.e., superficial "corrosion" of globules (because of the formation of liquid) occurs more quickly than total separation from the main dendritic arms as the temperature increases. This is expected since superficial "corrosion" requires less energy than the total separation of the dendritic arm.

The amount, distribution, morphology, and size of the $Al_{15}(FeMnCr)_3Si_2$ particles (Figure 11 and columns P_s and f_P in Table 6) in the B319+Fe alloy are similar for all of the soak times and temperatures analyzed, indicating that they are thermodynamically stable. Figure 12 shows these particles for two extreme heat treatment conditions (575 °C-0 s and 595 °C-90 s). The similarity between the particles at each extreme of the time/temperature range can be clearly seen.

The α-Fe phase is usually formed at the expense of the β-Al-Fe-Si phase in the presence of Mn. When the ratio of Fe content to Mn content is less than 2, all of the β phase is expected to be replaced by the polygonal compound [39]. The B319+Fe alloy analyzed here has an Fe-to-Mn ratio of ~4, so only part of the β phase should be replaced by the α-Fe phase. This agrees with the fact that both phases are observed in the as-cast microstructure of the alloy (Section 2).

The two distinct morphologies of the α-Fe phase observed in the as-cast microstructure of the B319+Fe alloy (Chinese script and polyhedrons) are formed at different stages of solidification. The Chinese script, or skeleton-like, α-Fe phase nucleates as a secondary phase at the grain boundaries of the already-solidified α-Al phase. The "arms" of the skeleton shape develop according to the direction in which the solute flows in the surrounding liquid, giving the characteristic Chinese script morphology. The polyhedral morphology is formed by the nucleation of the α-Fe phase directly from the liquid before the α-Al phase in regions with a higher solute concentration [22,37–39]. The rounder shape is formed because of the higher thermodynamic stability of a lower ratio of surface energy to volume energy. Thus, the α-Fe phase polyhedrons can be expected to be stable since they melt at a much higher temperature than those used in the isothermal heat treatments.

Quantitative analysis of G_s, GL_s, and C for the B319+Fe alloy under the conditions studied suffers from inaccuracy because of the high standard errors of the measurements. Considering (a) the large statistical population sizes used for each parameter (n, given in Section 2) and (b) the high values of the resulting standard errors (obtained using the standard deviation and n, as described in Section 2), the variations can be considered to be the result not of inadequate statistical sample sizes but rather of the dispersive nature of the variables analyzed. The Ostwald ripening mechanism, which is responsible for the formation of large agglomerates and smaller particles, makes the microstructure extremely heterogeneous, contributing to the spread of the measurements. G_s, GL_s, and C are therefore presented in the form of probability distributions in Figure 13a,b, Figure 13c,d, and Figure 13e,f, respectively, for temperatures of 591 °C (left) and 595 °C (right). The mean values of G_s/GL_s were obtained by an algebraic operation and thus cannot be represented as a statistical distribution.

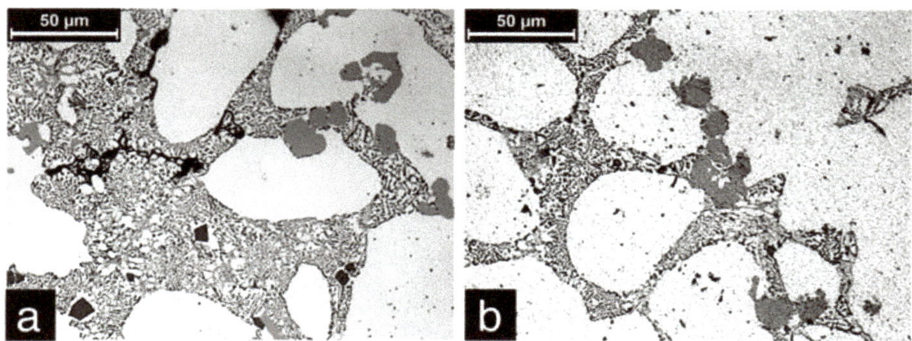

Figure 12. $Al_{15}(FeMnCr)_3Si_2$ particles in the B319+Fe alloy after heat treatments using two extreme combinations of temperature and soak time: 575 °C-0 s (**a**) and 595 °C-90 s (**b**).

The parameters G_s (Figure 13a,b) and GL_s (Figure 13c,d) have similar distributions, with a maximum frequency at smaller values. The maximum of the G_s distribution tends to be displaced to lower values as the soak time increases, reflecting the trend in Table 6. At 591 °C (Figure 13a) the progressive narrowing of the distribution of G_s as the duration of the isothermal treatment increases is particularly apparent, indicating that a longer soak time promotes a more homogeneous microstructure. At 595 °C (Figure 13b) a significant change in distribution is only seen when a soak time of 0 s (black columns) is compared with the other conditions, indicating that soaking for more than 30 s does not significantly affect G_s at this higher temperature. The distribution of GL_s (Figure 13c,d) also tends to be displaced to lower values with increased soak time. At 591 °C (Figure 13c) a significant reduction and narrowing of the distribution is observed only after 90 s of isothermal treatment (grey columns). However, at 595 °C (Figure 13d) this displacement and narrowing occurs earlier, after 30–60 s (blue and red columns), indicating the role played by temperature in increasing diffusion. Contrasting with the trend observed up to this point, the distribution of GL_s for the 595 °C-90 s condition (the grey columns in Figure 13d) is more spread out and extends to higher values, showing again that prolonged exposure of the microstructure to a high temperature leads to the excessive growth of α-phase particles, as discussed earlier. For some conditions, the G_s and GL_s distributions become slightly bimodal, with a discrete second maximum located to the right of the main maximum. This is a clear indication of the presence of large agglomerates along with smaller particles in the microstructure because of Ostwald ripening. In general, circularity behaves like a statistical random variable (Figure 13e,f). For some heat treatment conditions, there is a discrete transition to a bimodal distribution, indicating again the simultaneous presence of individual (convex) and interconnected (concave) particles in the microstructure (Figure 11). After 90 s of isothermal treatment at both temperatures (grey columns in Figure 13e,f), the maximum clearly moves to higher values of circularity, showing that soak time helps to increase spheroidization.

Suitable grain sizes for semisolid slurries produced by conventional casting are usually of the order of ~70–150 μm [1,2], below those observed here (~130–280 μm). For thixoforming, the solid globules in the liquid should have a refined, spheroidal microstructure to allow the specific interaction between particles under shear that confers thixotropic characteristics on the semisolid slurry. As the present study is only the first stage in the evaluation of the suitability of B319+Fe alloy for use in SSM processing, the alloy was produced using a relatively cheap, conventional production process. This resulted in an initial, coarse microstructure that was present (albeit transformed by diffusion) after all heat treatments. Other forms of raw-material production should therefore be tested for this alloy to ensure adequate refinement and spheroidization.

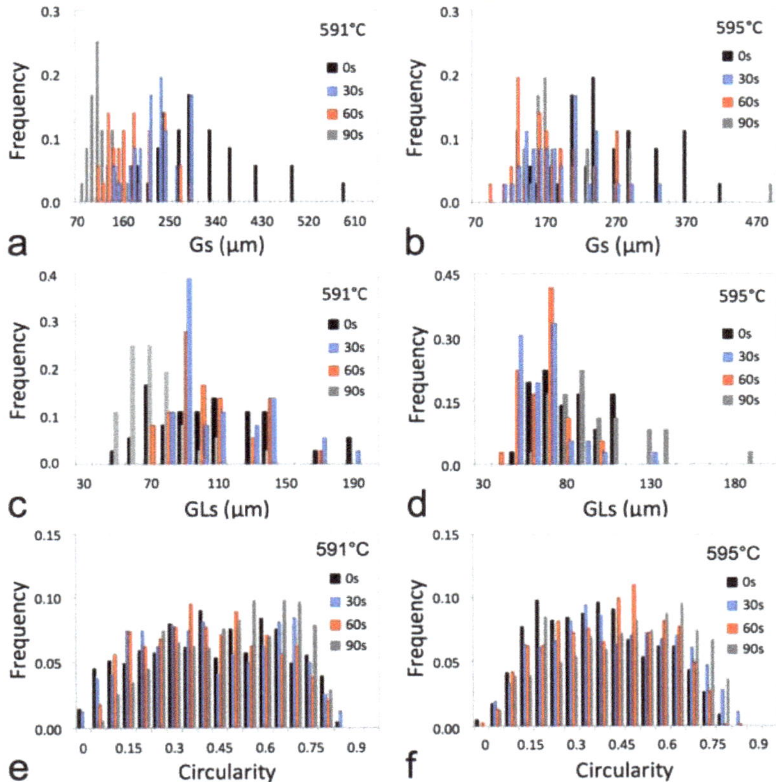

Figure 13. Distributions of B319+Fe alloy grain size (**a,b**), globule size (**c,d**) and circularity (**e,f**) measured after isothermal heat treatment (at 591 °C and 595 °C, respectively) for 0, 30, 60, and 90 s.

Another essential characteristic for thixoforming is that the microstructure is stable, i.e., the size and shape of the remaining globules must remain unchanged over a range of soak times in the semisolid state. This characteristic is important because industrial processes are time-controlled rather than temperature-controlled: thixoforming operations use induction-heating furnaces that provide a set power for a fixed time to achieve the required temperature [20]. Significant changes in the microstructure during reheating and partial melting will therefore change the semisolid rheological behavior, adversely affecting the reproducibility of the process. Although the microstructure of the B319+Fe alloy studied here is coarse, the results show that it is stable for all of the soak times studied, indicating that this alloy is a promising raw material for thixoforming if adequate grain refinement is performed before processing.

3.5. Rheological Behavior in the Semisolid State

Figure 14 shows the engineering stress vs. engineering strain (Figure 14a,b) and apparent viscosity vs. shear rate (Figure 14c,d) curves of B319+Fe alloy samples subjected to isothermal heat treatment at 591 °C (left) and 595 °C (right) for 0, 30, 60, and 90 s. As the heat-treated alloy has a coarse dendritic-like microstructure, three regions can be identified in the curves in Figure 14: (a) a region of dendritic stability—the beginning of thixoforging consists of the compression of a highly interconnected dendritic network. As the formation progresses, the solid structure becomes more interconnected and intricate, leading to an increase in stress and apparent viscosity until a peak is observed in the latter (μ_{max} in Figure 14c,d), when maximum interconnectivity is achieved; (b) a region of dendritic instability—the dendritic structure can no longer withstand the pressure applied and the dendritic arms break. The progressive movement of the particles causes the apparent viscosity to decrease. The stress continues to increase until it reaches a maximum (σ_{max} in Figure 14a,b) and all of the dendritic "skeleton" has collapsed; (c) an unrestricted flow region—broken particles flow "freely" through the liquid, producing a higher shear rate and, consequently, a rapid decrease in the apparent viscosity, which in turn results in smaller values of stress as the strain continues to increase. After the decrease, only a small change in apparent viscosity is observed as the shear rate increases; the midpoint (in time) of the unrestricted flow region, i.e., the interval extending from the maximum stress to the end of the test (the last measurement on the right of the graph), is generally considered a representative value of the apparent viscosity of the thixotropic slurry. The tests performed here were stopped when a strain of 0.8 was reached for all samples; had the strain been increased further, a fourth region would be seen in the curves, in which the liquid would be fully expelled to the outer radius of the thixoforming disk, leading to a central region of the sample with an extremely high solid fraction. This situation would result in a rapid rise in stress and apparent viscosity until the end of the test, as if solid metal were being formed.

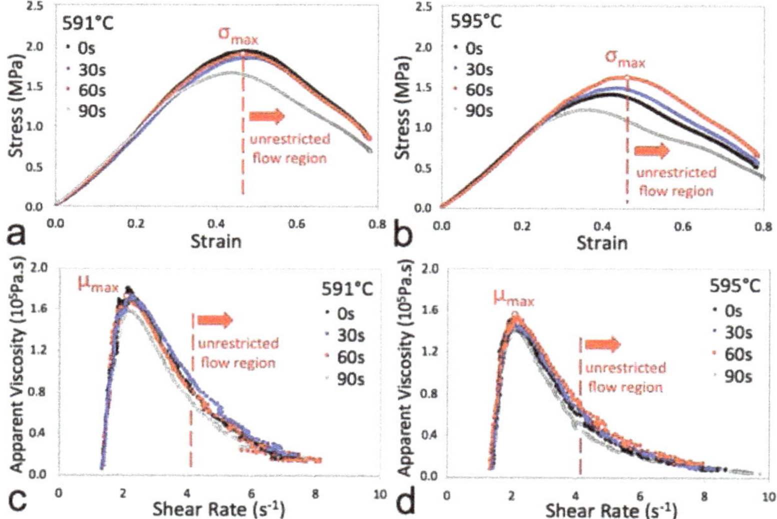

Figure 14. Engineering stress vs. engineering strain (**a,b**) and apparent viscosity vs. shear rate (**c,d**) curves measured during thixoforming (open-die forging) of B319+Fe alloy after heat treatment at 591 °C and 595 °C for 0, 30, 60, and 90 s. The maximum apparent viscosity, maximum stress, and unrestricted flow region are highlighted for a soak time of 60 s in all curves for illustrative purposes.

Table 7 shows the rheological parameters of the heat-treated samples at three moments during the thixoforming tests: the points of maximum apparent viscosity (μ_{max}) and maximum stress (σ_{max}) and the point that is representative of the unrestricted flow region. In Figure 14, the apparent viscosity and stress values for a given temperature are similar for soak times of 0, 30, and 60 s (black, blue, and red curves), while for a soak time of 90 s (grey curves) they are somewhat smaller. Thixoforming at 595 °C results in slightly smaller values of maximum stress and apparent viscosity (Figure 14b,d) than thixoforming at 591 °C (Figure 14a,c), which agrees with the higher liquid fraction of the alloy at 595 °C and consequent reduced viscosity and strength of the slurry. These results indicate that the liquid fraction (which is directly related to temperature) is more effective in changing the rheological behavior of the B319+Fe alloy than the slight changes in microstructure caused by different soak times. It should be borne in mind, as mentioned earlier (Section 3.4), that these changes were analyzed using mean values with large standard errors, i.e., distributions with large spreads, which were reflected in the homogeneous rheology of the semisolid slurry.

Table 7. Rheological parameters at the point of maximum apparent viscosity (beginning of dendritic breakdown) and point of maximum stress (end of dendritic breakdown) and in the unrestricted flow region (representative point) for heat-treated samples of B319+Fe alloy during thixoforming.

Point	Condition		μ (10^5 Pa.s)	$\dot{\gamma}$ (s^{-1})	σ (MPa)	e	t (s)
μ_{max}	591 °C	0 s	1.78	2.11	1.27	0.26	0.076
		30 s	1.73	2.25	1.36	0.28	0.085
		60 s	1.72	2.05	1.13	0.23	0.067
		90 s	1.59	2.08	1.05	0.22	0.064
	595 °C	0 s	1.49	1.95	0.89	0.20	0.057
		30 s	1.48	1.96	0.88	0.20	0.058
		60 s	1.55	2.06	1.02	0.24	0.068
		90 s	1.44	1.96	0.82	0.19	0.053
σ_{max}	591 °C	0 s	0.87	4.07	1.93	0.46	0.15
		30 s	0.85	4.06	1.85	0.47	0.14
		60 s	0.84	4.07	1.90	0.47	0.15
		90 s	0.92	3.57	1.66	0.43	0.13
	595 °C	0 s	0.74	3.72	1.40	0.43	0.14
		30 s	0.81	3.58	1.49	0.44	0.14
		60 s	0.72	4.13	1.62	0.46	0.15
		90 s	0.89	3.19	1.21	0.36	0.10
Unrestricted Flow (midpoint)	591 °C	0 s	0.30	6.10	1.59	0.60	0.22
		30 s	0.32	6.38	1.87	0.62	0.23
		60 s	0.35	6.04	1.66	0.57	0.20
		90 s	0.32	5.78	1.39	0.56	0.19
	595 °C	0 s	0.19	6.21	1.02	0.60	0.21
		30 s	0.24	6.07	1.19	0.57	0.20
		60 s	0.26	6.00	1.30	0.60	0.23
		90 s	0.18	6.33	0.86	0.56	0.19

The maximum apparent viscosity obtained by thixoforging the B319+Fe alloy was of the order of 10^5 Pa.s. This compares favorably with values considered suitable for this kind of process in the literature, i.e., 10^4–10^6 Pa.s [1,2]. Thus, although the microstructure of the alloy is coarse (Section 3.4), its influence over rheology seems to be less than that of other processing parameters, such as the liquid fraction, higher values of which facilitate the movement of solid particles in the slurry. The rheological stability, i.e., the homogeneous flow of the thixotropic slurry during thixoforming after heat treatment under different conditions (temperatures and soak times), shows again that this alloy may be easy to control in the semisolid state and could thus potentially be a suitable raw material for use in SSM processing.

Coarse particles are known to be detrimental to the rheology of the semisolid slurry. However, the effect of the Fe-rich precipitates on the rheological behavior of the B319+Fe alloy cannot be evaluated because the amount, distribution, size, and morphology of these particles were similar for all isothermal treatments applied (Section 3.4). If these precipitates do indeed have a detrimental effect, then they affect all of the heat treatments similarly. Further investigation of this issue would require a comparison of the rheological results presented here with those for conventional B319 alloy (with a lower Fe content).

4. Conclusions

The suitability for thixoforming of an Al-Si-Cu alloy with a high Fe content (here referred to as B319+Fe) was analyzed. The results reveal good agreement between the thermodynamic behavior predicted by the DM and the data obtained from semisolid forging of samples of the alloy. Although the alloy has a coarse microstructure, its rheological behavior is appropriate for thixoforming, indicating that the microstructure only plays a secondary role in determining viscosity when enough liquid is present in the slurry to enhance the flow. The microstructural and rheological stability of the alloy at the temperatures and soak times studied here also make it a promising raw material for thixoforming. The similar amount, distribution, size, and morphology of Fe-rich precipitates after the various isothermal treatments indicate that these compounds are thermodynamically stable in the temperature range analyzed, but also prevent an evaluation of their effect on the rheological behavior of the alloy.

Author Contributions: G.L.B., performed most part of this investigation in her PhD that include the raw material preparation, thermodynamic, microstructure and rheological characterization. She performed the experimental procedure as well as the data acquisition and analysis and wrote the initial version of this paper. C.T.W.P. help with the experimental procedure regarding the rheological analysis as well as the microstructure characterization. E.J.Z. was responsible for conceptualization, methodology, formal analysis, validation, funding acquisition, as well as project administration and the final writing and reviewing.

Funding: This research was funded by [Sao Paulo Research Foundation (Fapesp) and Nation Scientific Council (CNPq)] grant numbers [2015/22143-3 and PQ 306896-2013-3].

Acknowledgments: The authors would like to thank the Brazilian research funding agencies FAPESP (São Paulo Research Foundation—Projects 2013/09961-3 and 2015/22143-3), CNPq (National Council for Scientific and Technological Development—Project CNPq PQ 304921-2017-3), and CAPES (Federal Agency for the Support and Improvement of Higher Education) for providing financial support for this study. The authors are also indebted to the Faculty of Mechanical Engineering at the University of Campinas for the practical support very kindly provided.

References

1. Kirkwood, D.H.; Suéry, M.; Kapranos, P.; Atkinson, H.V.; Young, K.P. *Semisolid Processing of Alloys*, 1st ed.; Springer: Berlin/Heidelberg, Germany, 2010; p. 172. ISBN 978-3-642-00706-4.
2. Flemings, M.C. Behavior of metal alloys in the semisolid state. *Metall. Trans. A* **1991**, *22*, 957–981. [CrossRef]
3. Zoqui, E.J. Alloys for semisolid processing. In *Comprehensive Materials Processing*, 1st ed.; Hashmi, S., Ed.; Elsevier: Amsterdam, The Netherlands, 2014; Volume 5, pp. 163–190. ISBN 978-0-08-096533-8.
4. Atkinson, H.V. Modelling the semisolid processing of metallic alloys. *Prog. Mater. Sci.* **2005**, *50*, 341–412. [CrossRef]
5. Jung, H.K.; Kang, C.G. Reheating process of cast and wrought aluminum alloys for thixoforging and their globularization mechanism. *J. Mater. Process. Tech.* **2000**, *104*, 244–253. [CrossRef]
6. Chen, Y.; Wei, J.; Zhao, Y.; Zheng, J. Microstructure evolution and grain growth behavior of Ti14 alloy during semisolid isothermal process. *Trans. Nonferr. Metal. Soc.* **2011**, *21*, 1018–1022. [CrossRef]
7. Reisi, M.; Niroumand, B. Growth of primary particles during secondary cooling of a rheocast alloy. *J. Alloys Compd.* **2009**, *475*, 643–647. [CrossRef]
8. Lashkari, O.; Ghomashchi, R. The implication of rheology in semisolid metal processes: An overview. *J. Mater. Process. Tech.* **2007**, *182*, 229–240. [CrossRef]
9. Perez, M.; Barbé, J.C.; Neda, Z.; Bréchet, Y.; Salvo, L. Computer simulation of the microstructure and rheology of semisolid alloys under shear. *Acta Mater.* **2000**, *48*, 3773–3782. [CrossRef]
10. Liu, D.; Atkinson, H.V.; Jones, H. Thermodynamic prediction of thixoformability in alloys based on the Al–Si–Cu and Al–Si–Cu–Mg systems. *Acta Mater.* **2005**, *53*, 3807–3819. [CrossRef]

11. Uggowitzer, P.P.J.; Uhlenhaut, D.I. Metallurgical aspects of SSM processing. In *Thixoforming: Semi-Solid Metal Processing*, 1st ed.; Hirt, G., Kopp, R., Eds.; Wiley-VCH: Weinheim Germany, 2009; pp. 31–42. [CrossRef]

12. Zoqui, E.J.; Benati, D.M.; Proni, C.T.W.; Torres, L.V. Thermodynamic evaluation of the thixoformability of Al–Si alloys. *CALPHAD* **2016**, *52*, 98–109. [CrossRef]

13. Chiarmetta, G. Why Thixo? In Proceedings of the 6th International Conference on Semi-Solid Processing of Alloys and Composites, Turin, Italy, 27–29 September 2000; pp. 15–21.

14. Zoqui, E.J.; Gracciolli, J.I.; Lourençato, L.A. Thixo-formability of the AA6063 alloy: Conventional production processes versus electromagnetic stirring. *J. Mater. Process. Tech.* **2008**, *198*, 155–161. [CrossRef]

15. Proni, C.T.W.; Torres, L.V.; Haghayeghi, R.; Zoqui, E.J. ECAP: An alternative route for producing AlSiCu for use in SSM processing. *Mater. Charact.* **2016**, *118*, 252–262. [CrossRef]

16. Paes, M.; Zoqui, E.J. Semisolid behavior of new Al–Si–Mg alloys for thixoforming. *Mater. Sci. Eng. A* **2005**, *406*, 63–73. [CrossRef]

17. Nadal, R.L.; Roca, A.S.; Fals, H.D.C.; Zoqui, E.J. Mechanical properties of thixoformed hypoeutectic gray cast iron. *J. Mater. Process. Tech.* **2015**, *226*, 146–156. [CrossRef]

18. Haghayeghi, R.; Zoqui, E.J.; Timelli, G. Enhanced refinement and modification via self-inoculation of Si phase in a hypereutectic aluminium alloy. *J. Mater. Process. Tech.* **2018**, *252*, 294–303. [CrossRef]

19. Roca, A.S.; Fals, H.D.C.; Pedron, J.A.; Zoqui, E.J. Thixoformability of hypoeutectic gray cast iron. *J. Mater. Process. Tech.* **2012**, *212*, 1225–1235. [CrossRef]

20. Campo, K.N.; Zoqui, E.J. Thixoforming of an ECAPed Aluminum A356 Alloy: Microstructure Evolution, Rheological Behavior, and Mechanical Properties. *Metall. Mater. Trans. A* **2016**, *4*, 1–11. [CrossRef]

21. Irizalp, S.G.; Saklakoglu, N. Effect of Fe-rich intermetallics on the microstructure and mechanical properties of thixoformed A380 aluminum alloy. *Eng. Sci. Tech. Int. J.* **2014**, *17*, 58–62. [CrossRef]

22. Evolution of Phases in a Recycled Al-Si Cast Alloy During Solution Treatment. Available online: https://www.intechopen.com/books/scanning-electron-microscopy/evolution-of-phases-in-a-recycled-al-si-cast-alloy-during-solution-treatment (accessed on 8 March 2018).

23. Brollo, G.L.; Proni, C.T.W.; Paula, L.C.; Zoqui, E.J. An alternative method to identify critical temperatures for semisolid materials process applications using differentiation. *Thermochim. Acta* **2017**, *651*, 22–33. [CrossRef]

24. Brollo, G.L.; Paula, L.C.; Proni, C.T.W.; Zoqui, E.J. Analysis of the thermodynamic behavior of A355 and B319 alloys using the Differentiation Method. *Thermochim. Acta* **2018**, *659*, 121–135. [CrossRef]

25. *NADCA Standards for Semi-Solid and Squeeze Casting Processes*; Section 3; North American Die Casting Association: Arlington Heights, IL, USA, 2006.

26. Nafisi, S.; Emadi, D.; Ghomashchi, R. Semi solid metal processing: The fraction solid dilemma. *Mater. Sci. Eng. A* **2009**, *507*, 87–92. [CrossRef]

27. Kazakov, A.A.; Luong, N.H. Characterization of semisolid materials structure. *Mater. Charact.* **2001**, *46*, 155–161. [CrossRef]

28. Zoqui, E.J.; Shehata, M.T.; Paes, M.; Kao, V.; Es-Sadiqi, E. Morphological evolution of SSM A356 during partial remelting. *Mater. Sci. Eng. A* **2002**, *325*, 38–53. [CrossRef]

29. ASTM International E112-13. Standard Test Methods for Determining Average Grain Size. Available online: https://www.astm.org/Standards/E112.htm (accessed on 8 March 2018).

30. Proni, C.T.W.; Zoqui, E.J. The effect of heating rate on the microstructural breakdown required for thixoformability. *Int. J. Mater. Res.* **2017**, *108*, 228–236. [CrossRef]

31. Laxmanan, V.; Flemings, M.C. Deformation of semi-solid Sn-15 Pct Pb alloy. *Metall. Trans. A* **1980**, *11A*, 1927–1937. [CrossRef]

32. Warmuzek, M. Metallographic Techniques for Aluminum and Its Alloys. In *Metallography and Microstructures*, 1st ed.; Voort, G.F.V., Ed.; ASM International: Materials Park, OH, USA, 2004; Volume 9, pp. 711–775. ISBN 978-0-87170-706-2.

33. Prados, E.F.; Sordi, V.L.; Ferrante, M. The effect of Al_2Cu precipitates on the microstructural evolution, tensile strength, ductility and work-hardening behaviour of a Al–4 wt.% Cu alloy processed by equal-channel angular pressing. *Acta Mater.* **2013**, *61*, 115–125. [CrossRef]

34. Tabibian, S.; Chakaluk, E.; Constantinescu, A.; Szmytka, F.; Oudin, A. TMF–LCF life assessment of a Lost Foam Casting A319 aluminum alloy. *Int. J. Fatigue* **2013**, *53*, 75–81. [CrossRef]

35. Jorstad, J.L.; Pan, Q.Y.; Apelian, D. Solidification microstructure affecting ductility in semi-solid-cast products. *Mater. Sci. Eng. A* **2005**, *413–414*, 186–191. [CrossRef]

36. Gao, T.; Hu, K.; Wang, L.; Zhang, B.; Liu, X. Morphological evolution and strengthening behavior of a-Al(Fe,Mn)Si in Al–6Si–2Fe–xMn alloys. *Results Phys.* **2017**, *7*, 1051–1054. [CrossRef]

37. Warng, P.S.; Liauh, Y.J.; Lee, S.L.; Lin, J.C. Effects of Be addition on microstructures and mechanical properties of B 319.0 alloys. *Mater. Chem. Phys.* **1998**, *53*, 195–202. [CrossRef]

38. Martinez, R.; Russier, V.; Couzinier, J.P.; Guillot, I.; Cailletaud, G. Modeling of the influence of coarsening on viscoplastic behavior of a 319 foundry aluminum alloy. *Mater. Sci. Eng. A* **2013**, *559*, 40–48. [CrossRef]

39. Tzimas, E.; Zavaliangos, A. Materials selection for semisolid processing. *Mater. Manuf. Process.* **1999**, *14*, 217–230. [CrossRef]

40. Wang, N.; Zhou, Z.; Lu, G. Microstructural Evolution of 6061 Alloy during Isothermal Heat Treatment. *J. Mater. Sci. Technol.* **2011**, *27*, 8–14. [CrossRef]

41. Modigelli, M.; Pola, A. Modeling of shear induced coarsening effects in semi-solid alloys. *Trans. Nonferr. Metal. Soc.* **2010**, *20*, 1696–1701. [CrossRef]

42. Werz, T.; Baumann, M.; Wolfram, U.; Kril, C.E., III. Particle tracking during Ostwald ripening using time-resolved laboratory X-ray microtomography. *Mater. Charact.* **2014**, *90*, 185–195. [CrossRef]

43. Jiang, J.; Wang, Y.; Atkinson, H.V. Microstructural coarsening of 7005 aluminum alloy semisolid billets with high solid fraction. *Mater. Charact.* **2014**, *90*, 52–61. [CrossRef]

Article

Microstructure and Mechanical Properties of Thixowelded AISI D2 Tool Steel

M. N. Mohammed [1,*], M. Z. Omar [2], Salah Al-Zubaidi [3], K. S. Alhawari [2] and M. A. Abdelgnei [2]

[1]   Department of Engineering & Technology, Faculty of Information Sciences and Engineering,
    Management & Science University, Selangor 40100, Malaysia
[2]   Department of Mechanical and Materials Engineering, Faculty of Engineering and Built Environment,
    Universiti Kebangsaan Malaysia, Selangor 43600, Malaysia; zaidiomar@ukm.my (M.Z.O.);
    alhawary.khaled@yahoo.com (K.S.A.); mnel_abdelgnei@yahoo.com (M.A.A.)
[3]   Department of Automated Manufacturing Engineering, Al-Khwarizmi college of Engineering,
    University of Baghdad, Baghdad 10071, Iraq; salah_mfeng@yahoo.com
*   Correspondence: mohammed.ukm@gmail.com; Tel.: +60-111-123-5993

Received: 9 March 2018; Accepted: 19 April 2018; Published: 4 May 2018

Abstract: Rigid perpetual joining of materials is one of the main demands in most of the manufacturing and assembling industries. AISI D2 cold work tool steels is commonly known as non-weldable metal that a high quality joint of this kind of material can be hardly achieved and almost impossible by conventional welding. In this study, a novel thixowelding technology was proposed for joining of AISI D2 tool steel. The effect of joining temperature, holding time and post-weld heat treatment on microstructural features and mechanical properties were also investigated. Acceptable joints without defect were achieved through the welding temperature of 1300 °C, while the welding at lower temperature resulted in a series of cracks across the entire joint that led to spontaneous fracture after joining. Tensile test results showed that maximum joint tensile strength of 271 MPa was achieved at 1300 °C and 10 min holding time, which was 35% of that of D2 base metal. Meanwhile, tensile strength of the joined parts after heat treatment showed a significant improvement over the non-heat treated condition with 560 MPa, i.e., about 70% of that of the strength value of the D2 base metal. This improvement in the tensile strength attributed to the dissolution of some amounts of eutectic chromium carbides and changes in the microstructure of the matrix. The joints are fractured at the diffusion zone, and the fracture exhibits a typical brittle characteristic. The present study successfully confirmed that by avoiding dendritic microstructure, as often resulted from the fusion welding, high joining quality components obtained in the semi-solid state. These results can be obtained without complex or additional apparatuses that are used in traditional joining process.

Keywords: thixowelding; thixojoining; semisolid joining; cold-work tool steel

---

1. Introduction

Recently, AISI D2 tool steel get significant attention in the wide applications of industrial sector owing to its attractive properties. Unfortunately, a high-quality joint of this type of material can be hardly achieved and almost impossible by conventional welding methods owing to its high carbon and alloying elements content with an enormous amount of carbides. The existing fusion welding technique is the most commonly used process and has been widely practiced in industries. In this process, the welding is used with preheating and post-weld heat treatment (PWHT) to avoid solidification cracking and residual stresses that induced by the phase transformations throughout the welding. The most critical part in the process is the heat affected zone (HAZ), which can be very hard and brittle, along with susceptive to cracking if the tools are not heat-treated correctly.

In addition, fusion welding technologies are characterized by high temperature gradients that results in high thermal stresses and a fast solidification, which lead to the occurrence of segregation phenomena. Furthermore, the interface morphology is characteristically dendritic and the natural progression of solidification commonly leads to interior structural defects, such as shrinkage porosities and non-homogenous microstructure [1]. Therefore, an innovative route that can avoid the above mentioned problems inherent in fusion welding is required. Thixojoining is a possible technique that can weld the AISI D2 cold-work tool steel free from the above mentioned problems, as the process is carried at the semisolid state where the mechanisms of solidification and heat flow are different from other welding processes [2–4].

During the last 30 years, several authors have shown that the thixojoining process has the potential to be used for a wide range of metal production processes. Notwithstanding some technical and technological differences between the available semi-solid joining processes, they can be classified into four categories: (1) joining metals by using semisolid slurries, (2) addition of functional features, (3) semisolid stir joining, and (4) semisolid diffusion joining [2].

The first method involves the utilization of a thixotropic metal as filler applied into the joint groove for the purpose of joining materials. This kind of joining has the distinctive advantage of allowing a controlled flow during deposition process. Moreover, the applied temperatures and thermal gradients are considered smaller if compared with fusion welding. Earlier research work in the 1990s has demonstrated the feasibility of the process when Mendez and Brown used Sn-5 wt. % Pb as a filler and applied it on the joint groove to join bars of Sn-15 wt. % Pb model alloy. The microstructure of the joint cross-sectional area showed excellent metallurgical joining with the interface connection and smooth and homogeneous welding between the bars and filler metal without defects or adverse effects in the heating zone [5].

Other researchers have more investigations in the field of semisolid joining revealed the possibility of producing prototype components by combining the forming operation with the simultaneous insertion of additional components [6]. One special approach that takes advantage of the material's high flowability is the addition of functional features to a forged part. With this technology, it is possible to shorten conventional process chains and to create a new generation of components. Kiuchi et al. [6] studied the possibility of joining aluminum with aluminum and steel alloys. They found that thixojoining of simple insert shape with bulk material is promising approach. With decreasing of liquid fraction, there was increasing in joint strength. This clarifies the significant merits compared with conventional casting due to formation of intermetallic compounds and diffusion between base and insert materials. The big challenge of improvement this technique for other materials is accessible inserting component in mushy matrix with destruction of joint geometry by remelting or deformation.

Further investigations in Semisolid stir joining (SSSJ) showed the applicability of this type of joining process to improve and overcome the difficulties accompanied with current joining techniques which were based on friction stir welding. Recently, there was trend toward carrying out a vacuum-free SSSJ process that enables using low joining temperature when applied to a semisolid base or filler metal. Hosseini et al. [7] have also carried out extensive experiments on semisolid stir joining (SSSJ) for AZ91 alloy by using mechanical stirring and a Mg-25 wt. % Zn interlayer. The results showed the significant effect of mechanical stirring rate on the joint-strength. This supposes that the formed oxide layer has been disrupted due to stirring effect and consequently improve metallurgical bonds. They claimed that this method can be used as an alternative joining for Mg-based alloys.

More investigations to achieve joint with high-quality globular structure were undertaken by Mohammed et al. [8–10] in semisolid diffusion joining (SSDJ). Their research explored the possibilities of joining semisolid AISI D2 tool steel with functional elements of the same material as well as with other materials such as AISI 304 stainless steel by using a peculiar characteristic of the metals' liquid components. In their proposed method, when the interfaces of two metals are pasted together, the liquid components present in both sides of these "semisolid metals" penetrate and diffuse through the interface and then solidify together as one body. The experimental results showed that, the use of this technique can make homogeneous properties with high surface quality without any evidence of porosity or microcracking.

The semisolid metal joining (SSMJ) process can be used for processing different net near net shape components with complicated shape and geometries where other joining processes cannot be utilized for many reasons such as: low melting temperature, poor mechanical properties, and insufficient plastic deformation. Several applications have been considered based on this new joining method like heating radiator components, car shock absorber, slider system components and roller way elements. In addition, this joining technology can be used for joining craft glass and stones to plates and bars with different shapes. The products can be utilized in decoration various wall plates and floor panels. Furthermore, metal/glass joining products have great capabilities to add other functional properties to materials to produce lubricated, corrosion-proof, antiwear or heatresistant items [2].

The thixojoining is new semi-solid-based process that should take their place among other joining processes. Nevertheless, it takes little considerations by researcher and investigators. This new specific area needs continuous studies and works to overcome the related problem and propose solutions and suggestions especially for high grade alloys and materials. In this work, the thixowelding process was applied to join AISI D2 cold work tool steels by using a direct partial remelting method (DPRM) method. The objectives of the present research are to study the microstructure evolution and mechanical properties of the joints under different welding conditions.

2. Experimental Section

2.1. Materials and Sample Preparation

AISI D2 cold-work tool steel used in this study which was supplied after a soft annealing process, i.e., heating to 850 °C followed by cooling at 10 °C/h to 650 °C and finally air cooling. The chemical composition of the studied AISI D2 cold-work tool steel was determined by using X-ray fluorescence (XRF) (S8 TIGER ECO, Bruker, Karlsruhe, Germany) as shown in Table 1. The base blank of AISI D2 was cut Ø 16× 100 mm while the length of the pin insert was 18 mm with diameter of 7 mm as shown in Figure 1. Each specimen was disposed of rust and scale by submerging in pickling solution, washed with acetone, and finally rinsed using distilled water.

The temperatures of liquidus and solidus, in addition to liquid fraction, were estimated by carrying out differential scanning calorimetry test (DSC). Small pieces samples (less than 20 mg) were cut from the as-received material for testing with The Netzsch-STA 449 F3 simultaneous thermogravimeter (TGA-DSC) (STA 449F3, NETZSCH, Bavaria, Germany). The samples were heated at a rate of 10 °C/min in nitrogen atmosphere to prevent oxidation. The common tangent procedure is adopted to estimate the liquid fraction. Finding out the fraction of liquid was done by integrating of partial areas under the heating curves as shown in Figure 2.

Table 1. Chemical composition of AISI D2 cold-work tool steel (in Weight Percent).

C	Si	Mn	P	S	Cr	Ni	Mo	W	Cu	V	Fe
1.50	0.258	0.24	0.02	0.01	11.2	0.21	0.78	0.2	0.08	0.781	Balance

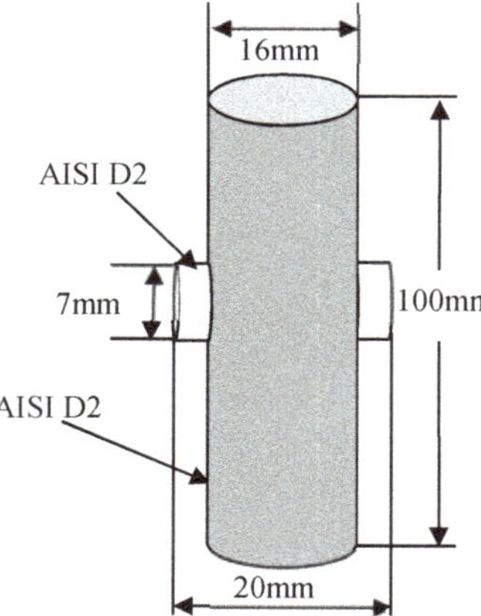

Figure 1. Schematic drawing of the AISI D2 parts combination.

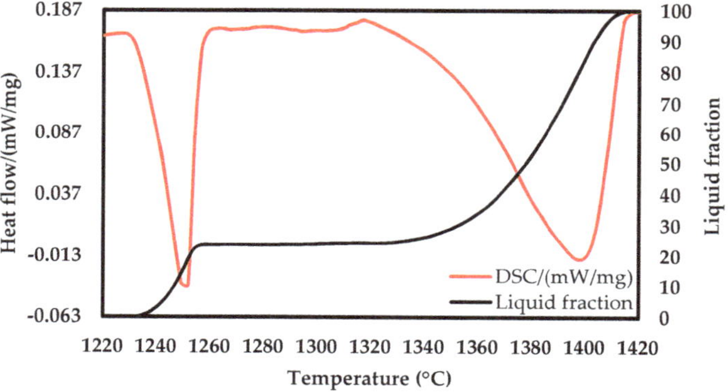

Figure 2. Liquid fraction vs. temperature and heating flow vs temperature curves of the AISI D2 tool steel calculated by differential scanning calorimetry (DSC).

2.2. Thixo Welding Process

High temperature carbolite furnace was utilised to carry out the thixo welding of AISI D2 tool steel as shown in Figure 3. The furnace is vertical type and protected with argon gas. The moment that furnace reached the pre-selected temperature (1250, 1275, and 1300 °C), the sample was lowered into the furnace by use chromehromel wire for 0, 5 and 10 min respectively then it was allowed to cool slowly at room temperature as shown in Figure 4. To investigate the influence of post welding heat treatment on the microstructure and mechanical properties of the welded plates, specimens were heat treated by heating the specimens to 850–900 °C followed by slow rates of furnace cooling (22 °C/h), when a steel cools to 650 °C, air cooling can be used to bring the steel to room temperature.

2.3. Metallography

The thixowelded joint samples were polished and then etched by using Villela reagent (1 g picric acid, 5 mL hydrochloric acid, and 95 mL ethyl alcohol) to reveal their microstructures. The microstructures of the joints were characterized by Optics microscopy (BX51-P, Olympus, Tokyo, Japan), scanning electron microscopy (SEM) (LEO 1455 VP SEM, Zeiss, Oberkochen, Germany) equipped with energy dispersive X-ray spectroscopy (EDX) (Oxford 7353, Oxford Instruments, Wiesbaden, Germany). X-ray diffraction (XRD) (Bruker D8 Discover, Bruker, Karlsruhe, Germany) analyses were carried out to analyze the phases present in the steel specimen before and after the thixojoining. It also gives an idea about the characteristics of joints and if there is any harmful phase or brittle phase have formed during thixojoining.

2.4. Test of Mechanical Properties

The mechanical characterization of the joints was measured using tensile strength and hardness tests. The tensile tests were carried out on using Zwick universal testing machine (ProLine Z005/Z100, Zwick, Ulm, Germany) with a cross-head speed of 0.5 mm/min at room temperature according to ASTM E8M Standards. The fracture morphologies of failed specimens were then observed by scanning electron microscopy (SEM). Vickers micro-hardness testing (FH5, Tinius Olsen, Horsham, PA, USA) was performed on the polished specimens of the transverse section of joints as well as the base material and insert to compare the phases present in the joint and the joining parameters on the base metal. The profiles of micro-hardness were constructed across the sample's horizontal section of the D2 tool steel base metal, the weld zone and the insert by using a load of 10 g for 30 s along the lines perpendicular to the joint interface. Hardness measurements were carried out by using Vickers micro-hardness tester FH5.

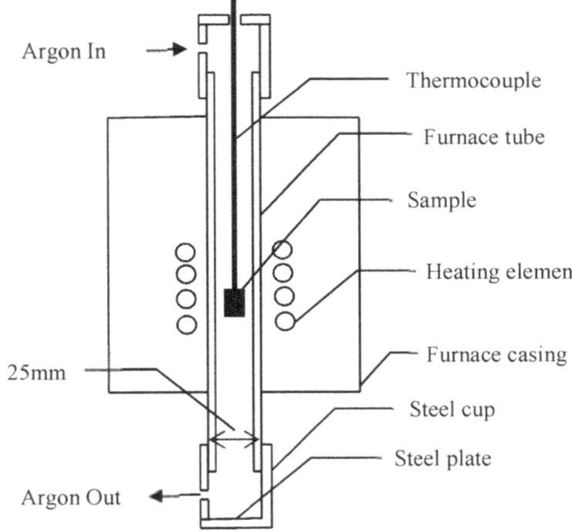

Figure 3. Schematic drawing of the furnace set-up.

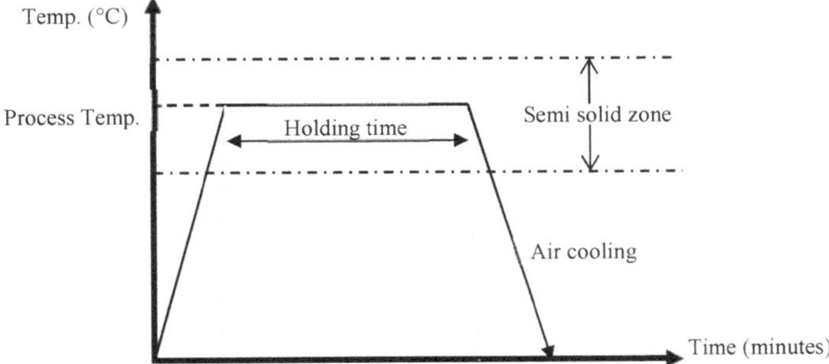

Figure 4. Temperature profile of direct partial remelting experiment.

3. Results and Discussion

3.1. Microstructure and Joint Quality

The microstructure of as-received AISI D2 consists of ferritic matrix parallel to working direction in which different surplus of large-coarse and fine carbides are distributed as shown in Figure 5. The revealed microstructure refers to annealed tool steel [11] and indicates that the heat treatment was carried out as recommended. The optical microscopy examination performed on the welded joints at various conditions is shown in Figure 6. From the complete weld section at 1250 °C heating temperature with 0, 5 and 10 min holding time, large gap can been seen and big cracks at the joint interface can be observed. With the increase of welding temperature to 1275 °C with 0 and 5 min holding time, a series of cracks resulted and the weld seam could not be well joined, as shown in Figure 6S4,S5. In addition, big fractures across the entire joint were shown with little and local interactions were lactated at different zones in the interface resulting in spontaneous fracture after joining. However, when the holding time is adequate (10 min) the weldability is relatively good and there is no macroscopic voids and fractures across the entire joint which is occupied by eutectic liquid, as shown in Figure 6S6. Meanwhile, by further increasing the welding temperature to 1300 °C, strong thixojoint can be produced. Further, fine and small boundary can be observed as transition zone between two alloys. No presence of porosity and micro-cracking at this zone and the joint shape was smooth and not corrugated, as shown in Figure 6S7–S9. The thixoweld joint was characterised by visible full penetration that indicate perfect thixojoning between base blank and the insert along the bounding border. Higher amount of the interfacial reactions has been produced between two base metals [12]. The structure is distinguished by two features that can be determined by grain direction, as illustrated in arrows direction in Figure 7. The direction of the grains are in bands in the forming direction and as a single pin of AISI D2 was inserted perpendicular into a base blank of AISI D2 to illustrate two distinct features. The interfacial reactions during heating have accelerated the diffusion of elements through the interface and propagated the eutectic liquid across the joined surfaces that resulted in good thixojoint of the interface at both side of steels.

Additionally, the properties of the thixojoining can be improved by subsequent heat-treatment. It is clearly visible that the changes in the microstructure patterns of the thixojoining occur after the heat treatment as shown in Figure 8. It is obviously to see that the primary distribution of grain boundary is decreased substantially and some amounts of eutectic carbides dissolve in the matrix. It means that the primary grain boundaries have distinctly pinning effect on growth of grain size in the heat treated condition [12]. The heat treated structures of the thixojoined parts only show improved contrast between solid and liquid phases, whereby the big solids appeared in little liquid matrix.

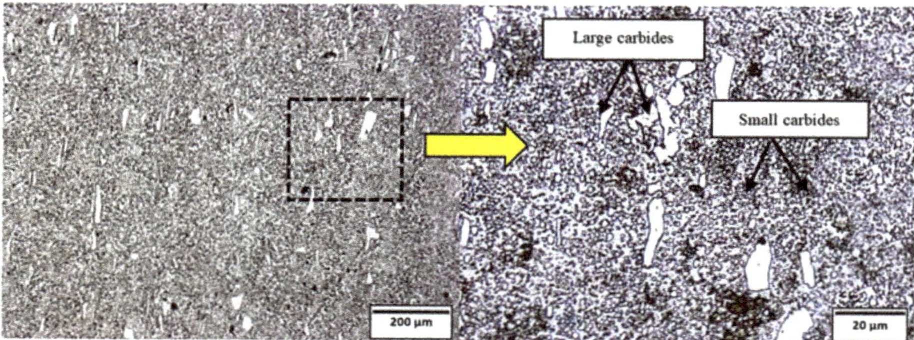

Figure 5. Optical microstructure of as received AISI D2 tool steel.

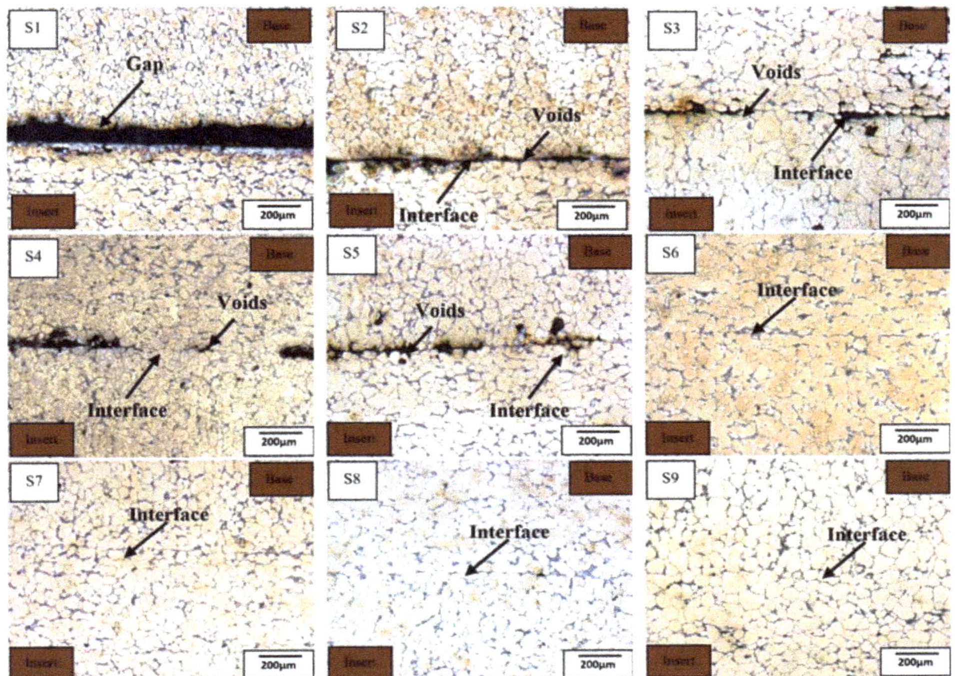

Figure 6. Microstructure of the thixojoined samples processed at different conditions: (**S1**) 1250 °C/0 min, (**S2**) 1250 C/5 min, (**S3**) 1250 °C/10 min, (**S4**) 1275 °C/0 min, (**S5**) 1275 °C/5 min, (**S6**) 1275 °C/10 min, (**S7**) 1300 °C/0 min, (**S8**) 1300 °C/5 min, (**S9**) 1300 °C/10 min.

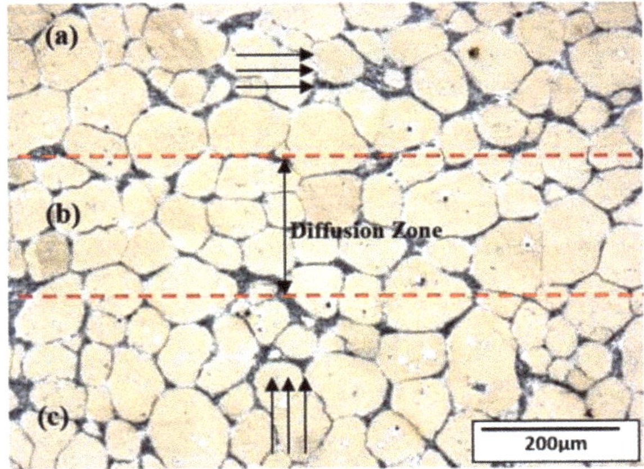

Figure 7. Microstructure of thixo-weld-joint of (**a**) base metal, (**b**) diffusion zone (**c**) insert, after subjecting to 1300 °C heating temperature for 10 min.

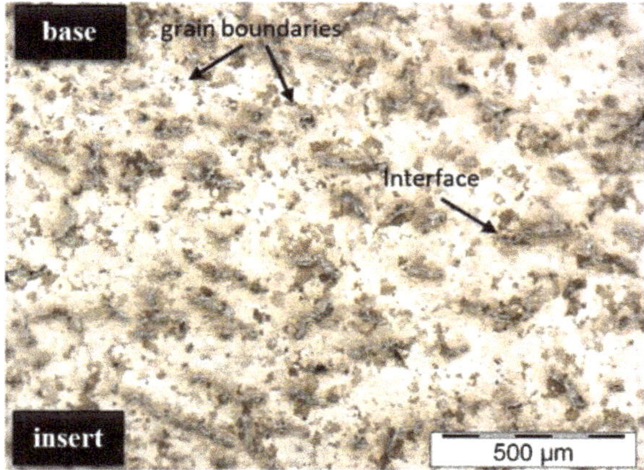

Figure 8. Microstructure of thixojoint parts after application of 1300 °C for 10 min and subjected to heat treatment.

The samples were also examined using SEM in order to get a complete assessment if there is any harmful or brittle phase have formed during thixojoining. Figure 9 shows the Scanning electron images of the microstructures of the thixojoint parts and the EDS spectra corresponding to the analyzed phases. While, SEM-EDX elemental mapping techniques was also performed on both sides at the periphery of the joint in order to determine the composition of the region near the diffusion zone of the joined part, as shown in Figure 10. Regarding to Figures 9 and 10, the EDX results confirm that the contact area was identical in microstructure, composition, and properties with the base and insert. On the basis of element distribution, it can be deduced that the both sides at the periphery of the joint display a uniform and homogeneous distribution in the structure due thermal diffusion driving force [12]. Along with this result, it was observed that the intermetallic compounds were not incorporated into the joining region, providing further evidence of the reliability of this process.

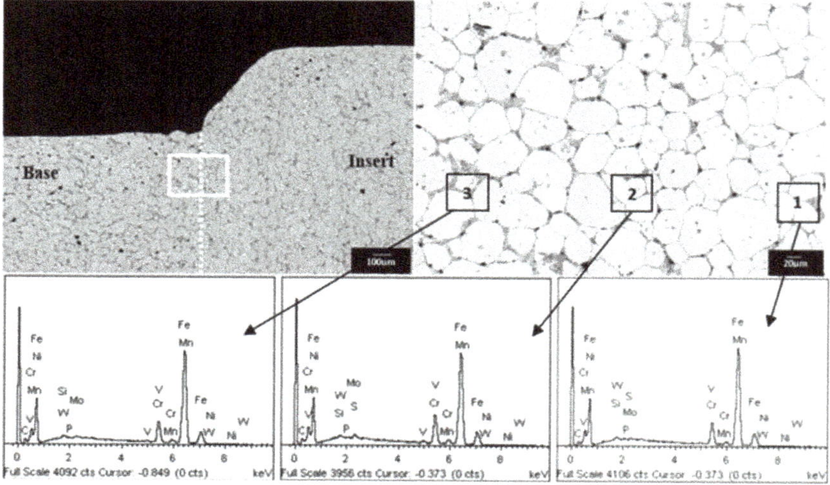

Figure 9. SEM micrographs of thixojoined part at 1300 °C with EDX of point (**1**) for base metal, (**2**) for diffusion area and (**3**) for insert.

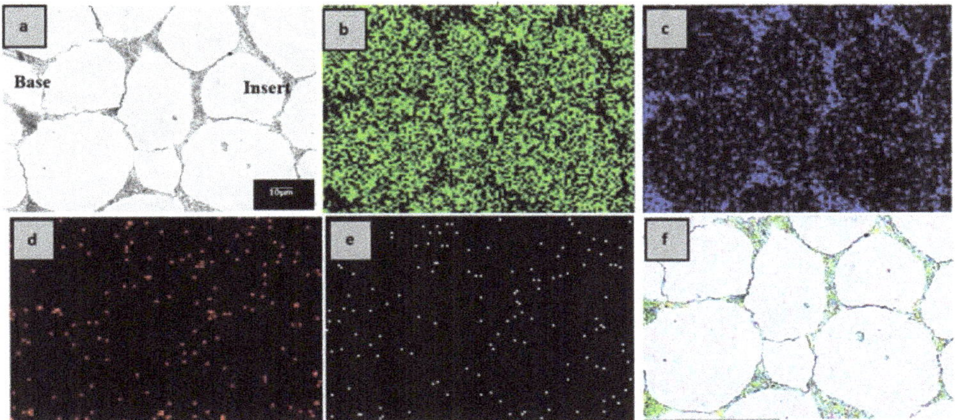

Figure 10. SEM-EDX elemental mapping of thixojoined part at 1300 °C of results near joint: (**a**) SEM micrograph, (**b**) Fe, (**c**) Cr, (**d**) C, (**e**) V, (**f**) all four elements.

X-ray diffraction test has been carried out for more characterisation of thixojoint as displayed in Figure 11. The analysis of the diffraction patterns of as-received tool steel found peaks corresponding to the ferrite phase and iron-chromium carbide (M_7C_3). From the other side, the thixoweld at 1300 °C is characterised by three peaks refer to austenite and ferrite phases in addition to M_7C_3 carbides. The presence of austenite is primarily connected with the increasing of carbon in the solid solution that stabilizing the austenite [13]. Higher contents of these elements have a significant effect on the martensite temperature and consequently the transformation of austenite to martensite is very difficult. In addition, when relatively fast cooling from the solidus–liquidus range is applied there is relatively better stability for the austenitic phase at room temperature [14].

Meanwhile, analysis of the XRD patterns of the joined parts after post welding heat treatment condition revealed a change of the structure character over the non-heat treated condition as shown in Figure 10. As can be observed that, the microstructures are composed of different amounts of retained austenite, martensite and chromium carbides. Retained austenite was recognised using crystallographic data for austenite, while martensite was recognised with data for ferrite because of similarity in the crystal structure and slight differences in lattice constants. The decomposition of metastable austenite in the structure was probably caused by the changes of carbon content. The effect of slow cooling give rise to dissolution of a large number of carbide particles causing in an augmentation of retained austenite after post welding heat treatment condition. During the thermal exposure, carbon migrates and diffused deeper to the grain boundaries while some amounts of eutectic chromium carbides dissolve in the matrix. It means that the primary grain boundaries have distinctly pinning effect on growth of grain size in the heat treated condition [15]. Based on these findings, no harmful or brittle phase was detected by XRD patterns at the thixoweld zone and these observations were in good agreement with microstructure examinations.

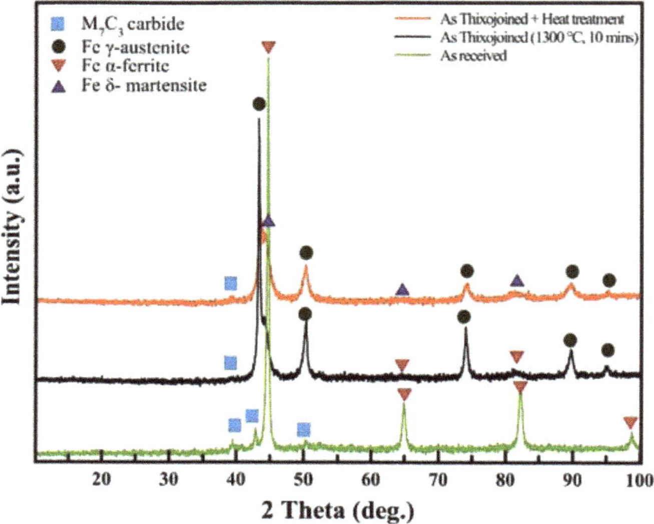

Figure 11. XRD patterns for AISI D2-AISI D2 joining processed at different conditions.

3.2. Mechanical Properties

3.2.1. Tensile Test and Fracture-surface Investigation

For the tensile strength, six samples were prepared using different joining parameters (the first three specimens with 1250 °C heating temperature was neglected). It is believed that the supplied heat input not providing sufficient diffusion within the joint interface even when the holding time was long. Typical stress-strain curves for the different test conditions are shown in Figure 12. With high thixowelding temperature, the produced interface is free from voids and pores due to high diffusion rates of different elements across this area. Consequently, when this high temperature accompany with long holding time, high tensile strength could be achieved. It is well known that the strength of a joint primarily depends on the quality of its microstructure. As can be observed, the tensile strength graph shows a relatively identical shape of the applied heating temperature to 1300 °C at 0, 5 and 10 min to give 241 MPa, 253 MPa and 271 MPa respectively. A minimum tensile strength of 241 MPa is recorded for the applied heating temperature of 1300 °C at 0 min. At 5 min a slight increment in tensile strength was recoded. Meanwhile, with 10 min. holding time there was remarkable improvement in tensile strength due to fact that diffusion rate is highly affected by both of temperature and time for non-steady state diffusion. The tensile value of the D2 as-supplied material (791 MPa) is more than two times if compared with the as-thixojoind value of 271 MPa. This is actually an interesting result which was achieved by a process which is simpler than conventional processes. Meanwhile, tensile strength of the joined parts after post welding heat treatment condition showed a significant improvement over the non-heat treated condition which is 560 MPa, i.e., about 70% the strength value of D2 base metal. This improvement of strength is linked to such factors as the size, distribution and dissolvement of carbides, relative proportions of austenite and martensite phases, and grain size. This explanation is also corroborated by microscopic studies which resulted in improved contrast between solid and liquid phases, whereby the big solids appeared in little liquid matrix where primary distribution of grain boundary is decreased substantially and some amounts of eutectic carbides dissolve in the matrix.

The mechanical properties of similar substrates will have an important effect on the joints properties because the temperature attained by each substrate markedly depends on the properties of the materials and on the selected joining parameters. According to the Arrhenius equation, an increase in the temperature will occasion an exponential enhancement in the diffusivity of solute [16]. Consequently, the heating temperature for the material will have a substantial impact on the joint properties. Weak joint may be produced if heating temperature is set on low level, even though if it is combined with high holding time. This resulted in inadequate diffusion along the interface because the elements do not have enough energy to migrate and diffuse within diffusion transition zone.

$$D = D^* \exp\left(-Qd \: / \: RT\right) \tag{1}$$

where D is the diffusivity, D^* is temperature-independent preexponential (m^2/s), Qd is the activation energy for diffusion (J/mol or eV/atom), R is the gas constant (8.31 J/(mol·K) or 8.62×10^{-5} eV/(atom·K)) and T is absolute temperature (K).

It is worth mentioning that the joining strength increases with the increase of holding time at the initial stage of the joining process due to the better coalescence of mating surfaces to ensures complete diffusion of atoms and attains a maximum strength value [17]. In general, in non-steady state diffusion, the concentration of diffusion species is function of both of time and position (Fick's second law). It is thus expected with the aid of more suitable holding time, joints with highly desirable microstructure can be obtained and in turn, highly reliable bonded joints with good performance can be anticipated. Moreover, holding time has an effect on the quantity of atomic diffusion and the creep of the protrusions (i.e., elemental diffusion of the joined alloy can be improved by increasing the holding time) [17].

$$\partial C / \partial t = D \left(\partial^2 C / \partial x^2\right) \tag{2}$$

where $\partial C / \partial t$ is the change in solute concentration with time at a given position in the base metal, which could provide an indication of isothermal solidification rate, and D is the diffusion coefficient, and $\partial^2 C / \partial x^2$ is the change in concentration gradient with distance.

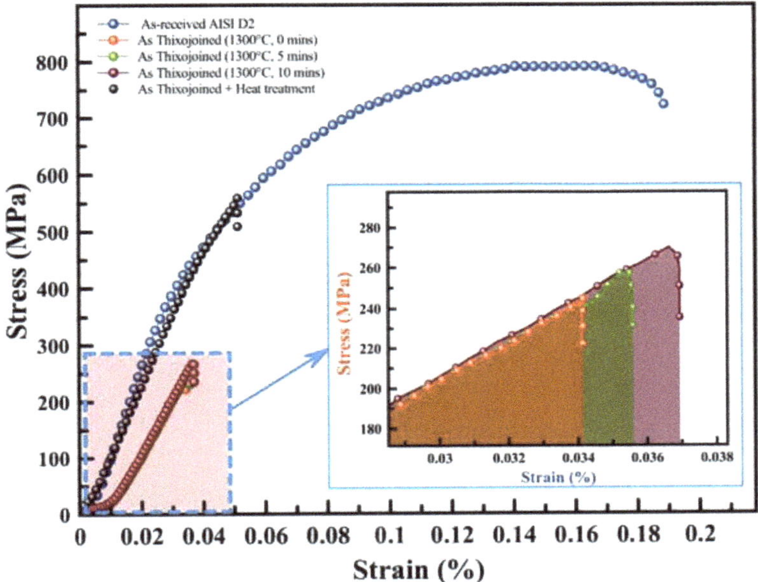

Figure 12. Typical stress-strain curves for the different test conditions.

Figure 13 displays the SEM morphology of the thixojoint that achieved with various joining conditions. Through observing the fracture surfaces created by the tensile test, it was found that the fracture appearance of the welded joint at 1300 °C was not completely along the interface between base metal and insert. The fractures were characterised by brittleness, where intergranular fracture mode is illustrative. The fracture path, which is typically dimpled, is located around liquid-solid zone and grain boundary. The close-up of the fracture path reveals a "cobbled" surface appearance, which could indicate that during solidification the fracture occurred through the liquid with the re-solidified grains then becoming exposed as "cobbles", a phenomenon also seen in 7075 Al alloy [18]. The brittleness was a result of coarse fracture with big solid lump that refer to unbroken structure. The solid γ (Fe) particles, mainly with globular shape, surrounded by the carbide networks which is a typical characteristic with respect to semi-solid microstructure of D2 tool steel. The fracture then appears to have quickly propagated along the grain boundaries and the weak liquidus–solidus interface of the lumps [19]. From the results it is believed that there is a good metallurgical joining at interface; this kind of compound fracture mechanism means that the rupture is not along the interface.

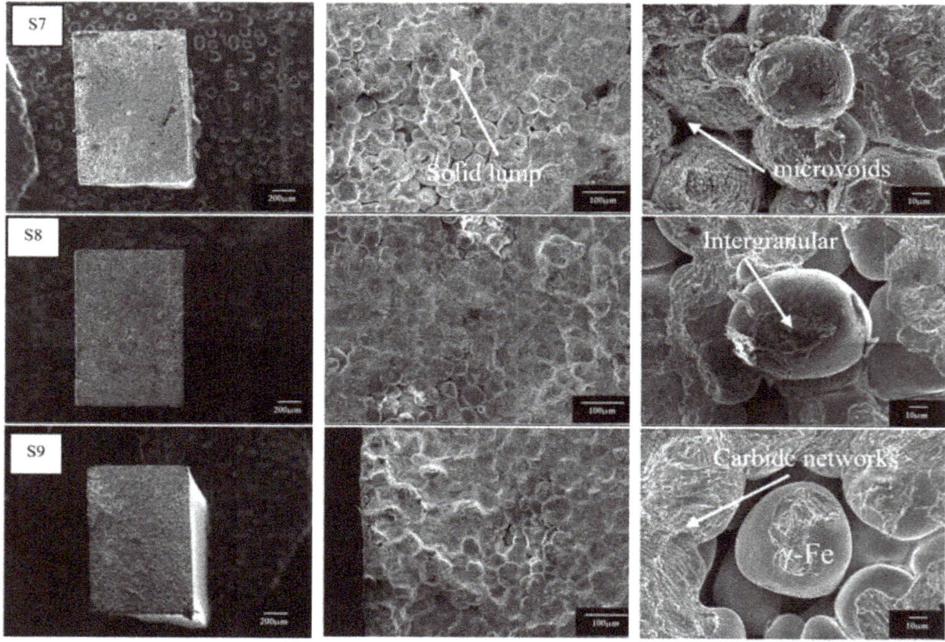

Figure 13. Fractured surface of welded joints of samples processed at various conditions: (**S7**) 1300 °C/0 min, (**S8**) 1300 °C/5 min, (**S9**) 1300 °C/10 min.

Figure 14 shows fractured surface of welded joints obtained after post welding heat treatment. As can be understood from this figure, post welding heat treatment changes the fracture mode to predominantly transgranular fracture, with few intergranular facets. On the contrary, in the tensile test of the non treated condition, disbanding of the solid particles has occurred as well as several microcracks with nano wide which are found out in fracture surface. Stress applied on the sample during the tensile tests led to the formation of big cracks out of small pores and in turn, fractures occurred when these cracks grew. It is worth mentioning that the main parameter responsible for the low ductility is the grain boundary. The liquid phase appearance in grain boundaries, results in formation of porosities after solidification, which reduces ductility. It is thus expected with the presence of grain boundary leads to long soft zones which distort discriminatory during plastic deformation [18,20]. It is attractive to note that post-joining heat treatment is one of the key parameters for improving the mechanical properties of thixojoined parts (although this would have cost-implications and could reduce the attractiveness of the process).

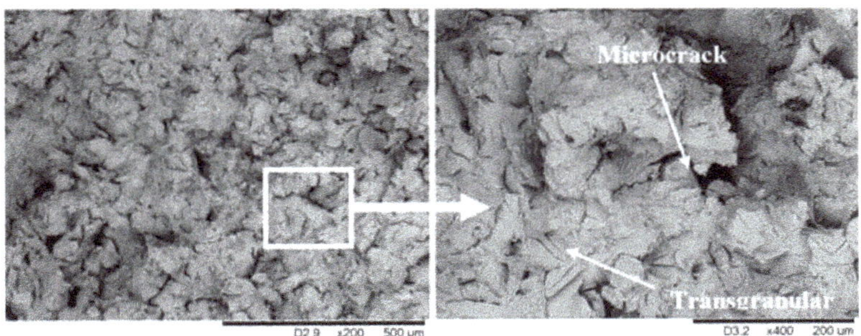

Figure 14. Fractured surface of welded joints after post welding heat treatment.

3.2.2. Hardness Measurements

Micro-hardness profiles were obtained for the AISI D2 tool steel base metal, the weld zone and the insert by using a load of 10 g for 30 s. Figure 15 presents the micro-hardness profile against distance from the interface and illustrates the variations of the HV value along the lines perpendicular to the joint interface. The hardness measurement of the base metal, the weld zone and the insert is fluctuated in the range 350 Hv to 370 Hv. This indicated that the hardness value of the welded joints was almost similar to those of base metal and insert. This indicated that the hardness profile gives a good impression of joint microstructure and the degree of homogenization, this looks very promising. However, the thixoweld-joint of AISI D2 exhibits different hardness values if compare to as supplied material. This exposure to thixojoining temperature also results in the phase change to metastable austenite. While the remaining interspaces are filled by precipitated eutectic carbides on the grain boundaries and in the lamellar network. As can be seen from this figure, almost the similar trend is observed in the microhardness profiles of all samples. Figure 15 clarifies small decrement in hardness values at same temperature with high holding time. The behaviour may be attributing to grain growth with increasing holding time during evolution of microstructures. [21,22]. These results evidenced that diffusion has occurred and weld zone micro-hardness was free from harmful or brittle phases. In the case of post welding heat treatment condition (S10), hardness measurements were somewhat higher than for the non-heat treated samples. This higher hardness can be attributed to the some amounts of eutectic chromium carbides dissolve in matrix during slow cooling. It means that the primary grain boundaries have distinctly pinning effect on growth of grain size in the heat treated condition [23]. In addition, the phase transformation is probably the reason of hardness increase to the (490 Hv). This value is higher in comparison with material hardness in the initial state after thixojoining processing. During the isothermal heat treatment at higher temperatures, particular decomposition of the austenite and ferrite phases to the martensitic phases was achieved and the carbides of chromium and retained austenite were included in the structure, as shown in XRD analysis. In tool steels with high alloy content, high hardenability may cause martensite to form during air cooling. Hardenability is the topic which relates alloying to the phase transformations that occur on cooling [24].

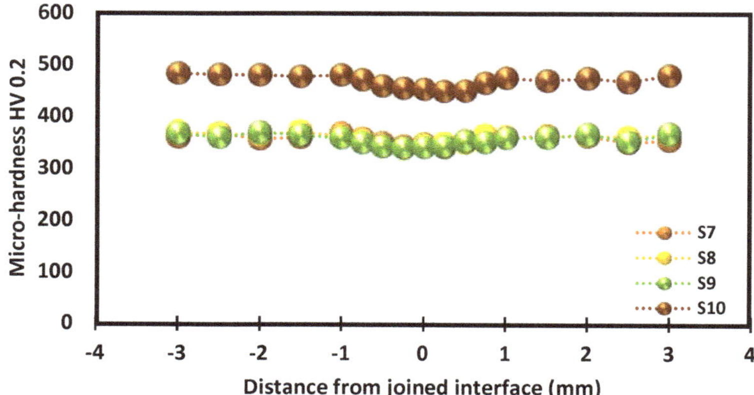

Figure 15. Hardness distributions of D2–D2 tool steel thixowelded joints at various conditions: (**S7**) 1300 °C/0 min, (**S8**) 1300 °C/5 min, (**S9**) 1300 °C/10 min, (**S10**) 1300 °C/10 min + post welding heat treatment.

4. Conclusions

The novel thixojoining process of AISI D2 similar tool steel was carried out by using the direct partial re-melting technique, and this was investigated through this research. The effect of joining temperature, holding time and post-weld heat treatment on microstructural characteristics and mechanical properties of AISI D2 tool steel joints have been studied. The summarised conclusions can be listed as below:

1. Adequate and defectless thixoweld-joint were successfully produced with thixowelding temperature of 1300 °C, while lower welding temperature caused a series of cracks across the entire joint resulting in spontaneous fracture after joining. While post weld heat treatment, resulted in improved contrast between solid and liquid phases, whereby the big solids appeared in little liquid matrix where primary distribution of grain boundary is decreased substantially and some amounts of eutectic carbides dissolve in the matrix.

2. According to the X-ray line scan analyses across the joint interface confirms a uniform elemental distribution in the structure while the X-ray diffraction pattern of the thixowelded area was free from any brittle or harmful phases.

3. Maximum tensile streghth of 271 MPa was achieved at 1300 °C and 10 min holding time, which was 35% that of D2 base metal. Meanwhile, tensile strength of the joined parts after heat condition showed a significant improvement over the non-heat treated condition which was 560 MPa, i.e., about 70% that of as received AISI D2 tool steel base metal.

4. The fracture of joints was located along the interface between base metal and insert with typical brittle characteristics (intergranular). Meanwhile, the joints after heat condition are fractured at the diffusion zone, and the fracture exhibits typical brittle characteristic (tranasgranular).

5. The hardness measurement of the base metal, the weld zone and the insert is fluctuated in the range of 350 Hv to 370 Hv. This indicated that the hardness value of the welded joints was almost similar to those of base metal and insert. While the hardness measurements of the (thixojoining + heat treated) condition (490 Hv) was somewhat higher than for the non-heat treated samples. This is more related to the phase transformation (austenite and ferrite phases to the martensitic phases). Moreover, some amounts of eutectic chromium carbides dissolve in the matrix.

Acknowledgments: The authors would like to thank Universiti Kebangsaan Malaysia (UKM) and the ministry of Higher Education (MoHE), Malaysia, for the financial support under Research Grants AP-2012-014.

Author Contributions: M.N.A. and M.Z.O. conceived and designed the experiments; M.N.A. performed the experiments; M.N.A., M.Z.O., S.A.-Z., K.S.A., M.A.A. analyzed the data; M.N.A. wrote the paper. All authors discussed the result and contributed to the final manuscript.

Conflicts of Interest: The authors declare no conflict of interest.

References

1. Sekharbabu, R.; Rafi, H.K.; Rao, K.P. Characterization of D2 tool steel friction surfaced coatings over low carbon steel. *Mater. Des.* **2013**, *50*, 543–550. [CrossRef]

2. Mohammed, M.N.; Omar, M.Z.; Sajuri, Z.; Salleh, M.S.; Alhawari, K.S. Trend and Development of Semisolid Metal Joining Processing. *Adv. Mater. Sci. Eng.* **2015**, *2015*. [CrossRef]

3. Chegeni, A.A.; Kapranos, P. An Experimental Evaluation of Electron Beam Welded Thixoformed 7075 Aluminum Alloy Plate Material. *Metals* **2017**, *7*, 569. [CrossRef]

4. Mohammed, M.N.; Omar, M.Z.; Salleh, M.S.; Alhawari, K.S.; Kapranos, P. Semisolid Metal Processing Techniques for Nondendritic Feedstock Production. *Sci. World J.* **2013**, *2013*. [CrossRef] [PubMed]

5. Mendez, P.F.; Brown, S.B. Method and Apparatus for Metal Solid Freeform Fabrication Utilizing Partially Solidified Metal Slurry. Patent number: 5,893,404, 13 April 1999.

6. Kiuchi, M.; Yanagimoto, J.; Sugiyama, S. Application of mushy/semi-solid joining—Part 3. *J. Mater. Process. Technol.* **2003**, *140*, 163–166. [CrossRef]

7. Hosseini, V.A.; Aashuri, H.; Kokabi, A.H. Characterization of newly developed semisolid stir welding method for AZ91 magnesium alloy by using Mg–25%Zn interlayer. *Mater. Sci. Eng. A* **2013**, *565*, 165–171. [CrossRef]

8. Mohammed, M.N.; Omar, M.Z.; Syarif, J.; Sajuri, Z.; Salleh, M.S.; Alhawari, K.S. Semi-Solid Joining of D2 Cold-Work Tool Steel. *Solid State Phenom.* **2015**, *217–218*, 355–360. [CrossRef]

9. Mohammed, M.N.; Omar, M.Z.; Alhawari, K.S.; Abdelgnei, M.A.; Saud, S.N. The Interface Morphology of Thixo-Joined Dissimilar Steels. *Solid State Phenom.* **2016**, *256*, 243–250. [CrossRef]

10. Mohammed, M.N.; Omar, M.Z.; Salleh, M.S.; Alhawari, K.S. Study on Thixojoining Process Using Partial Remelting Method. *Adv. Mater. Sci. Eng.* **2013**, *2013*, 1–8. [CrossRef]

11. Omar, M.Z.; Alfan, A.; Syarif, J.; Atkinson, H.V. Microstructural investigations of XW-42 and M2 tool steels in semi-solid zones via direct partial remelting route. *J. Mater. Sci.* **2011**, *46*, 7696–7705. [CrossRef]

12. Vigraman, T.; Narayanasamy, R.; Ravindran, D. Microstructure and mechanical property evaluation of diffusion-bonded joints made between SAE 2205 steel and AISI 1035 steel. *Mater. Des.* **2012**, *35*, 156–169. [CrossRef]

13. Püttgen, W.; Hallstedt, B.; Bleck, W.; Uggowitzer, P.J. On the microstructure formation in chromium steels rapidly cooled from the semi-solid state. *Acta Mater.* **2007**, *55*, 1033–1042. [CrossRef]

14. Aisman, D.; Jirkova, H.; Kucerova, L.; Masek, B. Metastable structure of austenite base obtained by rapid solidification in a semi-solid state. *J. Alloys Compd.* **2011**, *509*, S312–S315. [CrossRef]

15. Rogal, Ł.; Dutkiewicz, J. Heat Treatment of Thixo-Formed Hypereutectic X210CrW12 Tool Steel. *Metall. Mater. Trans. A* **2012**, *43*, 5009–5018. [CrossRef]

16. Wikstrom, N.P.; Egbewande, A.T.; Ojo, O.A. High temperature diffusion induced liquid phase joining of a heat resistant alloy. *J. Alloys Compd.* **2008**, *460*, 379–385. [CrossRef]

17. Cao, J.; Liu, J.; Song, X.; Lin, X.; Feng, J. Diffusion bonding of TiAl intermetallic and Ti_3AlC_2 ceramic: Interfacial microstructure and joining properties. *Mater. Des.* **2014**, *56*, 115–121. [CrossRef]

18. Mohammadi, H.; Ketabchi, M. Investigation of microstructural and mechanical properties of 7075 al alloy prepared by SIMA method. *Iran. J. Mater. Sci. Eng.* **2013**, *10*, 32–43.

19. Guo, J.F.; Chen, H.C.; Sun, C.N.; Bi, G.; Sun, Z.; Wei, J. Friction stir welding of dissimilar materials between AA6061 and AA7075 Al alloys effects of process parameters. *Mater. Des.* **2014**, *56*, 185–192. [CrossRef]

20. Mohammed, M.N.; Omar, M.Z.; Syarif, J.; Sajuri, Z.; Salleh, M.S.; Alhawari, K.S. Microstructural evolution during DPRM process of semisolid ledeburitic D2 tool steel. *Sci. World J.* **2013**, *2013*. [CrossRef] [PubMed]

21. Omar, M.Z.; Atkinson, H.V.; Kopranos, P. Semi-solid metal processing—A processing method under low flow loads. *J. Kejuruter.* **2007**, *19*, 137–146.

22. Mohammed, M.N.; Omar, M.Z.; Syarif, J.; Sajuri, Z.; Salleh, M.S.; Alhawari, K.S. Microstructural properties of semisolid welded joints for AISI D2 Tool Steel. *J. Kejuruter.* **2014**, *26*, 31–34. [CrossRef]

23. Aisman, D.; Jirkova, H.; Masek, B. The influence of deformation and cooling parameters after transition through semi-solid state on structure development of ledeburite steel. *J. Alloys Compd.* **2012**, *536*, S204–S207. [CrossRef]
24. Roberts, G.; Krauss, G.; Kennedy, R. *Tool Steels*, 5th ed.; ASM International: Geauga County, OH, USA, 1998.

Article

Microstructure and Properties of Semi-solid ZCuSn10P1 Alloy Processed with an Enclosed Cooling Slope Channel

Yongkun Li [1], Rongfeng Zhou [1,2,*], Lu Li [1,2], Han Xiao [1] and Yehua Jiang [1]

[1]   Faculty of Material Science and Engineering, Kunming University of Science and Technology, Kunming 650093, China; liyongkun@kmust.edu.cn (Y.L.); lilukust@126.com (L.L.); kmxh@kmust.edu.cn (H.X.); jiangyehua@kmust.edu.cn (Y.J.)

[2]   Research Center for Analysis and Measurement, Kunming University of Science and Technology, Kunming 650093, China

*   Correspondence: zhourfchina@hotmail.com; Tel.: +86-137-0886-8341

Received: 14 March 2018; Accepted: 16 April 2018; Published: 17 April 2018

Abstract: Semi-solid ZCuSn10P1 alloy slurry was fabricated by a novel enclosed cooling slope channel (ECSC). The influence of pouring length of ECSC on the microstructures of ZCuSn10P1 alloy semi-solid slurry was studied with an optical microscope an optical microscope (OM), scanning electron microscope (SEM), X-ray diffraction (XRD) and energy dispersive spectrometer (EDS). Liquid squeeze casting and semi-solid squeeze casting were performed under the same forming conditions, and the microstructure and properties were compared. The results show that primary α-Cu phase gradually evolved from dendrites to worm-like or equiaxed grains under the chilling action of the inner wall of the ECSC. The mass fraction of tin in the primary α-Cu phase increased from 5.85 to 6.46 after the ECSC process, and intergranular segregation was effectively suppressed. The finest microstructure can be obtained at 300 mm pouring length of ECSC; the equivalent diameter is 46.6 μm and its shape factor is 0.73. The average ultimate tensile strength and average elongation of semi-solid squeeze casting ZCuSn10P1 alloy reached 417 MPa and 12.6%, which were improved by 22% and 93%, respectively, as compared to that of liquid squeeze casting.

Keywords: enclosed cooling slope channel; ZCuSn10P1; semi-solid; microstructure refinement; properties

---

1. Introduction

Tin bronze has excellent flexibility, wear resistance, corrosion resistance, and high strength, and is widely used in the ship-building industry and is important in gears, valves, worm gears, etc. [1–3]. However, the primary α-Cu phase is a coarse dendritic structure and the segregation of tin in traditional tin-bronze ZCuSn10P1 is significant [4], which leads to microstructural heterogeneity and poor properties, limiting its uses in industry. Therefore, it is very necessary to find a technology that can improve the uniformity of microstructures and properties.

Semi-solid process is widely studied due to its many technical and economic advantages; it includes rheo-casting and thixo-casting [5]. In recent years, rheo-casting has attracted the interest of many researchers, and there are several ways to prepare non-dendrite microstructures [6–8]. The cooling slope plate method is a simple way to prepare semi-solid slurry. Abdelsalam et al. [9] fabricated A356/Al$_2$O$_3$ metal matrix nano-composites by a combination of stir casting and cooling slope casting typically. The results show that the average grain size of samples fabricated is fine, the finer α-Al grains were obtained at the bottom of the ingots. However, the addition of Al$_2$O$_3$ nanoparticles did not significantly influence the shape factor of the primary α-Al grains. Das et al. [10] studied the microstructural evolution of semi-solid slurry generation of Al-Si-Cu-Fe alloy and the effect of microstructural morphology of the slurry on its rheological behavior. Yoshida et al. [11] studied the formation of hollow AZ31B magnesium alloy pipe made by vertical continuous casting using semi-solid slurry prepared on an inclined cooling plate. The results show that it was possible to make the pipe using both a taper mold and a core rod. However, holding at 650 °C of molten alloy has no tapering of the mold, and the core rod 1 mm crack appears on both outside and inside surface of the hollow material. Cit et al. [12] prepared the semi-solid slurry of tin bronze alloy by the slope plate method, and the effect of tin mass fraction on the microstructure of semi-solid slurry was studied. The results show that the size of primary α-Cu phase decreases with the increase of tin mass fraction. The size of the primary α-Cu phase is about 60 μm when the tin mass fraction reaches 3%. Kose et al. [13] prepared the semi-solid slurry of ZCuSn10P0.1 alloy by Shearing Cooling Roll (SCR). The effect of mold temperature on the tensile properties was studied. The results show that ZCuSn10P0.1 alloy was suitable for semi-solid process and an excellent semi-solid slurry can be prepared. The elongation of direct-squeeze casting parts was increased by 60% due to the improvement of tin segregation in the microstructure, the mold temperature at this time was 700 °C. According to the above study, it is known that there are few studies on semi-solid rheo-casting of ZCuSn10P1 copper alloys.

In this investigation, we attempted to make semi-solid slurry of ZCuSn10P1 with fine solid grains for rheo-casting, the influence of the pouring length of ECSC on the microstructures of ZCuSn10P1 alloy semi-solid slurry was studied. Subsequently, a shaft sleeve was obtained from the semi-solid slurry with a favorable solid morphology by squeeze casting. Effect of melt processing by ECSC on microstructure and properties of squeeze casting ZCuSn10P1 alloy were investigated.

2. Materials and Methods

2.1. Materials

The chemical composition of the ZCuSn10P1 alloy used in this study is listed in Table 1. The alloy had liquidus and solidus temperatures of 1020.7 °C and 830.4 °C, respectively. Measurements by differential scanning calorimetry (STA 449F3, NETZSCH, Bavaria, Germany) were carried out on about 10 mg samples obtained from the as-cast alloys under an argon at a heating rate of 10 K·min^{-1}, as shown in Figure 1.

Table 1. Chemical composition of ZCuSn10P1 alloy (mass %).

Cu	Sn	P	Other
88.90	10.22	0.71	0.17

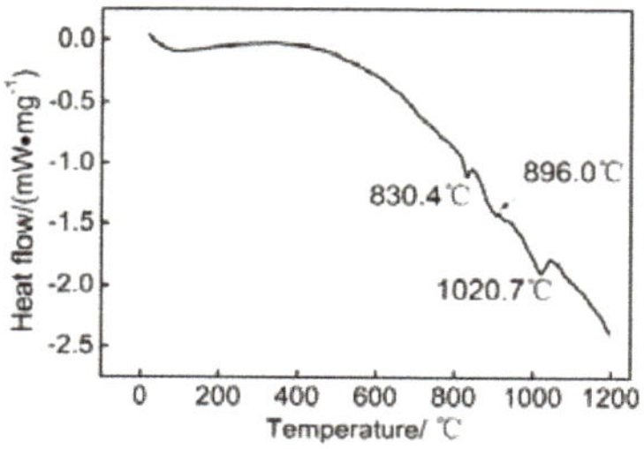

Figure 1. DSC heating curve of as-cast ZCuSn10P1 alloy.

2.2. Fabrication of the ZCuSn10P1 Alloy Semi-Solid Slurry

A schematic illustration of the ECSC equipment and forming process curve is shown in Figure 2. The slurry-making equipment consists of two parts: an enclosed cooling slope channel and an angle adjuster (Figure 2a). The channel thickness of metal flow was 5 mm [14]. The enclosed cooling slope channel was made from stainless steel and fixed at 45° with respect to the horizontal plane and was cooled by water circulation in the gap. The pouring length was set to different values in this investigation.

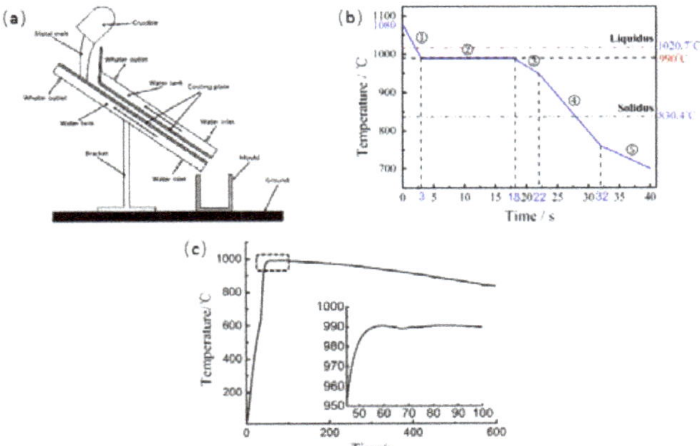

Figure 2. Schematic illustration of the: (**a**) ECSC equipment; (**b**) forming process curve; and (**c**) temperature recording curve during slurry transfer ① ECSC process, ② slurry transfer, ③ mold filling,④ holding pressure, ⑤ parts air cooling.

Fabrication of the ZCuSn10P1 alloy semi-solid slurry was carried out according to the following procedures (Figure 2b): about 6.5 kg of ZCuSn10P1 alloy in a graphite crucible was melted at 1200 ± 10 °C in a medium frequency induction furnace. After complete melting, the temperature dropped to 1080 °C and it was poured immediately into the ECSC. The ECSC casting of the ZCuSn10P1 involved pouring the melt over an enclosed cooling slope channel and its width is 100 mm. The prepared slurry was collected by a high-purity graphite crucible preheated to 950 °C to ensure that the temperature of the slurry was almost constant during the transfer process (Figure 2b).

By changing the pouring length of ECSC, the semi-solid slurry of ZCuSn10P1 alloy under different pouring length conditions was obtained. A 10 mm × 10 mm × 10 mm cube was removed from the slurry and water quenching immediately. The microstructure of these specimens was observed using an optical microscope. The number of primary α-Cu phases in the metallographic images under five different fields of view was counted to calculate the average value under each field of view. Then, the total number of per unit area based on the area of the field of view and the average number of primary α-Cu phases was calculated. Measurements of the equivalent diameter (*D*) and shape factor (*F*) of the primary α-Cu phase were carried out using image-analysis techniques [15]. The equivalent diameter (*D*) and shape factor (*F*) were determined from the following equations [15]:

$$D = 2\sqrt{\frac{A}{\pi}} \tag{1}$$

$$F = \frac{4\pi A}{p^2} \tag{2}$$

where *A* is the area of primary α-Cu phase and *p* is the perimeter. We obtained them using the Image-pro plus (Media Cybernetics, Rockvile, MD, USA) software suite. At *F* = 1, the shape is a perfect circle.

2.3. Squeeze Casting

Mold map and the product of ZCuSn10P1 alloy are shown in Figure 3. Squeeze casting of the semi-solid was at pressure 100 MPa and speed 21 mm/s to obtain a four-cavity part, as shown in Figure 3a. The semi-solid slurry was prepared at a pouring length of 300 mm. Liquid squeeze casting at casting temperature 1050 °C was also carried out for comparison with semi-solid squeeze casting. The forming process parameters of liquid squeeze casting and semi-solid squeeze casting were the same. The mold temperature was 450 °C. As shown in Figure 3c, samples were taken from different positions A, B and C for microstructure observation and phase analysis. To accurately reflect the properties of different position in the part, we took four tensile samples symmetrically on both sides of the parting surface (Figure 3b). The average value of four tensile samples as the final property index was calculated. Mechanical properties of ZCuSn10P1 alloy parts were measured using a CMT300 tensile machine.

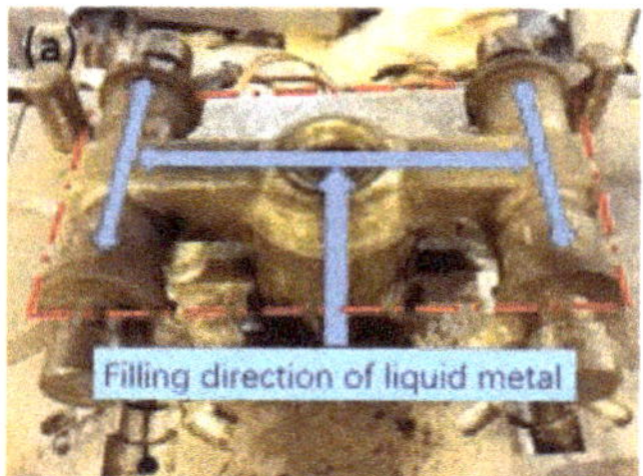

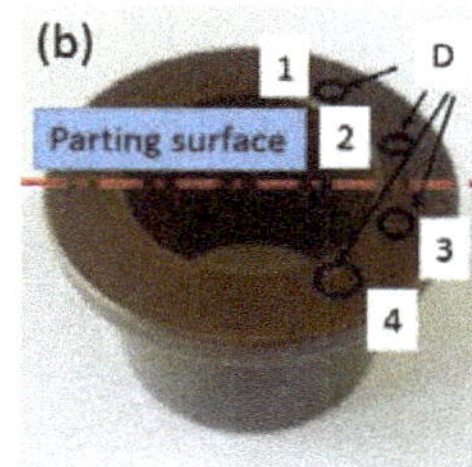

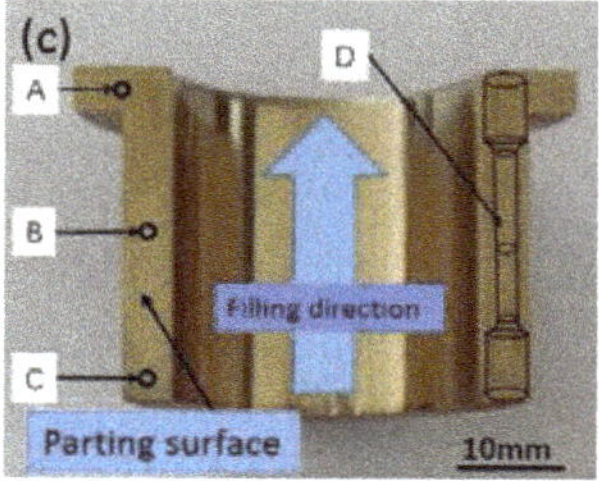

Figure 3. Mold map and the product of ZCuSn10P1 alloy: (**a**) mold map; and (**b**,**c**) squeeze part (A, B and C are sampling positions for microstructure observation, while D is sampling position and shape for tensile test, 1 to 4 are sampling positions for tensile test from above the parting surface to below the parting surface).

3. Results and Discussion

3.1. Influence of ECSC Process on the Microstructures of ZCuSn10P1 Alloy

Figure 4 shows the as-cast microstructures and semi-solid slurry microstructures of ZCuSn10P1 alloy. The as-cast sample was obtained by direct water quenching of the molten at 1050 °C. The processing parameters of ECSC were a pouring temperature of 1080 °C, and the pouring length and angle of 300 mm and 45°, respectively. As can be seen, the two types of samples are mainly composed of primary α-Cu and (α+δ +Cu3P) phases. The primary α-Cu phase is the coarse mesh dendritic structure and the (α+δ +Cu3P) phase is distributed in the dendrite clearance in as-cast samples (Figure 4a). The primary α-Cu become worm-like grains or equiaxed grains by ECSC process (Figure 4b). The primary α-Cu was surrounded by a liquid phase, the solid grain filling together with the liquid phase in the process of forming improved the synergistic effect of solid-liquid flows. Note that the liquid content shown in the micrographs is not suggested: the number present at temperature because the quench rate is not rapid enough for the content of liquid to be entirely "frozen in" [16]. During the quench, some liquid deposits onto the solid, appearing to be "solid" in the quenched microstructure [16,17].

Phase identification was accomplished by comparison of the observed XRD peaks with published crystallographic data (JCPDS database, International Centre for Diffraction Data, Powder Diffraction Standards Joint Committee, Newtown Square, PA, USA). Figure 5 shows the XRD traces from as-cast and semi-solid slurry of the ZCuSn10P1 alloy. Primary α-Cu and δ-Cu31Sn8 were detected in as-cast sample and semi-solid sample; the Cu3P phase was not detected, but the Cu3P phase was nevertheless present [18]. Some peaks become weaker and wider (as shown in Figure 5, marked "2"), indicating that the grains were more refined (see also Figure 4).

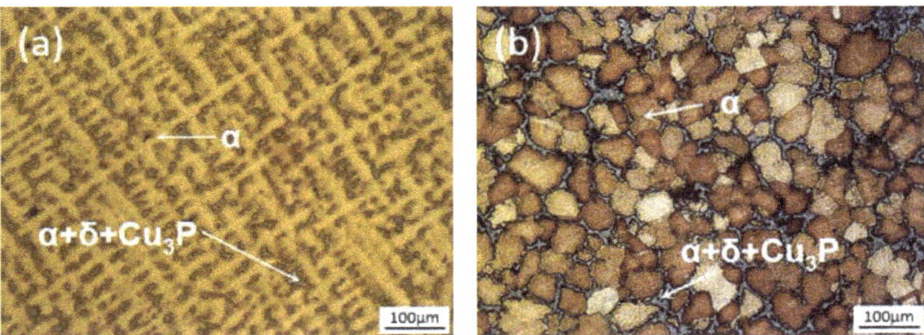

Figure 4. Microstructures of ZCuSn10P1 alloy: (**a**) as-cast; and (**b**) semi-solid.

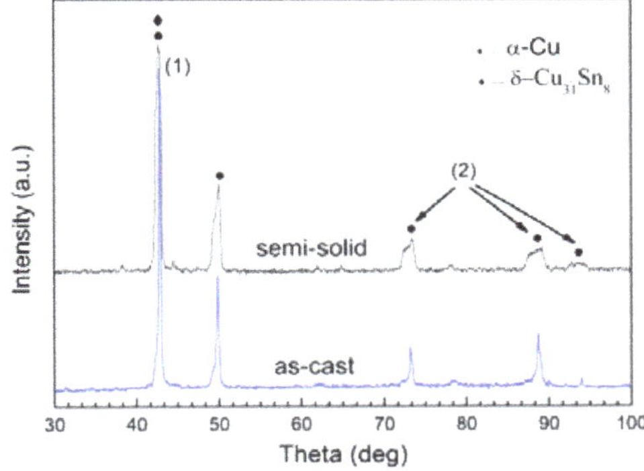

Figure 5. XRD spectrum of ZCuSn10P1 alloy.

Figure 6 shows the as-cast and semi-solid slurry SEM micrograph of ZCuSn10P1 alloy. They are composed of three regions due to the different tin mass fraction (dark grey areas, light grey areas, and bright white areas). The points showing EDS are marked in Figure 6 and the EDS results are listed in Table 2. The tin mass fraction gradually increases as the color becomes lighter, as indicated by the arrows in Figure 6. The EDS results contrast the as-cast and semi-solid slurry: the tin mass fraction in the α-Cu phase of semi-solid slurry is higher than the as-cast, and the tin mass fraction in the liquid phase of the semi-solid is lower than the as-cast, indicating that the distribution of tin is relatively uniform in the semi-solid slurry and less segregated. Some peak shift (as shown in Figure 5, marked "1") indicated that the solid solubility of tin has changed (consistent with Table 2). The Cu3P phase is lamellar structure and distributed on the edge of the primary α-Cu phase [18]. It may be affected by the Cu3P phase when measuring two- or three-point energy spectrum, and shows the accumulation of P near primary α-Cu phase.

Table 2. EDS results for the microstructure of the as-cast and semi-solid ZCuSn10P1 alloy.

| Element | As-Cast | | | | | | Semi-Solid | | | | | |
| | 1 | | 2 | | 3 | | 1 | | 2 | | 3 | |
	Wt. %	At. %	Wt. %	At. %	Wt. %	At. %	Wt. %	At. %	Wt. %	At. %	Wt. %	At. %
Cu	93.79	96.04	79.25	78.46	71.53	81.68	93.22	95.78	84.6	90.30	70.69	70.47
Sn	5.85	3.21	13.73	7.28	27.94	17.08	6.46	3.56	14.85	8.48	20.59	11.83
P	0.36	0.75	7.02	14.26	0.53	1.24	0.32	0.66	0.56	1.22	8.71	17.7

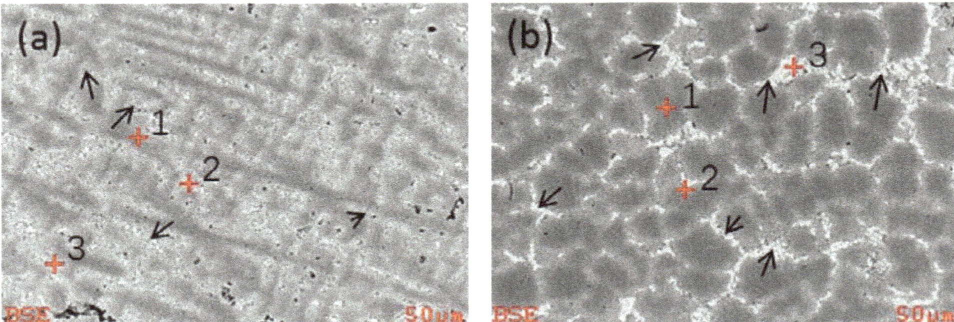

Figure 6. SEM micrograph of ZCuSn10P1 alloy: (**a**) as-cast; and (**b**) semi-solid.

3.2. Influence of Pouring Length on the Microstructure of the ZCuSn10P1 Semi-solid Slurry.

The pouring length directly affects the cooling length of the alloy on the cooling plate, so it is an important factor affecting the microstructure. When the pouring temperature is constant at 1080 °C and the pouring length is 200 mm, the alloy melt flows through the cooling channel rapidly, thus the cooling time is shorter. The number of nucleation sites decreases and the melt temperature is higher. The microstructure was mainly worm-like and dendrite-like, as shown in Figure 7a. With pouring length increased to 300 mm, the flow time of the melt in the cooling channel increases. The number of nucleation sites increases, and the temperature is more uniform. The microstructure is basically composed of worm-like grains or equiaxed grains due to suppression of the formation of dendrite structures, as shown in Figure 7b. However, with the increase of the pouring length to 400 mm, the flow time in the cooling channel is further increased. The primary α-Cu phase is finer and rounder due to the increase in pouring cooling length, as shown in Figure 7c. The temperature of the melt at the exit of the cooling channel is very low due to the increase of the pouring cooling length. It is easier to form crusts inside the cooling channel, which is not conducive to the preparation of a semi-solid slurry.

Figure 7. Microstructures of ZCuSn10P1 alloy at different pouring lengths at a pouring temperature of 1080 °C: (**a**) 200 mm; (**b**) 300 mm; and (**c**) 400 mm.

Figure 8 shows the relationship between equivalent diameter and shape factor of the primary α-Cu phase with pouring length (Figure 8a), and distribution of primary α-Cu phase equivalent diameter (Figure 8b) at different pouring lengths of semi-solid ZCuSn10P1 alloy. Figure 9 shows the relationship between the amount of primary α-Cu phase per unit area and pouring length in semi-solid ZCuSn10P1 alloy. In Figures 8 and 9, it can be found that the equivalent diameter of the primary α-Cu grains was 50 μm and the shape factor was 0.72 when the casting temperature was constant at 1080 °C and the pouring length was 200 mm. The equivalent diameters are mainly in the range of 30 μm to 60 μm, but there are also some larger grain diameters exceeding 100 μm. At this point, the number of grains per square centimeter of primary α-Cu phase is about 33,939 and the solid fraction is about

63.58% after quenching. The equivalent diameter of primary α-Cu grains is 46.6 µm and the shape factor is 0.73 when the pouring length is 300 mm. The number of grains with an equivalent diameter between 20 µm and 60 µm, accounting for 83% of the total number of grains. At this point, the number per square centimeter of primary α-Cu phase grains is maximum about 34,909 and the solid fraction was 66.34% after water-quenching. The equivalent diameter of primary α-Cu phase grains increased to 53.2 µm and the shape factor decreased to 0.68 when the pouring length was 400 mm. The size of primary α-Cu phases is mainly distributed within 40 µm and 70 µm, accounting for 78.8% of the total number grains. The number per square centimeter of primary α-Cu phase grains was 31,757 and the solid fraction was 75.3% after water-quenching. The actual solid fraction calculated is much higher than the theoretical solid fraction [16,17].

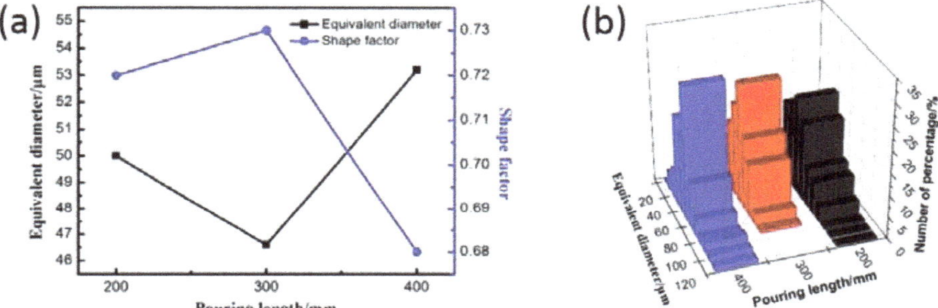

Figure 8. Relationship between equivalent diameter and shape factor of primary α-Cu phase with pouring length (**a**); and distribution of primary α-Cu phase equivalent diameter (**b**) at different pouring lengths of semi-solid ZCuSn10P1 alloy.

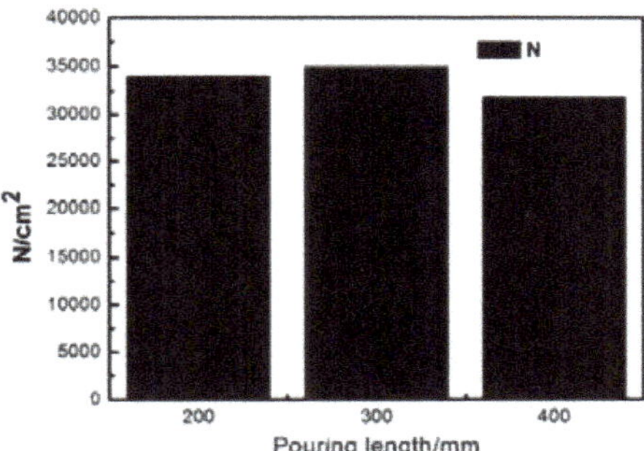

Figure 9. Relationship between number of primary α-Cu phase per unit area and pouring length in semi-solid ZCuSn10P1 alloy.

Figure 10a shows a schematic diagram of the Crystal Separation Theory [19] and Figure 10b shows the shear stress and velocity distribution of the melt during flow through the ECSC process. Many primary α-Cu phases are nucleated on the surface of the cooling plate due to the melt chilling effect when the liquid metal is poured onto a cooling slope plate [19]. The crystal nucleus and the cooling plate are detached into the interior of the solution under the action of the melt flow. The surface of the cooling slope plate forms new nuclei again so that the downstream liquid melt becomes the semi-solid slurry. Therefore, the formation of slurry is related to the temperature field and the velocity field. The cooling channel thickness of ECSC is only 5 mm, which means the temperature field and the velocity field are relatively uniform. The semi-solid slurry with equiaxed or worm-like grains is finally formed under the combined action of heterogeneous nucleation and homogeneous nucleation.

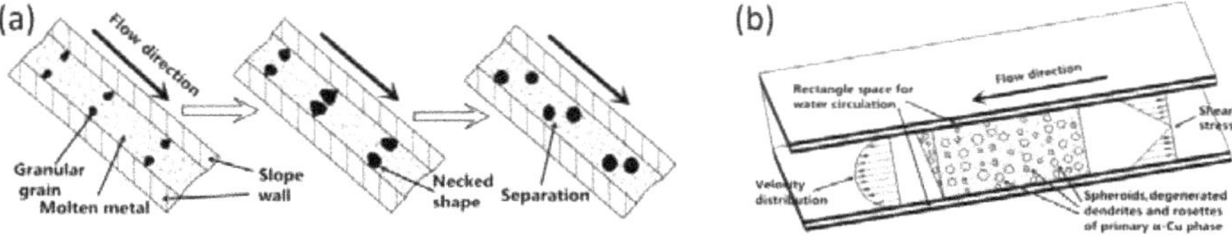

Figure 10. Schematic diagram of: (**a**) the mechanism of neck formation of α-Cu grains; and (**b**) principle of generation of new semi-solid slurry using the ECSC.

In summary, the pouring length determines the flow time through the cooling channel, too short or too long time cannot prepare excellent semi-solid slurry. There is not enough time for chilling when the pouring length is too short (200 mm). The melt chilling time is too long and it is not conducive to prepare of semi-solid slurry when the pouring length is too long (400 mm). To ensure that the melt has a certain chilling time and the fine semi-solid slurry can be prepared continuously, the recommended pouring length in this paper is 300 mm.

3.3. Rheo-Casting

Figure 11 shows the comparison of microstructures of ZCuSn10P1 alloy parts obtained by liquid squeeze casting and semi-solid squeeze casting. It can be found that liquid squeeze casting is made up of coarse dendrites or mesh and large area eutectoid phase segregation. This sort of microstructure would be detrimental to mechanical properties. During liquid squeeze casting, liquid metal was adopted. The squeeze casting mold and liquid metal had a huge temperature differential, which led to a large temperature gradient; therefore, coarse dendrites are easily formed in the alloy. At once, eutectoid phase could solidify rapidly due to high cooling rate. Moreover, the micro-segregation also occurred in the parts. It can be seen that dendritic at position A is relatively less than at position B, and minimum at position C. The different filling sequence of these positions leads to different solidification speed and cooling rate (Figure 3a). Filling mouth away from position A, the temperature of alloy was higher and the mold temperature was lower at position A. Thus, the solidification speed is fastest at position A, and slowest at position C.

However, it is discovered that the microstructures of ZCuSn10P1 alloy parts prepared by semi-solid squeeze casting is made up of worm-like or equiaxed grains and eutectoid structure. The average size of grains of semi-solid squeeze casting is finer than that of liquid squeeze casting. Microstructures at position A and B have almost no difference, therefore, the average size of grains and grain shape are almost the same. In addition, eutectoid phase segregation almost disappears. However, grain size at position C is bigger than at positions A and B, and solid fraction is higher than at positions A and B. The reason is that the filling sequence of these positions are different (Figure 3a), resulting in different solidification speed and cooling rate. The uniformity of semi-solid squeeze casting has been greatly improved compared to liquid squeeze casting, which reveals that melt processing by ECSC process has a significant effect. Semi-solid squeeze casting process and liquid squeeze casting have different solidification modes. Semi-solid squeeze casting process consists of two solidification stages, namely in the enclosed cooling slope channel and in the mold. However, liquid squeeze casting only has solidification in the mold. Semi-solid alloy had a relatively higher solid fraction than liquid alloy, so the cooling rate in the mold was slower, and the solidification was much closer to equilibrium solidification in the mold, thus suppressing microstructure segregation.

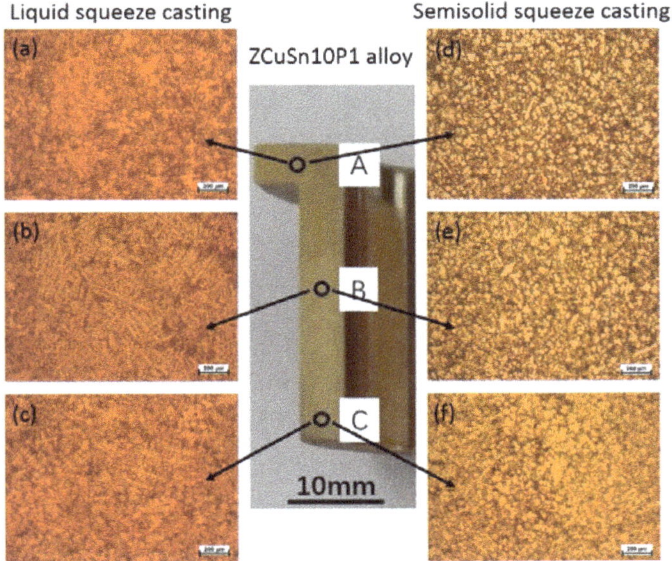

Figure 11. Comparison of microstructures of ZCuSn10P1 alloy parts obtained by liquid squeeze casting and semi-solid squeeze casting.

Table 3 shows ultimate tensile strengths and elongations of ZCuSn10P1 alloy parts produced by liquid squeeze casting and semi-solid squeeze casting for different positions. The average ultimate tensile strengths of ZCuSn10P1 alloy parts prepared by semi-solid squeeze casting reach 417 MPa and the average elongation reach 12.56%. The average ultimate tensile strengths of ZCuSn10P1 alloy parts prepared by liquid squeeze casting reach 342 MPa and average the elongation reach 6.5%. We can see that the average ultimate tensile strength and average elongation of ZCuSn10P1 alloy by semi-solid squeeze casting are higher than liquid squeeze casting. The average ultimate tensile strength and average elongation are improved by 22% and 93%, respectively, as compared to that of liquid squeeze casting. This is attributed to the grain refinement and homogeneity of microstructure. Therefore, the squeeze casting combined with the ECSC process is an effective technique to prepare tin bronze alloy sleeve with good mechanical properties.

Table 3. Ultimate tensile strengths and elongations of ZCuSn10P1 alloy parts produced by semi-solid squeeze casting and liquid squeeze casting for different positions.

Process	Mechanical Properties	Positions				Average Value
		1	2	3	4	
Semi-solid squeeze casting	Ultimate tensile strength/MPa	411	420	422	415	417
	Elongation/%	12.03	12.96	13.07	12.18	12.56
Liquid squeeze casting	Ultimate tensile strength/MPa	325	352	358	333	342
	Elongation/%	5.5	7.43	7.87	5.2	6.5

4. Conclusions

The conclusions obtained from the present study are described below:

1. The processing by ECSC can refine grain size and suppress tin segregation as well as regional segregation which usually appears in liquid squeeze casting. The mass fraction of tin in the primary α-Cu phase increased from 5.85 to 6.46 after the ECSC process.
2. The pouring length of ECSC process is an important factor affecting the microstructure and successful preparation of semi-solid slurry. The optimal pouring length is 300 mm; at this time, the equivalent diameter of primary α-Cu is 46.6 μm and its shape factor is 0.73.
3. The average ultimate tensile strength and average elongation of semi-solid squeeze casting ZCuSn10P1 alloy reached 417 MPa and 12.6%, which were improved by 22% and 93%, respectively, as compared to that of liquid squeeze casting. This is attributed to the grain refinement and homogeneity of microstructure.

Acknowledgments: The authors acknowledge funding for the research from National Science Foundation of China (51765026 and 51665024). This work is supported by National-local Joint Engineering Laboratory of Metal Advanced Solidification Forming and Equipment Technology of China.

Author Contributions: Rongfeng Zhou and Yongkun Li designed most of the experiments, analyzed the results and wrote this manuscript. Yongkun Li and Lu Li performed most experiments. Han Xiao and Yehua Jiang helped analyze the experiment data and gave some constructive suggestions about how to write this manuscript.

Conflicts of Interest: The authors declare no conflict of interest.

References

1. Charles, J.A. The development of the usage of tin and tin-bronze: some problems. In *The Search for Ancient Tin*; Smithsonian Institution: Washington, DC, USA, 1978; pp. 25–32.
2. Afshari, E.; Ghambari, M.; Farhangi, H. Effect of microstructure on the breakage of tin bronze machining chips during pulverization via jet milling. *Int. J. Min. Met. Mater.* **2016**, *23*, 1323–1332. [CrossRef]
3. Walsh, F.C.; Low, C.T.J. A review of developments in the electrodeposition of tin-copper alloys. *Surf. Coat. Technol.* **2016**, *304*, 246–262. [CrossRef]
4. Kodama, H.; Nagase, K.; Umeda, T.; Sugiyama, M. Microsegregation during dendritic growth in Cu-8% Sn alloys. *J. Jpn. Foundry Eng. Soc.* **1976**, *49*, 287–293.
5. Jorstad, J.L. Semi-Solid Metal Processing from an Industrial Perspective; The Best is yet to Come! *Solid State Phenom.* **2016**, *256*, 9–14. [CrossRef]
6. Hassas-Irani, S.B.; Zarei-Hanzaki, A.; Bazaz, B.; Roostaei, A.A. Microstructure evolution and semi-solid deformation behavior of an A356 aluminum alloy processed by strain induced melt activated method. *Mater. Des.* **2013**, *46*, 579–587. [CrossRef]
7. Yurko, J.A.; Martinez, R.A.; Flemings, M.C. Commercial development of the semi-solid rheocasting (SSRTM) process. *Metall. Sci. Technol.* **2013**, *21*, 10–15.
8. Hofmann, D.C.; Kozachkov, H.; Khalifa, H.E.; Schramm, J.P.; Demtriou, M.D.; Vecchio, K.S.; Johnson, W.L. Semi-solid induction forging of metallic glass matrix composites. *JOM* **2009**, *61*, 11. [CrossRef]
9. Abdelsalam, A.A.; Mahmoud, T.S.; El-Betar, A.A.; EI-Assal, A.M. A Study of Microstructures Characteristics of A356-Al2O3 Composites Produced by Cooling Slope and Conventional Stir Cast. *Int. J. Curr. Eng. Technol.* **2015**, *15*, 3560–3571.
10. Das, P.; Samanta, S.K.; Bera, S.; Dutta, P. Microstructure Evolution and Rheological Behavior of Cooling Slope Processed Al-Si-Cu-Fe Alloy Slurry. *Metall. Mater. Trans. A* **2016**, *47*, 2243–2256. [CrossRef]
11. Yoshida, R.; Motoyasu, G.; Motegi, T. Production of Continuous Casting Pipe Using Semi-solid Slurry of Magnesium Alloy. *Trans. Mater. Res. Soc. Jpn.* **2015**, *40*, 169–174. [CrossRef]
12. Cit, M.T.; Tanabe, F. New semi-solid casting of copper alloys using an inclined cooling plate. In Proceedings of the 8th International Conference on Semi-Solid Processing of Alloys and Composites, Limassol, Cyprus, 21–23 September 2004.
13. Kose, T.; Uetani, Y.; Nakajima, K.; Matsuda, K.; Lkeno, S. Effect of die temperature on tensile property of rheocast phosphor bronze. *Mater. Sci. Forum* **2012**, *706*, 931–936. [CrossRef]
14. Zhao, Z.Y.; Guan, R.G.; Wang, X.; Huang, H.Q.; Chao, R.Z.; Dong, L.; Liu, C.M. Boundary layer distributions and cooling rate of cooling sloping plate process. *Wuhan Univ. Technol. Mater. Sci. Ed.* **2013**, *28*, 701–705. [CrossRef]

15. Das, P.; Dutta, S.; Samanta, S.K. Evaluation of primary phase morphology of cooling slope cast Al-Si-Mg alloy samples using image texture analysis. *Proc. Inst. Mech. Eng. B J. Eng. Manuf.* **2013**, *227*, 1474–1483. [CrossRef]
16. Atkinson, H.V.; Liu, D. Microstructural coarsening of semi-solid aluminium alloys. *Mater. Sci. Eng. A* **2008**, *496*, 439–446. [CrossRef]
17. Tzimas, E.; Zavaliangos, A. Evolution of near-equiaxed microstructure in the semi-solid state. *Mater. Sci. Eng. A* **2000**, *289*, 228–240. [CrossRef]
18. Davis, J.R. *ASM Specialty Handbook: Copper and Copper Alloys*; ASM International: Geauga County, OH, USA, 2008; pp. 1–600.
19. Ohno, A.; Motegi, T.; Soda, H. Origin of the equiaxed crystals in castings. *Trans. Iron Steel Inst. Jpn.* **1971**, *11*, 18–23.

 *metals*

Article

Microstructure of Semi-Solid Billets Produced by Electromagnetic Stirring and Behavior of Primary Particles during the Indirect Forming Process

Chul Kyu Jin

School of mechanical engineering, Kyungnam University, 7 Kyungnamdaehak-ro, Masanhappo-gu, Gyeongsangnam-do, Changwon-si 51767, Korea; cool3243@kyungnam.ac.kr; Tel.: +82-55-249-2346; Fax: +82-505-999-2160

Received: 18 February 2018; Accepted: 12 April 2018; Published: 15 April 2018

Abstract: An A356 alloy semi-solid billet was fabricated using electromagnetic stirring. After inserting the semi-solid billet into an indirect die, a thin plate of 1.2 mm thickness was fabricated by applying compression. The microstructure of the semi-solid billets fabricated in various stirring conditions (solid fraction and stirring force) were analyzed. The deformation and behavior of the primary α-Al particles were analyzed for various parameters (solid fraction, die friction, compression rate, and compression pressure). In the stirred billets, a globular structure was dominant, while a dendrite structure was dominant in the unstirred billets. As the solid fraction decreased and the stirring current increased, the equivalent diameter and roundness of the primary α-Al particles decreased. The primary α-Al particle sizes were reduced as the compressing velocity increased, while a greater number of particles could move as the compressing pressure increased. As the path over which the motion occurred became smoother, the fluidity of the particles improved. Under compression, bonded primary α-Al particles became separated into individual particles again, as the bonds were broken. As wearing caused by friction and collisions between the particles during this motion occurred, the particle sizes were reduced, and the particle shapes become increasingly spheroid.

Keywords: semi-solid material; A356 alloy; electromagnetic stirring; compression; primary α-Al particle

1. Introduction

The semi-solid forming process, which incorporates the advantages of the casting and forging processes, uses semi-solid billet, in which both solid and liquid states coexist. Flemings and Metz (1970) described the rheological behavior by a series of experiments [1]. When a material is being fabricated into a semi-solid state, the primary α particle sizes can be reduced, and their shapes can be rendered almost spherical through the application of external factors such as stirring and vibration.

A number of methods have been introduced to control the crystal grain and particle growth processes when the liquid is solidified into a semi-solid state. The most common and effective techniques are material rotation methods, such as the mechanical stirring method, which uses an impeller and a screw. Spencer et al. (1972) investigated the rheological behavior of Sn-15 pct Pb alloy by using mechanical stirring. They announced that the rheological behavior of partially solidified sheared alloy has many characteristics of thixotropic slurries [2]. Fan et al. (2005) adopted the rheo-diecasting process with twin screw for manufacturing near-net shape components. They investigated the microstructure and mechanical properties of rheo-diecasting aluminum alloys under as-cast, and various heat treatment conditions [3]. Biswas et al. (2006) used an external magnetic field to achieve a strong stirring effect on the melt motion. They investigated the effect of melt convection during the solidification of $Ti_{45}Al_{55}$ alloys [4]. Nafisi et al. (2006) and Bae et al. (2007) fabricated a semi-solid metal slurry of Al–Si alloy by using electromagnetic stirring (EMS), which uses an electromagnetic field.

Nafisi et al. investigated the effects of process parameters such as cooling rate and superheat on the morphology and distribution of eutectic silicon- and iron-based intermetallics during EMS. Bae et al. fabricated knuckles using semi-solid slurry by EMS [5,6]. Tzimas and Zavaliangos (2000) used a stress-induced and melt-activated (SIMA) object, which reheated and recrystallized a cold-worked billet [7]. Haga and Suzuke (2001) used a cooling slope method, which generates nuclei by allowing a material to flow on a sloped plane plate [8]. Zhang et al. (2009) developed an ultrasonic treatment method which uses powerful ultrasonic waves to create intense convection through gas bubbles [9]. Canyook et al. (2012) developed a gas-induced semi-solid (GISS) method, which uses gas bubbles [10]. As regards chemical methods, Nafisi et al. (2006) proposed a chemical grain refining method which requires a grain refiner to be added to the material [11].

Of the approaches listed above, the electromagnetic stirring method can be the most effective technique, having more advantages than disadvantages. The stirring force is weak compared to that of the mechanical stirring method, and the problems that affect the mechanical stirring method do not occur in this technique. The other advantages of this method are that fabrication is relatively simple, and it can be used as an integrated technique by easily combining it with a forming device such as a press and a casting machine. To date, a number of studies have been conducted on stirring variables that can control the crystal grains of an aluminium semi-solid slurry using the electromagnetic stirring method. In the physical experiments of Bae et al. (2007) and Oh et al. (2008), the solid particle sizes and roundness were found to change in response to the stirring conditions [6,12]. The effect of the electromagnetic stirring was confirmed, since the results showed that smaller size and higher roundness of the solid particles in semi-solid slurry were obtained when electromagnetic stirring was applied, compared with the characteristics of unstirred semi-solid slurry. The majority of studies about semi-solid materials reported that the physical mechanism behind crystal grain control is the separation and remelting of the dendrite arms, which is caused by the destruction of the initial dendrite structure due to vigorous stirring. This fragmentation or remelting mechanism theory is based on a result obtained following the application of mechanical stirring, in which the shear deformation and shearing strain rate are quite large, and the solid and liquid phases coexist in the same state. In the electromagnetic stirring method, which has a relatively weak stirring force, it is almost impossible to operate stirring in a semi-solid state with coexisting solid and liquid phases. The electromagnetic stirring must be performed in a liquid state. No study has yet produced results that can explain the mechanism through which the crystal grains are controlled by the electromagnetic stirring from the initial solidification stage.

The microstructure of A356 alloy semi-solid billet was analyzed through the application of various stirring conditions (solid fraction of billet and stirring force) of the electromagnetic stirrer. After inserting the fabricated semi-solid billet into a compression forming die, a thin plate of 1.2 mm thickness was fabricated. The deformation and behavior of the primary α-Al particles were analyzed for various compression conditions (solid fraction of billet inserted into the die, die lubrication, velocity and pressure of the punch-compressing billet). The formability and mechanical properties of the thin plate according to the process conditions were analyzed.

2. Experimental Procedure

2.1. Fabrication of the Semi-Solid Billet

Semi-solid billets were fabricated through the application of electromagnetic field stirring, as aluminum in a liquid state was in the process of solidifying into a semi-solid state. Billets of 35, 45, and 55% solid fraction were fabricated under currents of 60 A. The billets with 45% solid fraction were fabricated under applied currents of 30, 60, and 90 A, respectively, so that the effects of stirring forces with varied magnitude could be examined.

2.1.1. Electromagnetic Stirrer

The electromagnetic stirrer was a horizontal electromagnetic stirrer fabricated with three-phase (T, S, R) three poles. The schematic diagram showing the electromagnetic stirrer setup is shown in Figure 1. A molten metal container is required for the semi-solid billet fabrication process using the electromagnetic stirrer. The container must not melt or become deformed at temperatures above 700 °C and must not be affected by the electromagnetic force. Stainless steel 304 with austenite structure, which is a nonmagnetic material, was used. A container of 100 mm height, Φ56 mm external diameter, and 2 mm thickness was fabricated.

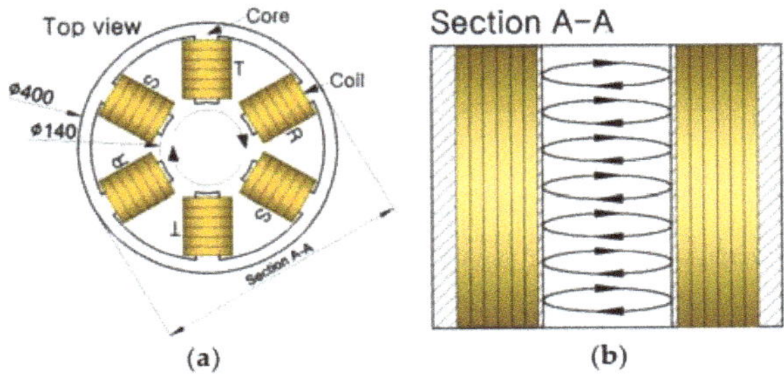

Figure 1. Schematic diagram of the electromagnetic stirring (EMS) device: (**a**) Top view; (**b**) Section A-A review.

2.1.2. Parameters of Electromagnetic Stirring

The material used in the semi-solid billet fabrication experiment was A356 alloy. The chemical composition of the A356 alloy is shown in Table 1. The solid fractions of the A356 material at different temperatures were obtained through differential scanning calorimetry (DSC). The liquidus and solidus temperatures were 617 and 547 °C, respectively, and the solidification range was 70 °C. The temperature at which a 35% solid fraction was obtained was 600 °C, while the temperatures for 45% and 55% solid fractions were 590 and 578 °C, respectively. An A356 alloy ingot was transformed into the molten state by being placed in a furnace and heated to 730 °C.

Table 1. Chemical composition of A356 alloy (wt %).

Si	Mg	Ti	Fe	Ni	Mn	Zn	Pb	Al
7.08	0.35	0.17	0.08	0.07	0.01	0.01	0.01	Bal.

With respect to the semi-solid billet fabrication sequence, first, the container was inserted into the electromagnetic stirrer. The container was then positioned at the center of the electromagnetic stirrer by placing a 100-mm-thick thermal insulator pad inside the stirrer and then placing the container on the pad. Next, using a ladle, the molten metal contained in the furnace was added to the container (filled to a height of 90 mm) and, immediately, stirring was initiated. Simultaneously, a thermocouple was inserted at the center of the molten metal in the container, and the temperature was measured. When the material was solidified (air cooling) at the necessary temperature for the required solid fraction, the fabricated semi-solid billet was removed from the stirrer and, as immediately as possible, cooled in water. After cooling, the primary α-Al particles and liquid inside the semi-solid billet formed primary α-Al particles and eutectic structures, respectively, at room temperature. The stirring conditions for the semi-solid billet fabrication process are shown in Table 2. The temperature at which the stirring was initiated was set to 620 °C, which is 3 °C higher than the liquidus temperature of the A356 alloy. Conditions 1–6 in Table 2 are different solid fractions. For conditions 1–3, billets of 35 (T = 600 °C), 45 (T = 590 °C), and 55% solid fraction (T = 578 °C) were fabricated by stirring under

60 A. Under conditions 4–6, billets with 35%, 45%, and 55% solid fraction, respectively, were fabricated without stirring. Conditions 7 and 8 are the different stirring currents. To that end, billets with 45% solid fraction were fabricated by applying 30 A (condition 7) and 90 A (condition 8).

Table 2. Fabrication conditions of semi-solid billets.

No.	Temperature (°C)/Solid Fraction (%)	Stirring	Stirring Current (A)
1	600/35	Yes	60
2 (standard)	590/45	Yes	60
3	578/55	Yes	60
4	600/35	No	-
5	590/45	No	-
6	578/55	No	-
7	590/45	Yes	30
8	590/45	Yes	90

2.2. Thin-Plate Forming Process

After putting a semi-solid billet into the die for thin plate, the compression was conducted using a punch. A thin plate of 1.2 mm thickness was fabricated by varying the solid fraction of the billet, die lubrication, and velocity and pressure of the punch for compressing the billet.

2.2.1. Die and Punch

To analyze the deformation and behavior of the primary α-Al particles when the semi-solid billet was compressed by the punch, an indirect closed die for thin plate was fabricated. A schematic diagram of the indirect die and punch used for the fabrication of the thin plate is shown in Figure 2. The left-hand side shows the structures of the die and punch, while the right-hand side shows the die cavity shape. The structure of the indirect die allows the semi-solid billet to be compressed by the punch when the top and bottom die are closed. A thin plate can be formed by filling the die cavity with the compressed billet. The punch compressing the semi-solid billet was 60 mm in diameter, while the die cavity size was 265 and 185 mm. A concavo-convex space 0.3 mm in depth was present at the center of the cavity. The die was installed on a 200 ton oil hydraulic press, and the punch was connected to the oil hydraulic cylinder. The load was controlled by the rate of oil discharge. Cartridge heaters were inserted in the die heating hole and preheated to approximately 270 °C. During the experiment, the die temperature was maintained at 250 °C.

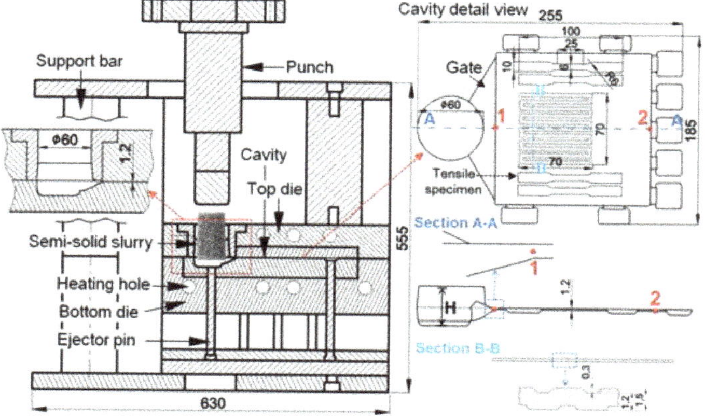

Figure 2. Schematic diagram of the compression die structure for forming thin plate and of the specimen position for investigating its microstructure and mechanical properties.

2.2.2. Parameters of the Forming Process

As regards the fabrication sequence of the thin plate using compression, firstly, the fabricated semi-solid billet was moved into the die. The semi-solid billet was then immediately compressed by dropping the punch. Next, the formed thin plate was removed from the die, and immediately cooled in water. The experiment conditions for thin plate fabrication through semi-solid billet compression with the punch are shown in Table 3. The compression experiment variables are classified into four categories: solid fraction of the semi-solid billet, cavity lubrication state, punch velocity, and punch pressure. A total of six experiments were configured. Conditions 1–3 involved different solid fractions of semi-solid billets. Graphite lubricant on the die cavity was used, with 300 mm/s of punch velocity and 200 MPa of punch pressure. For conditions 4–6, the parameters of condition 2 were used with the exception of one parameter. Specifically, the punch pressure was 100 MPa for condition 4, the punch velocity was 30 mm/s for condition 5, and, for condition 6, the graphite lubricant was not applied to the die cavity.

The friction factors (m) with and without the graphite lubricant were obtained through a ring compression test. A ring shape with external diameter of 16 mm, internal diameter of 8 mm, and thickness of 5.33 mm was used. The ring specimen was preheated to 590 °C, and compression was then performed. Through the ring compression test, it was found that the friction factor for the graphite lubricant was 0.4, whereas a factor of 0.9 was measured when the graphite lubricant was omitted.

Table 3. Compression conditions in indirect forming of thin plates.

No.	Temperature (°C)/Solid Fraction (%) of Billets	Velocity, V_P (mm/s)	Pressure, P_P (MPa)	Friction Factor (m)
1	600/35	300	200	0.4
2 (standard)	590/45	300	200	0.4
3	578/55	300	200	0.4
4	590/45	300	100	0.4
5	590/45	30	200	0.4
6	590/45	300	200	0.9

2.3. Analysis of Microstructure and Mechanical Properties

2.3.1. Microstructure

Since the semi-solid billet had a cylindrical shape of 55 mm in diameter and 90 mm in height, cooling occurred later at the center than at the edge when the billet was cooled in water. The billet microstructure was measured at a position close to the sample edge. The center of the semi-solid billet sample was cut in the horizontal direction, and a microstructure specimen was obtained from the cut billet, 2 mm from the edge.

To observe the microstructure of the thin plate, the gate of the thin plate was cut in the direction of section A-A, and the microstructure of the cut section (position 1) was measured (see Figure 2). To observe the behavior of the primary α-Al particles, which were compressed and caused to flow into the cavity, the microstructure was measured at position 2. Along with the microstructure measurement, the equivalent diameter D and roundness R of the primary α-Al particles were measured. The D value of a primary α-Al particle is defined in Equation (1), while R, which indicates how close the primary α-Al particle is to a circle, is defined in Equation (2). For a completely circular particle, R is 1, and this value increases with increased particle shape's irregularity,

$$D = \sqrt{\frac{4A}{\pi}} \tag{1}$$

$$R = \frac{p^2}{4\pi A} \tag{2}$$

where A is the area of a particle, and p is the perimeter of a particle.

2.3.2. Mechanical Properties

To analyze the formability of the thin plate in response to specific compression conditions, the height reduction ratio R_H was calculated using Equation (3). Note that a larger R_H indicates greater compression and shows that the formability is good.

$$R_H = \frac{H_0 - H}{H_0} \times 100 \tag{3}$$

In this formula, H_0 is the height of the semi-solid billet sample, and H is the height of the compressed thin plate, as shown in section B-B of Figure 2.

As can be seen in the cavity detail view of Figure 2, a tensile test was conducted by fabricating a tensile test specimen at the side of the fabricated thin plate. Vickers hardness was measured for the tip of the tensile test specimen. This specimen was fabricated according to the ASTM E 8M, subsize standard, with a gage length of 25 mm and width of 6 mm. The specimen thickness was the thickness of the fabricated thin plate. The tensile test was conducted on the formed state, without any additional treatment, such as surface and thermal treatments. For the tensile test, 25 ton Material Testing System (MTS) equipment was used, and the strain rate was set to 1 mm/min. For accurate elongation measurements, the elongation was measured using an extensometer. For the Vickers hardness test, the load was set to 200 g. The measurements in the tensile test and Vickers hardness test were taken five times for each specimen.

3. Results

3.1. Microstructures of the Semi-Solid Billet

The temperature curve of the A356 alloy as a function of solidification time is shown in Figure 3. It can be seen that the time required for solidification at 578 °C (solid fraction of 55%) from the liquid state at 620 °C decreased as the stirring current increased. The solidification rate increases because the stirring velocity increases when the stirring current is raised. Without electromagnetic stirring, the time required for the sample to solidify at 578 °C was 247 s, while for 30, 60, and 90 A stirrings, 238, 227, and 191 s were required, respectively. In the 60 A case, the time necessary for solidification at 600 °C was 106 s, and the required time to decrease the temperature from 600 to 590 °C was 58 s.

Note that, when a molten metal is solidified through electromagnetic stirring, the stirring is stopped during solidification. The stirring is not conducted until the target temperature value is reached because the solidification progresses more quickly at the top surface and at the edge. When a stirring current of 60 A was applied, stirring was performed until a temperature of 611 °C was reached (for 36 s). From 611 to 578 °C, stirring was not performed. For the 30 and 90 A, stirring was performed until reaching 617 °C (for 29 s) and 611 °C (for 41 s), respectively. In other words, for higher stirring currents, the stirring is performed for a longer period of time until a lower temperature is reached. For the 60 and 90 A cases, the stirring was stopped when a solid fraction of approximately 20% was obtained, whereas, for the 30 A case, the stirring was stopped at 617 °C at a liquid state. This is because of the weak stirring force and temperature difference between the positions due to heat transfer during stirring. The billet temperature was measured at the center of the container when the semi-solid billet was fabricated, and a temperature difference of approximately 2.5 °C was obtained between the center and the edges of the billet. The stirring was stopped because the edge was at approximately 615 °C (12% solid fraction). Note that the stirring process was stopped as solidification progressed more quickly at the edge, but vibration persisted because of the continued effect of the electromagnetic field.

The microstructure of the stirred billets (60 A) and unstirred billets are shown in Figure 4. The stirred and unstirred samples exhibited significant differences, which were particularly noticeable between the various solid fractions. For the unstirred semi-solid billet, a dendrite structure was dominant, while a fine globular structure was dominant for the stirred semi-solid billet. As the solid fraction increased, the primary α-Al particles became coarser and more irregularly shaped.

The unstirred samples are shown in the left-hand side of Figure 4. For the billet with 35% solid fraction, it can be seen that the irregularly shaped primary α-Al particles were arranged in a dendrite-type structure. In particular, it must be noted that the primary α-Al particles were not attached to each other. In the billet with 45% solid fraction, coarse particles formed by the bonding of primary α-Al particles could be seen, and the corners or tips of the irregular primary α-Al particles were connected with each other. Newly created small, spherical primary α-Al particles could also be seen. The billet with 55% solid fraction exhibited a coarse dendrite structure, in which irregular primary α-Al particles were completely bonded, and the irregularly shaped primary α-Al particles were arranged in a dendrite structure. As regards the microstructures of the unstirred billets, it could be seen that a dendrite structure was formed, as the irregularly shaped primary α-Al particles arranged in a dendrite-type structure became bonded with each other.

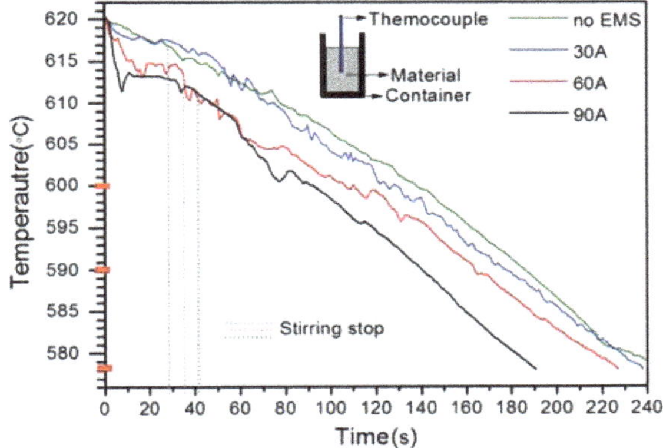

Figure 3. Solidification temperature according to the solidification time at different stirring currents.

The stirred samples are shown in the right-hand side of Figure 4. Only the small, spherical particles were distributed in the 35%-solid-fraction billet. The primary α-Al particles in the billet with 45% solid fraction were coarse and irregularly shaped compared with those in the former case because of the bonding between the primary α-Al particles and the particle growth. Further, the primary α-Al particle shapes were coarser and more irregular in the billet with 55% solid fraction than in the 45%-solid-fraction billet. Globular particles could be observed in the billet with 55% solid fraction. The majority of the particles were rosettes, and some exhibited dendrite shapes formed by the bonding of the rosette particles.

Thus, the unstirred and stirred billets had dendrite and globular structures, respectively. This was due to the bonding of primary α-Al particles in the unstirred case, and to the primary α-Al particle motion by the liquid fluid flow in the stirred case. It has been observed that, when the solidification begins below the liquidus temperature, nuclei are generated with a dendrite structure arrangement and are separated by certain gaps without being bonded with each other. Nuclei are also generated without dendrite arrangement [13].

When solidified without stirring, starting above the liquidus temperature, the particles, that are initially organized in a dendrite structure arrangement, irregularly grow as the solidification proceeds and form a dendrite structure, as they bond with each other through diffusion. In the dendrite structure arrangement, because the angle grain boundary between the primary α-Al particles is small, bonding between particles occurs more easily.

When solidified by stirring starting above the liquidus temperature, the primary α-Al particles are moved by the liquid fluid flow, and the dendrite structure arrangement is broken. The primary α-Al particles are deformed to spheroids by the liquid fluid and by wearing due to collisions between the particles. The electromagnetic field which continues to be applied even after stirring is stopped, generates vibration, and causes the primary α-Al particles to move. Because of this vibration, primary α-Al particle growth into irregular shapes is hindered [14].

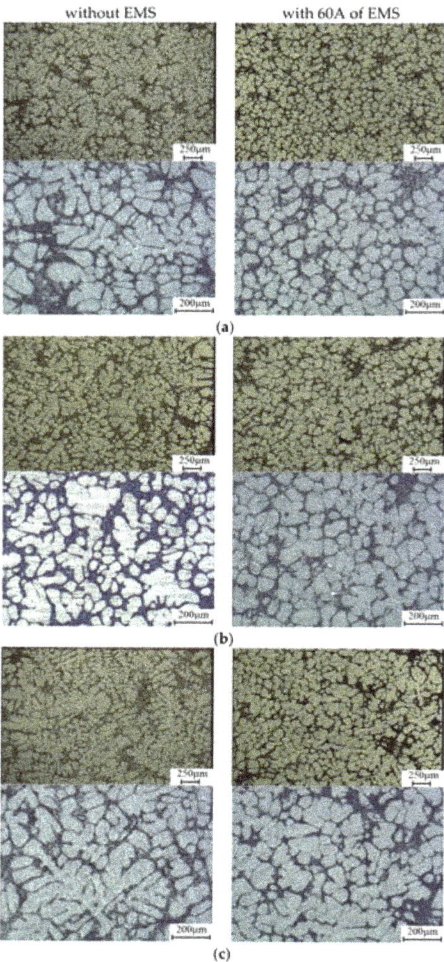

Figure 4. Microstructure of semi-solid billets with and without stirring: (**a**) 35% solid fraction; (**b**) 45% solid fraction; (**c**) 55% solid fraction.

The microstructures of the 45%-solid-fraction billets fabricated by stirring with 30 and 90 A, respectively, are shown in Figure 5. It can be seen that the primary α-Al particles in the billet fabricated using 30 A were coarser than those in the billet fabricated with 60 A. The majority of the primary α-Al particles were spheroids, but some dendrite shapes were present, in which rosette and irregularly shaped particles were bonded. From the microstructures presented in Figure 5b, it can be seen that the primary α-Al particles were smaller and more globular in the billet fabricated with 90 A than in that fabricated with 30 and 60 A (see Figure 4b). Rosette and dendrite-shaped particles were not present in the 90 A billet case. It can be seen that the particles with relatively irregular shapes were bonded with other particles at the corners. Because the stirring force and stirring velocity (shear strain rate) increased as the stirring current increased, the deformation of the primary α-Al particles and the number of collisions between the particles increase and, thus, the primary α-Al particles were deformed to become more globular. On the other hand, it can be seen that the bonding between the particles increased as a result of more frequent collisions between the particles.

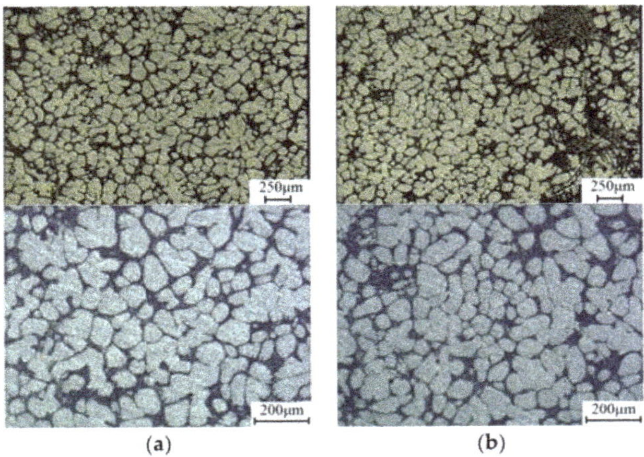

Figure 5. Microstructure of semi-solid billets with 45% solid fraction fabricated by different stirring currents: (**a**) 30 A of EMS; (**b**) 90 A of EMS. For 60 A of EMS A, see Figure 4b.

The D and R values of the primary α-Al particles in the billets are shown in Figure 6. The D and R of the primary α-Al particles were significantly smaller in the stirred billets than in the unstirred billets. As the solid fraction decreased and the stirring current increased, the D and R values of the primary α-Al particles decreased.

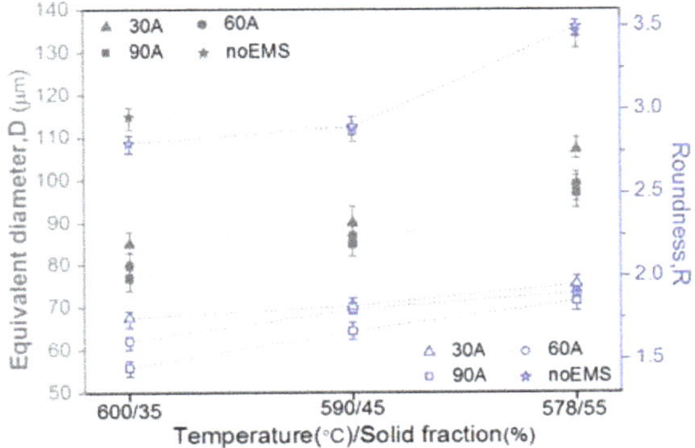

Figure 6. Equivalent diameter and roundness of primary α-Al particles in semi-solid billets.

3.2. Microstructure of the Thin Plates

The microstructure of the thin plates fabricated by insertion of 60 A stirred billets into the die with graphite lubricant, compressing them with 200 MPa pressure at 300 mm/s velocity (conditions 1–3 of Table 3) is shown in Figure 7. The primary α-Al particles were significantly smaller with much rounder shapes in all three thin plates, compared to the billets before compression (Figure 4). This is because a considerably large shear strain rate was applied by the punch compression, which caused the particles irregularly bonded with each other in the billet to become detached. The bonded particles, detached particles, single particles, and the residual liquid phase began to move because of the turbulent flow that occurred when they passed through the gate. At this point, the cross section was narrow, and the flow velocity increased significantly. Wearing of the particles occurred because of the collisions between them and as a result of the friction generated by the motion. Because of a large shear strain rate, bonding by collisions between particles did not occur. Similar to the observed billet microstructure (Figure 4), the microstructure of the thin plate fabricated with the 35%-solid-fraction billet had smaller and rounder particles than in the two other conditions. Considering the microstructure of the thin plate compressed with the 55%-solid-fraction billet shown in Figure 7c, a trace of rapidly expelled liquid phase could be confirmed at the bottom part of the gate (position 1). This phenomenon was due to the large bonding force between the particles and the low fluidity of the coarse particles. In the 55%-solid-fraction billet, because a large number of bonds existed between the particles, which were coarse, detachment and movement of the bonded particles were difficult even during compression. The same amount of liquid phase and solid phase could not move through the gate to the exterior simultaneously and, instead, a large amount of relatively good fluidity liquid phase exited the gate rapidly.

For all three conditions, the primary α-Al particles were rounder and smaller at position 2 than at position 1. The number of primary α-Al particles was small, and a large amount of eutectic structure could be confirmed, in which the liquid phase was filled. Because the distance over which the motion occurred was a little longer for the primary α-Al particles at position 2 than for those at position 1, a larger amount of wear due to friction and collisions between the particles occurred. Because small and round particles have superior fluidity than coarse particles [15,16], the round and small particles exhibited greater motion than the coarse particles at position 2. Hence, the primary α-Al particles of position 2 were smaller and rounder. The fluidity of the primary α-Al particles was decreased because of the friction occurring during the movement and was further obstructed in the concavo-convex space. Therefore, the liquid phase, which had relatively excellent fluidity, moved faster than the primary α-Al particles, and the eutectic structure occupied more space at position 2.

For the effect of the punch pressure, the microstructure of thin plates fabricated by insertion of 60 A stirred 45%-solid-fraction billets into the die with graphite lubricant, compressing them with 100 MPa pressure at 300 mm/s velocity (condition 4 of Table 3), is shown in Figure 8. The number of primary α-Al particles was small at both positions 1 and 2, compared to the thin plate fabricated with 200 MPa shown in Figure 7b. Porosities were apparent at position 2. It could be determined that it was difficult for many of the primary α-Al particles to move if a low punch pressure was applied during the compression of the semi-solid billet. The pressure compressing the semi-solid billet results in kinetic energy at the gate. The applied pressure of 100 MPa primarily caused the detachment of the bonded primary α-Al particles in the semi-solid billet, while the force required to initiate the primary α-Al particles motion became small. Therefore, many primary α-Al particles could not pass the gate, and more liquid phase was moved, which filled the cavity.

For the effect of the punch velocity, the microstructure of thin plates fabricated by insertion of 60 A stirred 45%-solid-fraction billets into the die with graphite lubricant, compressing them with 200 MPa pressure at 30 mm/s velocity (condition 5 of Table 3), is shown in Figure 9. Compared to the thin plate fabricated at 300 mm/s velocity shown in Figure 7b, the sizes of the primary α-Al particles were larger at both positions 1 and 2. At the position 1, primary α-Al particles arranged in a dendrite shape could be seen. When the compression velocity applied to the semi-solid billet was low, the deformation was reduced, and the kinetic energy at the gate was reduced. The coarse and irregularly shaped primary

α-Al particles were almost unable to exit the gate at the narrow cross section. Relatively smaller primary α-Al particles moved. For the moving primary α-Al particles, round deformation occurred as a result of wearing due to friction during motion and collisions between the particles.

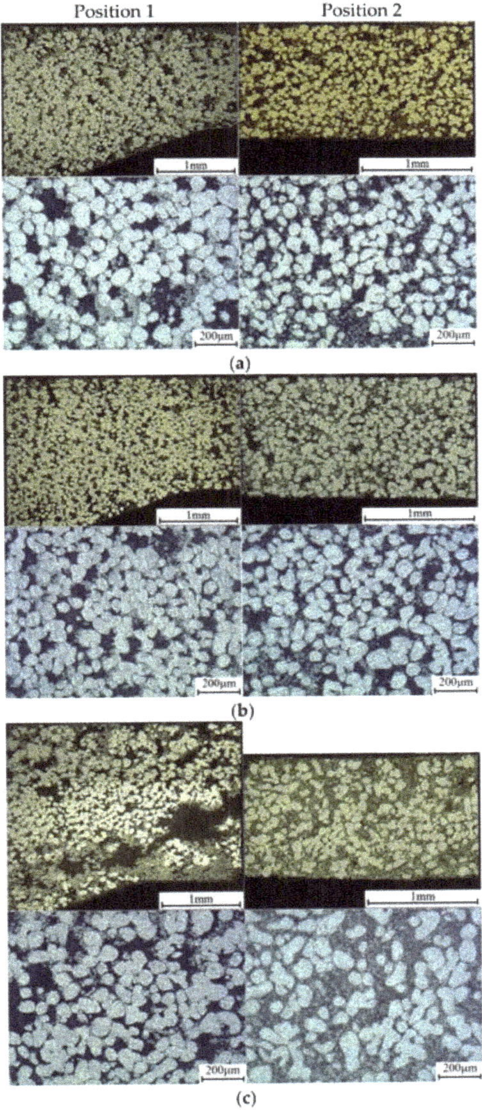

Figure 7. Microstructure of thin plates fabricated using 60 A stirred billets (graphite lubricant, 200 MPa of pressure and 300 mm/s of velocity): (**a**) Using a billet with 35% solid fraction; (**b**) Using a billet with 45% solid fraction; (**c**) Using a billet with 55% solid fraction.

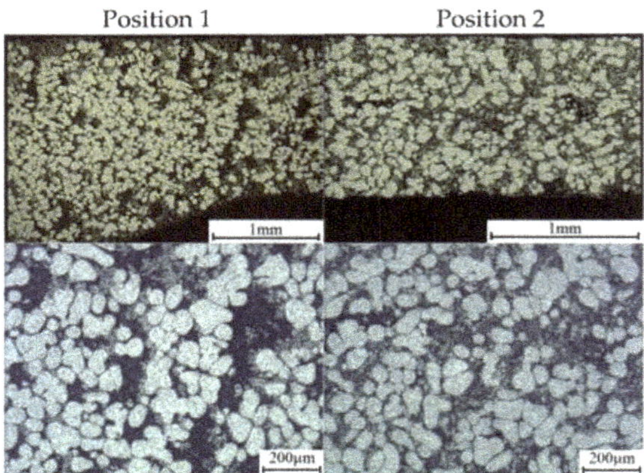

Figure 8. Microstructure of a thin plate fabricated using 60 A stirred 45%-solid-fraction billet, graphite lubricant, 100 MPa of pressure, and 300 mm/s velocity.

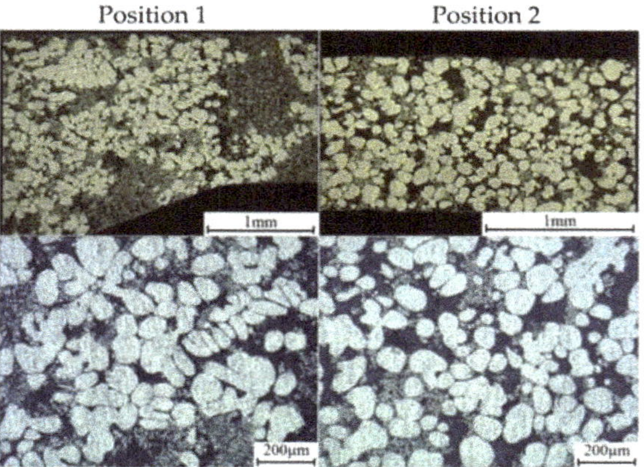

Figure 9. Microstructure of a thin plate fabricated using 60 A stirred 45%-solid-fraction billet, graphite lubricant, 200 MPa of pressure, and 30 mm/s velocity.

For the effect of the die friction, the microstructure of thin plates fabricated by insertion of 60 A stirred 45%-solid-fraction billets into the die, compressing them with 200 MPa pressure at 300 mm/s velocity (condition 6 of Table 3), is shown in Figure 10. At position 1, irregularly shaped primary α-Al particles were condensed, and approximately half the substance was a eutectic structure of liquid phase. The microstructure at position 2 exhibited a primarily eutectic structure, and only several primary α-Al particles were condensed. The friction factor of no graphite lubricant was 0.9, which was over two times greater than that of the graphite lubricant (m = 0.4). If the friction of the die is large, almost no primary α-Al particle can move, and only the liquid phase, which has relatively good fluidity, exits the gate. It must be noted that, because the primary α-Al particles had an inter-particle bonding force, they were condensed in a lump shape in one position, despite the high flow velocity.

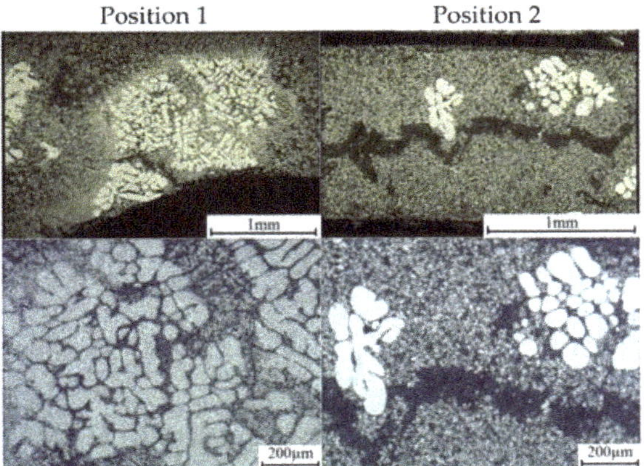

Figure 10. Microstructure of a thin plate fabricated using 60 A stirred 45%-solid-fraction billet, no graphite lubricant, 200 MPa of pressure, and 300 mm/s of velocity.

3.3. Formability and Mechanical Properties of the Thin Plates

The fabricated thin plate samples are shown in Figure 11. In the cases of conditions 1–3, where only the solid fractions of the billet were varied and the remaining compression conditions were identical, the H values were 17, 21, and 24 mm for condition 1 (35% solid fraction), condition 2 (45% solid fraction), and condition 3 (55% solid fraction) billets, respectively, while the height reduction ratio R_H was 81.1%, 76.6%, and 73.3%, respectively. In the sample fabricated under condition 3, an unformed part existed, and a flash phenomenon occurred at the compressed part. It is apparent that the compression was not well-performed, as the solid fraction of the billet increased, and a large number of solid particles were evident. Hence, the formability was degraded. The H and R_H of the sample fabricated under condition 4 were 27 mm and 70%, respectively. The H and R_H of the sample fabricated under condition 5 were 30 mm and 66.6%, respectively. Nonforming occurred in this sample. In the samples fabricated under conditions 4 and 5, the H was the highest obtained value for all conditions. It can be seen that velocity and pressure were the factors with the greatest effect on compressibility (formability). As regards compression condition 6, the H and R_H of the compressed part were 23 mm and 74.4%, respectively. Although a large amount of material was compressed in this case compared to the other conditions (as shown in Figure 10), this was filled with the liquid phase that flowed out from the billet only. Also, it was confirmed that a flash phenomenon occurred in the compressed part in this case, because of the rapid discharge of the liquid phase.

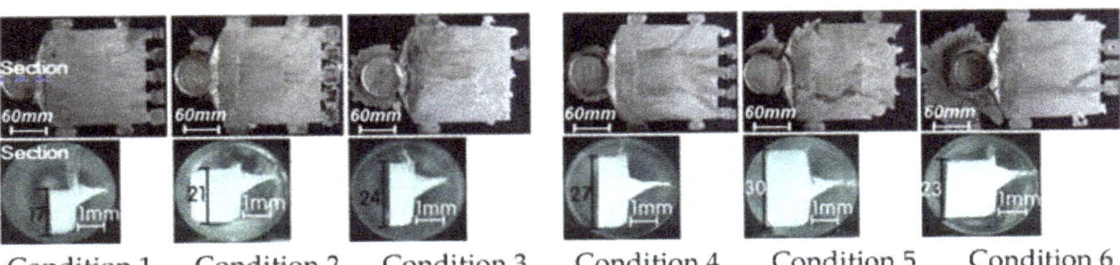

Figure 11. Thin plate samples fabricated in different compression conditions.

The mechanical properties of the thin plates fabricated under the six compression conditions are shown in Table 4. The tensile strength and elongation of conditions 1 and 2 were higher than those of the other conditions. The tensile strength of condition 2 was 205 MPa and the elongation was 8%, which was the highest obtained value. For condition 6, tensile strength and elongation were the lowest, and Vickers hardness was the highest because there were almost no primary α-Al particles, as shown in Figure 10, and there was only a eutectic structure filled with a liquid phase; thus, high brittleness was exhibited.

Table 4. Mechanical properties of thin plates fabricated in different compression conditions.

No.	Ultimate Tensile Strength (MPa)			Elongation (%)			Vickers Hardness (HV)		
	Mean	Max.	Min.	Mean	Max.	Min.	Mean	Max.	Min.
1	190	202	186	7	9	6	64	66	63
2	205	210	198	8	10	7	65	66	64
3	188	192	185	5	6	4	66	68	65
4	175	181	170	6	7	5	67	69	65
5	160	165	154	6	7	6	62	65	61
6	125	130	122	3	3	2	89	90	87

4. Conclusions

Semi-solid billets were fabricated by electromagnetic stirring, and experiments were conducted involving the fabrication of thin plates through compression of the billets. From these experiments, the effect of stirring on particle growth and the behavior of the particles by the compression were examined.

(1) For unstirred semi-solid billets, a dendrite structure was dominant, while a fine globular structure was dominant for the stirred semi-solid billets. As the solid fraction increased, the primary α-Al particles became coarser and more irregularly shaped.

(2) The equivalent diameter and roundness of the primary α-Al particles were significantly smaller in the stirred billets than in the unstirred billets. As the solid fraction decreased, and the stirring current increased, the equivalent diameter and roundness of the primary α-Al particles decreased.

(3) The primary α-Al particle sizes were reduced as the compressing velocity was increased, while a greater number of particles could be moved if the compressing pressure was increased. As the path over which the motion occurred became smoother, the fluidity of the particles improved.

(4) As a group of primary bonded α-Al particles was compressed under a large strain rate, the bonds were broken, and the group separated into individual particles. When the primary α-Al particles and residual liquid phase passed through a gate, at which the cross section was narrow, the flow velocity increased significantly, and the particles experienced turbulent flow. As wearing caused by friction and inter-particle collisions during this motion occurred, the particle shapes became increasingly spherical, because the particle sizes decreased, and the corner curvatures were increased. As the distance over which the particles moved increased, the particles were reduced in size and became more spherical.

Acknowledgments: This work was supported by Kyungnam University Foundation Grant, 2017.

Author Contributions: Chul Kyu Jin designed the experiment tools, performed the experiment, analysed the experimental results, and has contributed to the discussions as well as revisions.

References

1. Metz, S.A.; Flemings, M.C. *Fundamental Study of Hot Tearing*; American Foundrymen's Society: Schaumburg, IL, USA, 1970; Volume 78, pp. 453–460.

2. Spencer, D.B.; Mehrabian, R.; Flemings, M.C. Rheological Behavior of Sn-15 Pct Pb in the Crystallization Range. *Metall. Mater. Trans. A* **1972**, *3*, 1925–1932. [CrossRef]

3. Fan, Z.; Fang, X.; Ji, S. Microstructure and mechanical properties of rheo-diecast (RDC) aluminium alloys. *Mater. Sci. Eng. A* **2005**, *412*, 298–306. [CrossRef]

4. Biswas, B.; Hermann, R.; Das, J.; Priede, J.; Gerbeth, G.; Acker, J. Tailoring the microstructure and mechanical properties of Ti–Al alloy using a novel electromagnetic stirring method. *Scr. Mater.* **2006**, *55*, 1143–1146. [CrossRef]

5. Nafisi, S.; Emadi, D.; Shehata, M.T.; Ghomashchi, R. Effects of electromagnetic stirring and superheat on the microstructural characteristics of Al–Si–Fe alloy. *Mater. Sci. Eng. A* **2006**, *432*, 71–83. [CrossRef]

6. Bae, J.B.; Kim, T.W.; Kang, C.G. Experimental investigation for rheology forming process of Al–7% Si aluminum alloy with electromagnetic system. *J. Mater. Process. Technol.* **2007**, *191*, 165–169. [CrossRef]

7. Tzimas, E.; Zavaliangos, A. A comparative characterization of near-equiaxed microstructures as produced by spray casting, magnetohydrodynamic casting and the stress induced, melt activated process. *Mater. Sci. Eng. A* **2000**, *289*, 217–227. [CrossRef]

8. Haga, T.; Suzuki, S. Casting of aluminum alloy ingot for thixoforming using a cooling slope. *J. Mater. Process. Technol.* **2001**, *118*, 161–172. [CrossRef]

9. Zhang, Z.; Li, J.; Yue, H.; Zhang, J.; Li, T. Microstructure evolution of A356 alloy under compound field. *J. Alloy. Compd.* **2009**, *484*, 458–462. [CrossRef]

10. Canyook, R.; Wannasin, J.; Wisuthmethangkul, S.; Flemings, M.C. Characterization of flow behavior of semi-solid slurries containing low solid fractions in high-pressure die casting. *Acta Mater.* **2012**, *60*, 3501–3510. [CrossRef]

11. Naflsi, S.; Lashkari, O.; Ghomashchi, R.; Ajersch, F.; Charette, A. Microstructure and rheological behavior of grain refined and modified semi-solid A356 Al–Si slurries. *Acta Mater.* **2006**, *54*, 3503–3511.

12. Oh, S.W.; Bae, J.B.; Kang, C.G. Effect of Electromagnetic Stirring Conditions on Grain Size Characteristic of Wrought Aluminum for Rheo-forging. *J. Mater. Eng. Perform.* **2008**, *17*, 57–63. [CrossRef]

13. Flemings, M.C. Solidification processing. *Metall. Trans.* **1974**, *5*, 2121–2134. [CrossRef]

14. Hunt, J.D.; Jackson, K.A. Nucleation of solid in an undercooled liquid. *J. Appl. Phys.* **1966**, *37*, 254–257. [CrossRef]

15. Seo, P.K.; Kim, D.U.; Kang, C.G. The effect of the gate shape on the microstructural characteristic of the grain size of Al–Si alloy in the semi-solid die casting process. *Mater. Sci. Eng. A* **2007**, *445–446*, 20–30. [CrossRef]

16. Matsumiya, T.; Flemings, M.C. Modeling of Continuous Strip Production by Rheocasting. *Metall. Trans. B* **1981**, *12B*, 17–31. [CrossRef]

Article

Study of Semi-Solid Magnesium Alloys (With RE Elements) as a Non-Newtonian Fluid Described by Rheological Models

Marta Ślęzak

Department of Ferrous Metallurgy, Faculty of Metals and Industrial Computer Science, AGH University of Science and Technology, Al. Mickiewicza 30, 30-059 Kraków, Poland; mslezak@agh.edu.pl; Tel.: +48-12-617-26-02

Received: 28 December 2017; Accepted: 23 March 2018; Published: 28 March 2018

Abstract: This paper includes the results of high-temperature rheological experiments on semi-solid magnesium alloys and the verification of different models describing the rheological behaviour of semi-solid magnesium alloys. Such information is key from the point of view of designing alloy forming processes in their semi-solid states. Magnesium alloys are a very attractive material, due to their light weight and good plastic properties; on the other hand, this material is very reactive in a liquid (semi-solid) state, which is challenging from a testing and forming perspective. Formulating/finding models for an accurate description of the rheological behaviour of semi-solid magnesium alloys seems to be key from the standpoint of developing and optimising forming processes for semi-solid magnesium alloys.

Keywords: rheological model; semi-solid state; Mg alloys; high-temperature rheology; rheological properties

1. Introduction

Magnesium alloys are currently growing in importance as materials for parts used in the automotive industry. In addition, materials made of magnesium alloys have been accepted by the European Civil Aviation Conference and NASA as materials for the production of parts which are not prone to corrosion. Magnesium, with its specific gravity of 1.8 g/cm^3, is the lightest structural material. It is over four times lighter than steel and 1.5 times lighter than aluminium; at the same time it maintains very good mechanical properties, including ductility, which can be modified by the addition of appropriate alloying elements. Magnesium ranks eighth as the most frequently-occurring element in the lithosphere. It is produced from seawater, brines or magnesium rock and, therefore, its resources are enormous. Moreover, it is 100% recyclable. At present, annual magnesium output is estimated at about 500,000 tonnes p.a.

On the other hand magnesium is a very reactive material in a liquid (semi-solid) state, which makes it challenging from a laboratory and industrial perspective.

Viscosity is a property of liquid metals which plays a key role in many effects occurring in high-temperature conditions. It is a very important parameter when controlling manufacturing processes in which liquid metal is present: casting, forming [1–5], also in the semi-solid state. Data from measurements taken at high-temperatures are necessary for engineering new processes, and for the optimisation of those that already exist [6,7]. Many mathematical models that can assist in describing the thermodynamics, kinetics, fluid flow, and heat exchange have been created in recent years [8–10]. Obtaining the correct measurement data has been the basis for the creation of accurate models. The above-mentioned models may be helpful in modelling/optimising processes with the participation of the liquid phase. Mathematical modelling and the control of molten metal processing operations require knowledge of the thermophysical properties of liquid metals. The accuracy of the measurements of these properties is the basic precondition for the development of processes in materials engineering.

The issue of the influence of rheological parameters on semi-solid metal forming processes (SSM) has been considered in the subject literature. The beginning of semi-solid metal forming (SSM—so-called thixotropic forming) goes back to 1970 [11]. At the moment, it is believed that knowledge of the rheological properties in the semi-solid metal alloy forming processes plays a key role in process engineering [12–14]. Current semi-solid metal forming processes have been applied primarily in light metals processing [15,16]. Viscosity is the main rheological parameter considered in the SSM processes [17–21]; it is an indicator defining the capability of the metal to fill a mould, and it determines the force required to deform a material.

Many authors have taken up the subject of analysing the value of the dynamic viscosity coefficient of magnesium alloys [22–28]; however, this data did not concern systems in which rare earth elements had been added. Additionally, authors usually make rheological investigations of semi-solid slurries of alloys [22–28]. At the same time, the most frequently tested magnesium alloy—AZ91—was analysed in a slightly different way: by analysing rest time and subjecting the system to the impact of forces [22,23], instead of gradually changing the shear rate. This paper supplements the research on the rheological characteristics of magnesium alloys containing rare earth metals and include the results of rheological analysis by using models which are often mentioned in papers about aluminum alloys [16–21].

This paper contains the selected results of rheological tests of semi-solid magnesium alloys of the Mg-Zn-Al, Mg-Zn-RE groups: three chemical compositions with applied shear rates from 10 to 150 s^{-1}. The results of rheological tests conducted on magnesium alloys have been used to verify rheological models by Herschel-Bulkley, Ostwald, Carreau, and Bingham, which are most often used in the subject literature to describe the rheological behaviour of semi-solid metallic systems (aluminium, magnesium alloys). These models may be used for modelling alloy-forming processes in semi-solid states, and for computing individual rheological parameters (dynamic viscosity coefficient, shear stress, etc.) without the need to conduct expensive, complicated, and time-consuming tests.

The research materials presented in this paper form part of the tests and analyses performed, which, due to the complexity of the topic, constitute a cycle of studies concerning broadly-understood rheological analyses of liquid and semi-solid magnesium alloys (with various shares of the solid and liquid phases).

2. Materials and Methods

The rheological tests were carried out with a high-temperature rheometer [29–34] designed by the Anton Paar company (Anton Paar GmbH, Graz, Austria). The FRS1600 rheometer is a very precise instrument, equipped with an air bearing, one of the few instruments of this type that enables measurements to be performed at high temperatures, testing a very broad range of liquids, characterised by both high and low viscosity values (thanks to the measurement range of torque from 0.05 mNm to 200 mNm). Basically, it consists of the head of a rheometer and a furnace which enables a temperature in the range 673–1805 K to be obtained. There is also the possibility of providing measurements at room temperature. One of the main advantages is an operating system based on pneumatic servomotors used to manipulate the crucibles and the rotating rods inside the furnace. Control of the furnace is also possible using the rheometer software (which is called Rheoplus), which allows experiments with changes of temperature to be programmed. It is not only possible to study the rheological properties of materials, in the liquid state in this type of rheometer, but also in the semi-solid state.

The measurement is performed in a fixed crucible into which a sample of the material tested is placed, then a rotating spindle is immersed within the material being tested. The crucible is then placed inside a ceramic shroud, being a component of a heating furnace. The furnace, which is comprised of four electrically-heated SiC-type heating elements, enables the maximum temperature of 1793 K to be obtained within the sample. The whole device is shielded from the outside with an insulating material.

The temperature inside the furnace is controlled by a change in feeder power in the measurement and control system. The heating rate, along with maintaining the temperature at a constant level, are set in the control panel of the Rheoplus software of the rheometer. Rotary movements of the spindle are controlled by a motorised measurement head—the spindle being suspended on a ceramic tube placed in an air bearing. The head is cooled with water and air in order to ensure a low temperature.

In this rheometer the torque values are measured by the head and then the software calculates the values of shear stress, viscosity, etc. [34]. The instrument features torque accuracy of 0.001 mNm. The parameters of the geometry of the measurement system used are implemented in the Rheoplus software before starting the experiments. This method of measurement, with adequate equations for viscosity calculation, is fully described in [34].

The measurements were provided in a Searle-type system [18,20,29–34]. Concentric cylinder systems are described by the standards ISO 3219 and DIN 53019. Bobs with perforated surfaces, with diameters of 16 mm, and cups with smooth inner surfaces and an internal diameter of 30 mm were used for the tests. Materials were selected for the tools that prevented the tool surface reacting with the sample tested. The measurement system was made of low-carbon steel.

The rheological tests were conducted for three magnesium alloys with different chemical compositions. Table 1 presents the chemical compositions of the magnesium alloys tested.

For each of the aforesaid grades (Table 1), the values of the liquidus temperatures and temperatures of the solid phase content of 50% (Table 2) were determined with DSC (Differential Scanning Calorimetry) analysis (Figure 1).

Table 1. Magnesium alloys tested.

Mg Alloy	Composition (wt %)				
AZ91	Al	Mn	Zn	Mg	-
	8.7	0.13	0.7	bal.	
WE43B	Nd	Y	TR	Zr	Mg
	2.2	4.2	3.3	0.53	bal.
E21	Gd	Nd	Zn	Mg	-
	1.4	3.0	0.31	bal.	

Table 2. The values of the liquidus temperature and the temperature of the solid phase content of 50% for the magnesium alloys analysed according to the DSC analysis.

Mg Alloy	Liquidus Temperature (K)	$T_{50\% \text{ fs}}$
AZ91	893	846
WE43B	955	915
E21	947	918

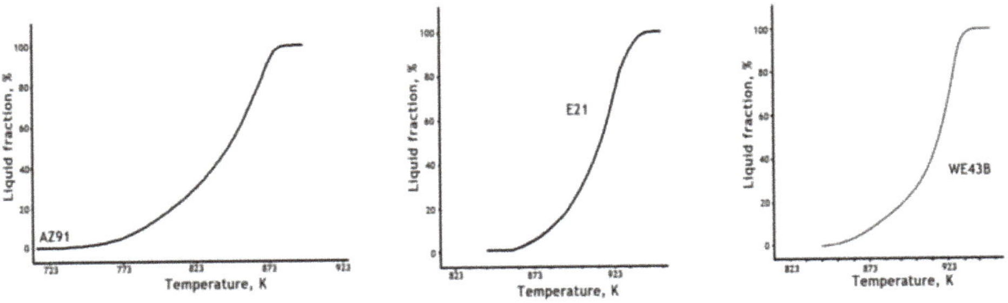

Figure 1. Graphs of DSC analysis of the magnesium systems analysed: AZ91, E21 and WE43B.

The determination of a fraction of the liquid phase as a function of temperature was calculated using data collected from the differential scanning calorimeter. It was assumed that a fraction of a liquid phase is proportional to the absorbed/released energy during the transformation (melting/solidification). The estimation of the liquid phase fraction changes was carried out by application of a partial peak area integration. Liquid fraction at a given temperature is determined by calculating the ratio of the area corresponding to the partial heat of melting over the total peak area. The former area is limited by the solidus temperature and the temperature range between solidus and liquidus lines (semi-solid range), the latter area is limited by solidus to liquidus temperature. It is expressed in volume percentage.

The amount of liquid phase was also determined as 50% on the basis of the content of eutectic phase in the sample after cooling from semi-solid temperature range. The discrepancy in the volume of liquid phase results from different heating rates, as well as an inhomogeneous chemical composition of the feedstock.

The samples were cylindrically shaped: 40 mm height, 25 mm in a diameter.

The rheological tests were conducted from the liquidus temperature to the temperature of a solid phase share of 50% (cooling rate was 2 K/min) for each alloy tested. The magnesium alloy tests were carried out in conditions of variable rheological parameters: shear rate values varied from 10 to 150 s^{-1}, and the objective of the tests was to find the influence of the aforementioned variables on the value of shear stress and, thus, to attempt to determine the rheological nature of the magnesium alloys tested.

The scheme of measurements (for each Mg alloy) was as follows:

- heating the sample;
- homogenising the temperature value in a whole volume of the sample—wait about 20 min after the assumed temperature value is obtained—while slowly shearing the sample;
- after stabilisation of the viscosity values of totally liquid alloys—decreasing the temperature up to f_s = 50% in each of alloys investigated (stirring with a shear rate 10 s^{-1});
- homogenise the temperature value in a whole volume of the sample;
- measurement with a shear rate 10–150 s^{-1};
- cooling the sample.

The findings are presented in the form of flow curves.

In the case of research on very reactive magnesium alloys, conducted on a wide range of solid phase shares (the upper range of the nominal torque operation of the measuring head), it was decided to carry out the measurements at the maximum shear rate of 150 s^{-1}. However, the description of data by rheological models was also assessed by paying special attention to the approximation of the shear rate to higher values, referring to thixoforming process conditions.

3. Results

Figure 2 presents a graph of changes in the shear stress value of AZ91 for variable shear rate values.

The values of shear stress of alloy AZ91, for a solid phase share of 50%, grow non-linearly as the shear rate grows, which shows non-Newtonian rheological behaviour of the body tested. The shear stress values grow from about 10 Pa to about 50 Pa, as the shear rate values grow from 10 to 150 s^{-1}.

The next figure (Figure 3) presents a graph of changes in the shear stress value of alloy E21 for variable shear rate values.

The values of shear stress of alloy E21, for 50% solid phase share, grow non-linearly as the shear rate grows, in a similar manner to alloy AZ91. However, for alloy E21, the shear stress values grow from about 3 to about 20 Pa.

Figure 4 presents a graph of changes in the tangential stress value of alloy WE43B for variable shear rate values.

The values of the shear stress of alloy WE43B, for a share of 50% solid phase, grow non-linearly as the shear rate increases. However, the values of shear stress obtained for alloy WE43B were the lowest, from about 4 to about 10 Pa.

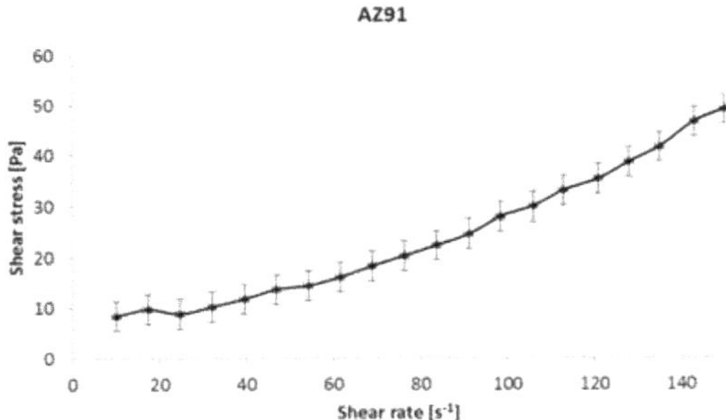

Figure 2. Flow curve of alloy AZ91, at a solid phase share of 50%.

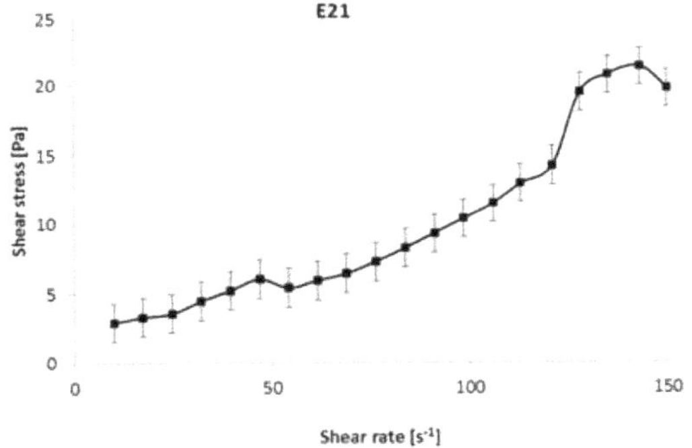

Figure 3. Flow curve of alloy E21 for 50% of the solid phase.

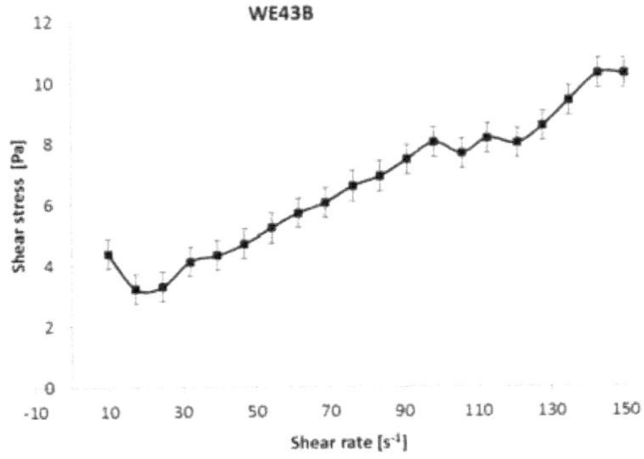

Figure 4. Flow curve of alloy WE43B for 50% of the solid phase.

Analysis of the results obtained allowed us to establish that the highest values of shear stress were obtained for alloy AZ91 (50 Pa), while the shear stress values for alloys E21 and WE43B achieved a maximum of 10–20 Pa. It should be borne in mind that alloys E21 and WE43B contain rare earth elements (yttrium, neodymium, gadolinium), which may influence changes in the shear stress values.

However, rheological tests of alloys with various contents of the above-mentioned rare earth metals would need to be performed to verify this.

Alloy E21 shows some deviation during the end of the test, it is probably the effect of shearing particles (the high amount of solid fraction). The author provided the wider rheological and microstructure measurements of these Mg alloy and did not observe the deviations of E21 alloy behaviour in comparison to the WE43B alloy (both contain RE elements).

In addition, all of the alloys tested showed a tendency to non-linear increase of the shear stress value as the shear rate increased. This may indicate a tendency towards shear-thinning (a decline in the value of the dynamic viscosity coefficient resulting from the forces applied). This is one of the key examples of non-Newtonian behaviour, characteristic for metals in a semi-solid state.

3.1. Data Description with Rheological Models

The rheological behaviour of a material is described by the relationships between stresses, strains, shear rates, and the time during which the material has been subjected to such strains. Such relationships are called rheological equations of the state of the material, or rheological equations, for short. The main task of rheology is to formulate models for describing the behaviours of bodies that have been subjected to an impact force.

In the subject literature, many attempts have been made to describe the flow curve with an appropriate rheological mathematical model [32–37]. The foregoing models are necessary for the analytical solution of problems related to non-Newtonian fluid flow [38–40].

Rheological models constitute a group of equations, which, apart from dynamic viscosity, also take into account other rheological parameters—shear rate, shear time, etc. The simplest mathematical rheological model, which describes a non-Newtonian fluid flow curve within a range of intermediate shear rates, is the so-called Ostwald-de Waele power law model in the form [33]:

$$\tau = k(\dot{\gamma})^n, \tag{1}$$

where k is the empirically-determined constant (Pa·s^n), n is the empirically-determined index exponent (-), τ is the shear stress (Pa), and $\dot{\gamma}$ is the shear rate (s^{-1}).

The power law model created by Ostwald and DeWaele is the simplest mathematical rheological model of a generalised Newtonian fluid, containing only two constants that need to be determined.

To better describe experimental data, numerous authors have proposed to use mathematical rheological models with more complex structures. This study attempted to approximate the results obtained with four selected rheological models, which are most often used in the subject literature to compute (approximate and describe) the values of the shear stress of aluminum and magnesium alloys intended to be formed in a semi-solid state [24,28,41,42].

The following models were used to describe the data:

- Ostwald (Equation (1))
- Herschel-Bulkley (Equation (2)):

$$\tau = \tau_{\text{HB}} + k(\dot{\gamma})^n \tag{2}$$

- Carreau (Equation (3)):

$$\eta = \frac{\eta_0 - \eta_\infty}{1 + (c \cdot \dot{\gamma})^{2p}} + \eta_\infty, \tag{3}$$

Form of the Carreau model which is used in the Rheoplus calculations:

$$\tau = \frac{\tau_0 - \tau_\infty}{1 + (c \cdot \dot{\gamma})^{2p}} + \tau_\infty \tag{3a}$$

- Bingham (Equation (4)):

$$\tau = \tau_{B} + k(\dot{\gamma}),\tag{4}$$

where τ is the shear stress (Pa), τ_{HB} is the Herschel-Bulkley shear stress (Pa). $\dot{\gamma}$ is the shear rate (s^{-1}), k is the Bingham constant (Pa·s), n is the empirically-determined index exponent (-), c is the Carreau constant (s), p is the Carreau exponent (-), η is the dynamic viscosity coefficient (Pa·s), η_0 is the dynamic viscosity coefficient for shear rates approaching 0 (Pa·s), η_∞ is the dynamic viscosity coefficient for shear rates approaching ∞ (Pa·s), τ_B is the Bingham shear stress (Pa), τ_0 is the shear stress for shear rates approaching 0 (Pa), and τ_∞ *is the* shear stress for shear rates approaching ∞ (Pa).

3.2. Rheoplus Calculations

Rheoplus V3.40 (Anton Paar GmbH, Ostfildern, Germany) is the integrated software for Anton Paar rheometers. By using Rheoplus it is possible to control instruments during measurement and analyse measurement data after testing. The different rheological models which were used to describe and fit the results obtained were implemented in the software.

Using Rheoplus software for each of the alloys tested over a range with variable shear rate values from 10 to 150 s^{-1}, an approximation of the results obtained was attempted with the four selected rheological models. The results obtained were presented in the form of a graph (with the calculated correlation coefficient R^2—the degree to which the model matched the actual data) with the actual flow curve obtained by measurements and the flow curves obtained from each model marked on the graph. The curves were presented in the shear stress τ—shear rate $\dot{\gamma}$ system (flow curve), as such relationships occur in three of the rheological equations presented. However, the Carreau equation is usually only defined for the dynamic viscosity coefficient η, so to enable models to be compared, the Carreau equation was also presented as shear stress versus shear rate.

Figure 5 presents the graphs of flow curves for alloy AZ91: actual and for three models.

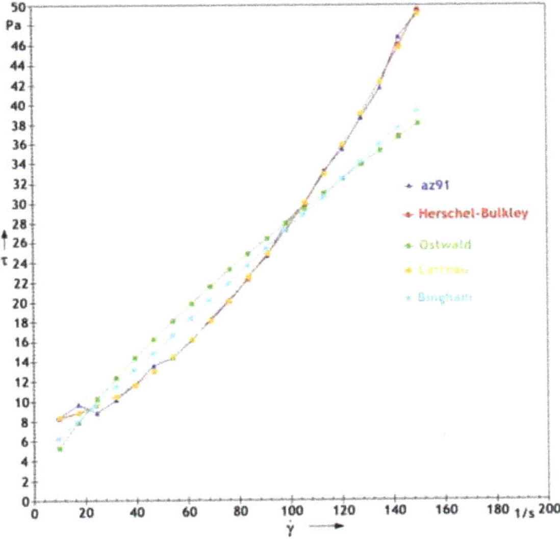

Figure 5. Flow curves for alloy AZ91: actual and computed from the Hershel-Bulkley ($R^2 = 0.99873$), Ostwald ($R^2 = 0.87918$), Carreau ($R^2 = 0.99875$), and Bingham ($R^2 = 0.91265$) models.

As we can observe, on graphs (Figure 5) the model and actual flow curves largely overlap, with the greatest deviations being seen for the flow curve described by the Ostwald model, which is reflected in the lowest value of the correlation coefficient R^2 out of all those computed.

Figure 6 presents the graphs of the flow curves for alloy E21: actual results and for three models.

The graphs (Figure 6) display the large overlap between the model and actual flow curves. The greatest deviations can be seen for the flow curves described by the Ostwald and Bingham models (the lowest values of coefficients R^2).

Figure 7 presents graphs of flow curves for alloy WE43B: actual and for three models.

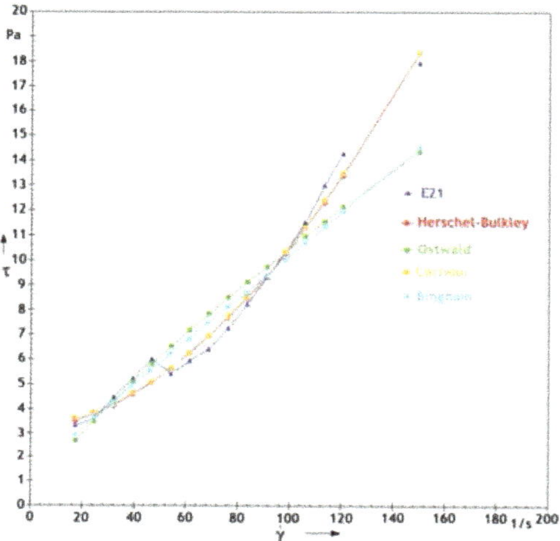

Figure 6. Flow curves for alloy E21: actual and computed from the Hershel-Bulkley ($R^2 = 0.9861$), Ostwald ($R^2 = 0.89346$), Carreau ($R^2 = 0.98654$), and Bingham ($R^2 = 0.90697$) models.

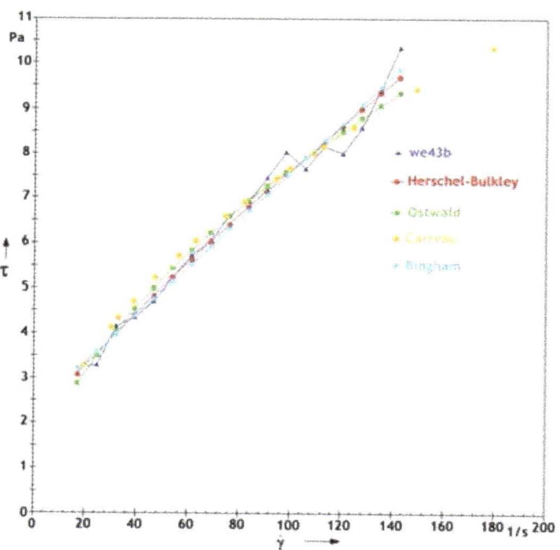

Figure 7. Flow curves for alloy WE43B: actual and computed from the Hershel-Bulkley ($R^2 = 0.97892$), Ostwald ($R^2 = 0.97174$), Carreau ($R^2 = 0.93052$), and Bingham ($R^2 = 0.97731$) models.

On the basis of the analysis of graphs on Figure 7, one may find that for flow curves describing the rheological behaviour of alloy WE43B, all four models represent the measurement results rather well; correlation coefficients over 0.93 were obtained for all four models.

As the Rheoplus software only enabled the quality of models to be assessed by analysing the correlation coefficient R^2, and the models were non-linear, the models were verified in Wolfram Mathematica. This analysis enabled the quality of the data description (mean prediction bands) to be more explicitly assessed by the selected rheological models, and allowed us to determine which non-linear model works best for the description of data from the measurements of semi-solid magnesium alloys with a 50% solid phase share.

3.3. Wolfram Mathematica Calculations

Wolfram Mathematica 11 (developed by The Wolfram Centre, Long Hanborough, United Kingdom) is a mathematical symbolic computation program used in many scientific, engineering, mathematical, and computing fields. The data from measurements where analysed in the Mathematica software implemented four different rheological models (Equations (1)–(4)). The results are presented as graphs and Equations (5)–(16), with calculated factors.

Below (Tables 3–5, Equations (5)–(16)), mathematical formulae of rheological models (calculated with Wolfram Mathematica software) are presented for each of the test alloys. In each of the cases the shear stress τ is given in Pa.

Table 3. Mathematical formulas of rheological models for the AZ91 alloy.

Model	Equation	
Herschel-Bulkley	$\tau = 8.05454 + 0.00473304 \cdot \dot{\gamma}^{1.81092}$	(5)
Ostwald	$\tau = 0.178404 \cdot \dot{\gamma}^{1.11004}$	(6)
Carreau	$\tau = \dfrac{8.43588 - 943.273}{\left(1 + \left((0.00339303 \cdot \dot{\gamma})^2\right)^{0.193341}\right)} + 943.273$	(7)
Bingham	$\tau = 0.434913 + 0.295291 \cdot \dot{\gamma}$	(8)

Table 4. Mathematical formulas of rheological models for the E21 alloy.

Model	Equation	
Herschel-Bulkley	$\tau = 2.97499 + 0.000466836 \cdot \dot{\gamma}^{2.15206}$	(9)
Ostwald	$\tau = 0.0116226 \cdot \dot{\gamma}^{1.52645}$	(10)
Carreau	$\tau = \dfrac{2.53371 - 963.647}{\left(1 + \left((0.00136759 \cdot \dot{\gamma})^2\right)^{0.572441}\right)} + 963.647$	(11)
Bingham	$\tau = 1.84953 + 0.158344 \cdot \dot{\gamma}$	(12)

Table 5. Mathematical formulas of rheological models for the WE43B alloy.

Model	Equation	
Herschel-Bulkley	$\tau = 3.15251 + 0.0171284 \cdot \dot{\gamma}^{1.20392}$	(13)
Ostwald	$\tau = 0.641002 \cdot \dot{\gamma}^{0.541978}$	(14)
Carreau	$\tau = \dfrac{3.46426 - 141.604}{\left(1 + \left((0.0129022 \cdot \dot{\gamma})^2\right)^{0.0314924}\right)} + 141.604$	(15)
Bingham	$\tau = 2.66302 + 0.0495468 \cdot \dot{\gamma}$	(16)

On the basis of the analysis of the above graphs and equations, we can conclude that the Herschel-Bulkley and Carreau models describe the results obtained well and, in addition, the mean prediction bands are relatively narrow for both cases. The mean prediction bands are the confidence bands for mean predictions and give functions of the predictor variables. This confirms the good quality of the description of the measurement data for these models. However, the Carreau model is much more complicated mathematically; therefore, the H-B model seems to be more appropriate for describing the rheological data obtained for the aforementioned magnesium alloys. Furthermore, the Carreau model is a function for calculations of flow behaviour including zero-shear and infinite-shear viscosity, thus, it dedicated for systems in which a wide range of shear rates are measured.

As a result of the analysis of the above models, it was found that the Herschel-Bulkley model best described the rheological behaviour of semi-solid magnesium alloys with a significant solid phase share in the alloy tested. For this model, the shear stress value was approximated for shear rates of 200 s^{-1}. The results obtained are presented in Figure 8.

As a result of the analysis of the approximation of the flow curves described with the Herschel-Bulkley model, one may find that this model is suitable for alloys AZ91, WE43B (large coverage with measurement data, small calculation uncertainty). This model describes data for alloy E21 as being slightly worse and, moreover, there is greater model uncertainty at higher values of the shear rate. However, this is likely to be related to the deviations of recorded measurement points that were not observed for the other two alloys.

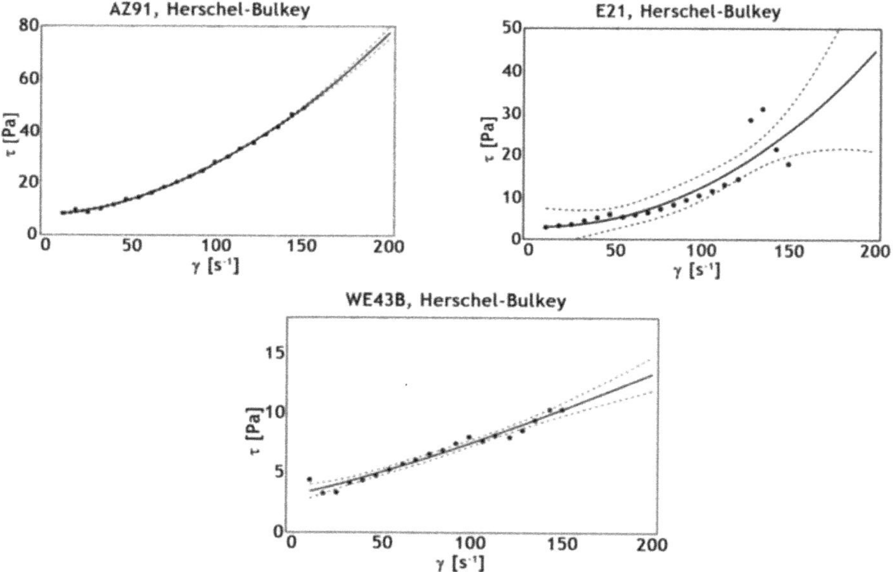

Figure 8. Approximated flow curves, along with mean prediction bands (functions of the predictor variables) calculated for the Herschel-Bulkley models, for each of the tested alloys: AZ91, E21 and WE43B (black continuous line—flow curve, grey dotted line—mean prediction bands).

4. Discussion

Flow curves obtained from measurements were compared with curves resulting from the use of four different rheological models: Herschel-Bulkley, Ostwald, Carreau, and Bingham. This allowed us to determine that all models provided a good level of accuracy of description, however, the measurement data were most accurately described by the models of Carreau and Herschel-Bulkley. Due to a simpler mathematical form, the model that is the most recommended for the description of data from rheological measurements of semi-solid magnesium alloys (with 50% solid phase share in the alloy) is the Herschel-Bulkley model.

The approximation conducted for higher values of the shear rate, for the Herschel-Bulkley model, showed that the model predicted shear stress values well, in particular for alloys AZ91 and WE43B (narrow mean prediction bands), and that it might be used to calculate shear stress values for Mg alloys under higher values of shear rate (according to the thixoforming process). The model performs slightly worse for alloy E21, but this is likely to be a result of the greater span of measurement points obtained.

5. Conclusions

- Rheological tests of magnesium alloys are challenging due to the high reactivity of the materials tested. They require considerable experience during measurements and analysis of the results obtained.
- Alloys E21 and WE43B contain rare earth elements (yttrium, neodymium, gadolinium), which may influence changes in the shear stress values. However, rheological tests of alloys with various contents of the above-mentioned rare earth metals would need to be performed to verify this.
- All of the alloys tested showed a tendency towards non-linear growth of the shear stress value as the shear rate grew. This is one of the key examples of non-Newtonian behaviour (shear-thinning), characteristic for metals in a semi-solid state.
- Four different rheological models: Herschel-Bulkley, Ostwald, Carreau, and Bingham provided a good level of accuracy of the description, however, the measurement data were most accurately described by the models of Carreau and Herschel-Bulkley.
- Due to a simpler mathematical form (and shear rate "range"), the model that is the most recommended for the description of data from rheological measurements of semi-solid magnesium alloys (with 50% solid phase share in the alloy) is the Herschel-Bulkley model.
- The approximation conducted for higher values of the shear rate, for the Herschel-Bulkley model, showed that the model predicted shear stress values well, in particular for alloys AZ91 and WE43B (narrow mean prediction bands). The model performs slightly worse for alloy E21, but this is likely to be as a result of the greater span of the measurement points obtained.
- The rheological models may be helpful for modelling and optimising the forming process of semi-solid magnesium alloys. Complementing thermodynamic databases with the results of rheological measurements will contribute to the development of the above mentioned processes and will facilitate modelling/engineering these processes without the need for conducting time-consuming and demanding measurements.
- During the thixocasting process the solid fraction might vary a great deal, thus, a wide range of rheological experiments have to be conducted to obtained reliable data for different amounts of liquid/solid fractions of Mg alloys. Authoritative data are also required from the optimal rheological model (for Mg alloys) point of view. The provided test with various shares of the solid and liquid phases show a similar rheological dependency as described in this paper.

Acknowledgments: Research financed through statutory funds of AGH University of Science and Technology in Krakow, No. 11.11.110.502. Special thanks to Bogusz Kania for his help in the execution of the computing.

Conflicts of Interest: The authors declare no conflict of interest.

References

1. Bakhtiyarov, S.I.; Overfelt, R.A. Measurement of liquid metal viscosity by rotational technique. *Acta Mater.* **1999**, *47*, 4311–4319. [CrossRef]
2. Budai, I.; Benko, M.; Kaptay, G. Comparison of different theoretical models to experimental data on viscosity of binary liquid alloys. *Mat. Sci. Forum* **2007**, *537*, 489–496. [CrossRef]

3.   Miłkowska-Piszczek, K.; Korolczuk-Hejnak, M. An analysis of the influence of viscosity on the numerical simulation of temperature distribution, as demonstrated by the CC process. *Arch. Metall. Mater.* **2013**, *58*, 1267–1274. [CrossRef]

4.   Qi, Y.; Cagin, T.; Kimura, Y.; Goddard, W.A. Viscosities of liquid metal alloys from nonequilibrium molecular dynamics. *J. Comput. Aided Mater.* **2001**, *8*, 233–243. [CrossRef]

5.   Zhao, P.; Wang, Q.; Liu, L.; Wei, Z.; Zhai, C. Fluidity of Mg-Al-Ca alloys in the high-pressure die casting proces. *Int. J. Mater. Res.* **2007**, *98*, 33–38. [CrossRef]

6.   Huppmann, M.; Reimers, W. Microstructure and mechanical properties of differently extruded AZ31 magnesium alloy. *Int. J. Mater. Res.* **2010**, *101*, 1264–1271. [CrossRef]

7.   Braszczyńska-Malik, K.N.; Froyen, L. Microstructure of AZ91 alloy deformed by equal channel angular pressing. *Int. J. Mater. Res.* **2005**, *96*, 913–917. [CrossRef]

8.   Chen, J.Y.; Fan, Z. Modelling of rheological behaviour of semisolid metal slurries Part 1—Theory. *Mater. Sci. Technol.* **2002**, *18*, 237–242. [CrossRef]

9.   Terzieff, P. The viscosity of liquid alloys. *J. Alloys Compd.* **2008**, *453*, 233–240. [CrossRef]

10.   Chhabra, R.P. A simple method for estimating the viscosity of molten metallic alloys. *J. Alloys Compd.* **1995**, *221*, L1–L3. [CrossRef]

11.   Flemings, M. Solidification processing. *Metall. Mater. Trans. B* **1974**, *5*, 2122–2134. [CrossRef]

12.   Atkinson, H.V. Modelling the Semisolid Processing of Metallic Alloys. *Prog. Mater. Sci.* **2005**, *50*, 341–412. [CrossRef]

13.   Kapranos, P.; Kirkwood, D.H. Thixoforming M2 tool steel—A study of different feedstock routes. *Metal. Ital.* **2010**, *9*, 17–21.

14.   Kapranos, P.; Kirkwood, D.H.; Sellars, C.M. Semi-Solid Processing of Aluminium and High Melting Point Alloys. *Proc. Inst. Mech. Eng. Part B J. Eng. Manuf.* **1993**, *207*, 1–8. [CrossRef]

15.   Sołek, K.P.; Rogal, Ł.; Kapranos, P. Evolution of Globular Microstructure and Rheological Properties of Stellite 21 Alloy after Heating to Semisolid State. *J. Mater. Eng. Perform.* **2017**, *26*, 115–123. [CrossRef]

16.   Rogal, Ł.; Dutkiewicz, J.; Atkinson, H.V.; Lityńska-Dobrzyńska, L.; Czeppe, T.; Modigell, M. Characterization of semi-solid processing of aluminium alloy 7075 with Sc and Zr additions. *Mat. Sci. Eng. A Struct.* **2013**, *580*, 362–373. [CrossRef]

17.   Brabazon, D.; Browne, D.J.; Carr, A.J. Experimental investigation of the transient and steady state rheological behaviour of Al-Si alloys in the mushy state. *Mater. Sci. Eng. A Struct.* **2003**, *356*, 69–80. [CrossRef]

18.   Salleh, M.S.; Omar, M.Z.; Syarif, J.; Mohammed, M.N. An Overview of Semisolid Processing of Aluminium Alloys. *ISRN Mater. Sci.* **2013**, *2013*, 1–9. [CrossRef]

19.   Das, P.; Samanta, S.K.; Dutta, P. Rheological Behavior of Al-7Si-0.3Mg Alloy at Mushy State. *Metall. Mater. Trans. B* **2015**, *46*, 1302–1313. [CrossRef]

20.   Lashkari, O.; Ajersch, F.; Charette, A.; Chen, X. Microstructure and rheological behavior of hypereutectic semi-solid Al-Si alloy under low shear rates compression test. *Mater. Sci. Eng. A Struct.* **2008**, *492*, 377–382. [CrossRef]

21.   Kim, W.Y.; Kang, C.G.; Lee, S.M. Effect of viscosity on microstructure characteristic in rheological behaviour of wrought aluminium alloys by compression and stirring process. *Mater. Sci. Techol.* **2010**, *26*, 20–30. [CrossRef]

22.   Chang, D.Y.; Kwang, S.S. Semi-Solid Processing of Magnesium Alloys. *Mater. Trans.* **2003**, *44*, 558–561. [CrossRef]

23.   Chen, H.I.; Chen, J.C. Thixotropic Behavior of Semi-Solid Magnesium Alloy. *Solid State Phenom.* **2006**, *116*, 648–651. [CrossRef]

24.   Kramer, M.; Jenning, R.; Lohmüller, A.; Hilbinger, M.; Randelzhofer, P.; Singer, R.F. Characterization of magnesium alloys for semi solid processing. In Proceedings of the 8th International Conference on Magnesium Alloys and Their Applications, Weimar, Germany, 26–29 October 2009; Kainer, K.U., Ed.; Wiley-VCH Verlag GmbH & Co. KGaA: Weinheim, Germany, 2009; pp. 376–383.

25.   Chen, H.-I.; Chen, J.-C.; Liao, J.-J. The influence of shearing conditions on the rheology of semi-solid magnesium alloy. *Mater. Sci. Eng. A Struct.* **2008**, *487*, 114–119. [CrossRef]

26.   Hu, Y.; He, B.; Yan, H. Rheological behavior of semi-solid Mg2Si/AM60 magnesium matrix composites at steady state. *Trans. Nonferr. Metal. SOC* **2010**, *20*, 883–887. [CrossRef]

27. Li, L.; Zheng, M. Theoretical research on rheological behavior of semisolid slurry of magnesium alloy AZ91D. *Comp. Mater. Sci.* **2015**, *102*, 202–207. [CrossRef]

28. Yan, H.; Rao, Y.; Chen, G. Rheological behavior of semi-solid AZ91D magnesium alloy at steady state. *J. Wuhan Univ. Technol.* **2015**, *30*, 162–165. [CrossRef]

29. Korolczuk-Hejnak, M.; Migas, P. Analysis of Selected Liquid Steel Viscosity. *Arch. Metall. Mater.* **2012**, *57*, 963–969. [CrossRef]

30. Korolczuk-Hejnak, M. Determination of Flow Curves on Selected Steel Grades in their Liquid State. *Arch. Metall. Mater.* **2014**, *59*, 1553–1558. [CrossRef]

31. Ślęzak, W.; Korolczuk-Hejnak, M.; Migas, P. High Temperature Rheometric Measurements of Mould Powders. *Arch. Metall. Mater.* **2015**, *60*, 289–294. [CrossRef]

32. Korolczuk-Hejnak, M. *Determination of the Dynamic Viscosity Coefficient Value of Steel Based on the Rheological Measurements*; Wydawnictwa AGH: Kraków, Poland, 2014; ISBN 978-83-7464-720-5.

33. Ostwald, W. Ueber die Geschwindigkeitsfunktion der Viskositat Disperser Systeme. *Kolloid Z.* **1925**, *36*, 99–117. [CrossRef]

34. Mezger, G.T. *The Rheology Handbook: For Users of Rotational and Oscillatory Rheometers*, 2nd ed.; Vincentz Network: Hannover, Germany, 2006; ISBN 3878701748.

35. Wiśniowski, R.; Skrzypaszek, K. Analiza modeli reologicznych stosowanych w technologiach inżynierskich. *Wiertnictwo Nafta Gaz* **2006**, *23*, 523–532.

36. Ślęzak, M. Mathematical Models for Calculating the Value of Dynamic Viscosity of a Liquid. *Arch. Metall. Mater.* **2015**, *60*, 581–589. [CrossRef]

37. Korolczuk-Hejnak, M. Empiric Formulas for Dynamic Viscosity of Liquid Steel Based on Rheometric Measurements. *High Temp.* **2014**, *52*, 667–674. [CrossRef]

38. Eskin, D. Modeling non-Newtonian slurry convection in a vertical fracture. *Chem. Eng. Sci.* **2009**, *64*, 1591–1599. [CrossRef]

39. Eskin, D.; Miller, M.J. A model of non-Newtonian slurry flow in a fracture. *Powder Technol.* **2008**, *182*, 313–322. [CrossRef]

40. Eskin, D. Modeling non-Newtonian slurry flow in a flat channel with permeable walls. *Chem. Eng. Sci.* **2015**, *123*, 116–124. [CrossRef]

41. Bührig-Polaczek, A.; Afrath, C.; Modigell, M.; Pape, L. Comparison of rheological measurement techniques for semi-solid aluminium alloys. *Solid State Phenom.* **2006**, *116*, 610–613. [CrossRef]

42. Modigell, M.; Pape, L.; Maier, H.R. Rheology of Semi-Solid Steel Alloys at Temperatures up to 1500 °C. *Solid State Phenom.* **2006**, *116*, 606–609. [CrossRef]

Article

Effect of Segregation and Surface Condition on Corrosion of Rheo-HPDC Al–Si Alloys

Maryam Eslami [1], Mostafa Payandeh [2], Flavio Deflorian [1], Anders E. W. Jarfors [2] and Caterina Zanella [2,*]

[1] Department of Industrial Engineering, University of Trento, 38123 Trento, Italy; maryam.eslami@unitn.it (M.E.); flavio.deflorian@unitn.it (F.D.)
[2] Department of Materials and Manufacturing, School of Engineering, Jönköping University, 55111 Jönköping, Sweden; mostafa.payandeh@ju.se (M.P.); anders.jarfors@ju.se (A.E.W.J.)
*   Correspondence: caterina.zanella@ju.se; Tel.: +46-36-10-1691

Received: 22 February 2018; Accepted: 23 March 2018; Published: 24 March 2018

Abstract: Corrosion properties of two Al–Si alloys processed by Rheo-high pressure die cast (HPDC) method were examined using polarization and electrochemical impedance spectroscopy (EIS) techniques on as-cast and ground surfaces. The effects of the silicon content, transverse and longitudinal macrosegregation on the corrosion resistance of the alloys were determined. Microstructural studies revealed that samples from different positions contain different fractions of solid and liquid parts of the initial slurry. Electrochemical behavior of as-cast, ground surface, and bulk material was shown to be different due to the presence of a segregated skin layer and surface quality.

Keywords: Al–Si alloys; rheocasting; HPDC; electrochemical evaluation

1. Introduction

High pressure die casting (HPDC) is one of the most used manufacturing process for light alloy components [1,2], due to its high productivity, capability to cast complex geometry, dimensional accuracy, reduced need for finishing operations, and producing component with fine grain microstructure and good mechanical properties [3–5].

Coupling semisolid metal (SSM) casting to HPDC (SSM-HPDC) is a promising technology to produce high quality components with sound microstructure. Higher viscosity of semisolid material in this process reduces air entrapment and consequent porosity in the component. Such a technology introduces a new opportunity to enhance the castability of a component, which is impossible to achieve by traditional manufacturing methods [2,3,6,7].

There are two kinds of semisolid processes: "rheocasting" and "thixocasting". In 1976, Flemings et al. [8] introduced rheocasting as an alternative manufacturing process to die casting and even forging, and suggested it can be used to prepare high quality parts. Rheocasting, the method which is used in the current research, involves shearing force during a first solidification to produce a slurry. The slurry is then transferred into a mold and solidifies with non-dendritic microstructure [9]. Rheocasting can be coupled with high pressure die casting and Rheo-HPDC parts have a globular microstructure and usually show low porosity. This leads to heat treatability and high performance [2,6]. Proper materials for Rheo-HPDC process are limited to those which have good castability with HPDC process and also have low sensitivity of the solid fraction to variations of temperature [10]. Hypoeutectic Al–Si alloys in the range of 5–8% silicon content are suitable choices for this process [11].

While Rheo-HPDC widens the composition range of castable alloys and allows for casting of thin sections, it is a process which increases the inhomogeneity of the microstructure at the macroscale in the final component, in comparison to the conversional HPDC process [5]. This phenomenon arises from the fact that the primary α-Al phase solidifies at higher temperature during slurry preparation, and is characterized with low solubility of alloying elements, therefore, these elements will be higher in the remnant liquid phase [11]. In addition, macrosegregation is formed during the filling process: solid and liquid fractions tend to separate in the gating system (longitudinal macrosegregation) and from the surface to the core of the component (transverse macrosegregation), increasing the microstructure inhomogeneity of the final component. This leads to variations of properties in different locations of the component either in microscopic or macroscopic scale [12–14]. The liquid will preferentially fill the furthest parts of the mold, or the thinner sections, while the solid fraction will concentrate in the core. This is due to the higher viscosity of the slurry compared to the liquid molten metal during casting [15]. As a consequence, higher yield strength and ultimate tensile strength are developed near the vent, where more liquid fraction and refined grains solidify compared to the region near the gate [11].

Park et al. [16] showed that the tensile elongation of thixoformed 357-T5 semi solid aluminum alloy is strongly affected by the α_1-Al volume fraction which changes in different locations of the component.

Masuku et al. [17–19] investigated the corrosion behavior of the surface layer of SSM-HPDC 7075-T6 and 2024-T6 alloys in sodium chloride solution. They observed a surface liquid segregated layer (mainly formed by the eutectic) in all the alloys, with higher amounts of alloying elements (such as copper). They did not report any difference between the pitting potential, however, they mentioned that pitting morphology is affected by the amount and distribution of the intermetallic particles and therefore differences are expected between the SSM-HPDC and wrought alloys [18]. How longitudinal macrosegregation affects the corrosion resistance of semisolid aluminum alloys and the behavior of the as-cast surface is still not studied.

Limitation of scientific and technical knowledge makes it essential to evaluate properties and behavior of aluminum alloys produced by means of Rheo-HPDC process under different operational circumstances. Corrosion resistance is a critical property for Al alloys, especially in outdoor applications, and therefore, is an interesting subject either for researchers or industries.

Many authors have investigated corrosion resistance of Al–Mg–Si (6xxx series) [20–28] and Al–Si alloys [29–35]. Regarding Al–Mg–Si alloys, pitting and intergranular corrosions (IGC) have been reported as localized corrosion features [23,27]. IGC susceptibility is especially influenced by the amount of copper, iron, and Mg/Si ratio in the alloy composition [23,24,27,36,37]. The localized corrosion occurs in the presence of phases such as β-phase (Mg$_2$Si), silicon (in alloys with excess Si) and copper-containing phases such as Q-phase (Al$_4$Cu$_2$Mg$_8$Si$_7$) (in alloys with Cu) [38]. Iron-rich intermetallics in Al–Mg–Si alloys are nobler compared to the matrix [21]. Nobler intermetallic particles (IMs) in grain boundaries form a microgalvanic couple with the adjacent precipitate free zones (PFZ), and result in IGC [37]. In chloride containing solutions, Mg$_2$Si undergoes Mg dealloying, before turning to an active cathodic site [21,22].

Al–Si alloys generally suffer from the localized corrosion (pitting) in the Al–Si eutectic, due to impurities, such as Fe [39]. Generally, corrosion behavior of these alloys depends on the amount and morphology of iron-rich IMs, such as β-AlFeSi and π-AlFeSiMg [34]. Both Fe and Si are cathodic with respect to the aluminum. Therefore, together they can form a microgalvanic couple, resulting in localized corrosion [34]. Silicon also increases the corrosion potential [23,27].

Previous corrosion studies of semisolid-cast Al-Si alloys such as those performed by Yu et al. [40], Park et al. [31], Tahamtan et al. [30,35] and Arrabal et al. [34], have mostly emphasized on the pitting corrosion in the eutectic regions of A356 and A357 alloys.

It is shown that semisolid-cast process, such as thixoforming, can effectively modify the Si morphology in 357 alloy, resulting in a higher corrosion resistance [40]. The acicular eutectic silicon phase in permanent mold cast 357 alloy has more contact area with the aluminum matrix in comparison to that of the globular eutectic silicon in the thixoformed alloy, and this can encourage the galvanic corrosion [40].

Rheocast process can increase the concentration of silicon in α-Al particles in A356 aluminum alloy. This leads to smaller potential differences between this phase and the eutectic silicon phase and β-AlFeSi IM particles, which result in higher resistance to pitting corrosion [34]. Although both of the eutectic silicon phase and iron-rich IMs contribute in the localized corrosion, some authors consider the contribution of IM particles to be more important [34,41].

The present study focused on the Al low Si alloys produced by Rheo-HPDC process and on the effect of the microstructure inhomogeneity on the corrosion resistance. In addition, the influence of silicon content and surface condition on the microstructure characteristics and the corrosion behavior is examined.

2. Materials and Methods

A telecom cavity filter (Figure 1) was rheocast by a 400 ton HPDC machine equipped with an automated RheoMetal™ slurry generator. The RheoMetal™ (Stockholm, Sweden) process uses an Enthalpy Exchange Material (EEM) as slurry generator [42].

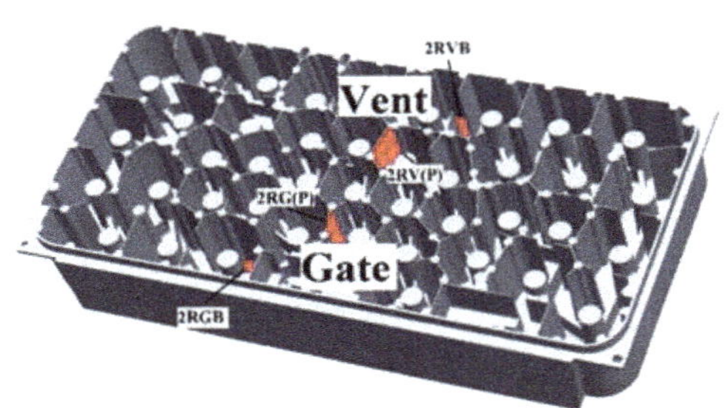

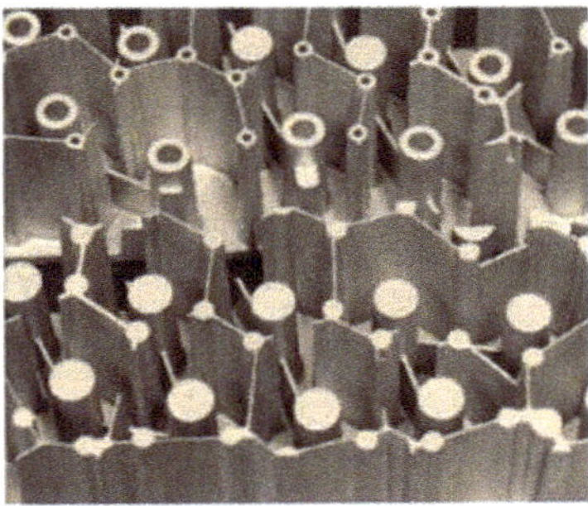

Figure 1. The experimental cavity filter used for Rheo-high pressure die cast (HPDC) process.

The alloys were prepared, and their composition were adjusted by adding pure silicon to a primary alloy in a resistance furnace. The chemical compositions of the alloys were measured by optical emission spectroscopy (OES) (SPECTRO Analytical Instruments GmbH, Kleve, Germany) and are presented in Table 1.

Table 1. Measured composition (wt %) of alloys.

Name	Si	Fe	Cu	Mn	Mg	Zn	Al
Alloy 2.5	2.41	0.462	0.131	0.019	0.58	0.038	96.338
Alloy 4.5	4.50	0.481	0.137	0.019	0.58	0.035	94.223

In the casting process, the temperature of the fixed half of the die was maintained at 230–250 °C, while the temperature of the moving half was set to 280–320 °C. Shots of about 5 kg were held at 675 °C in the ladle where 5% of EEM was added under stirring at 900 rpm. The final slurry temperature was 610 °C and had 40% of solid fraction. The die was filled in two stages with piston speed at 0.23 and 5.2 m/s, respectively, and the shot time was 31 ms. The solid fraction in the slurry was estimated by image analyses of a quenched sample of slurry after light etch using a 5% NaOH.

Specimens for corrosion tests were taken from the thin walls (thickness ≈ 1.5 mm), in as-cast or ground condition, or from the thick bottom plate (thickness ≈ 4 mm), underneath the component only in ground condition.

To investigate longitudinal segregation, both thin wall and thick bottom plate samples were taken from different locations: near the feeding gate (G) or near the die vent (V). The detailed samples designation is presented in Table 2.

Table 2. Designation of samples.

Name	Section	Near the Gate	Near the Vent
Alloy 2.5	Thin wall surfaces	2.5 RGP	2.5 RVP
	Plate bulk	2.5 RGB	2.5 RVB
Alloy 4.5	Thin wall surface	4.5 RGP	4.5 RVP
	Plate bulk	4.5 RGB	4.5 RVB

R = Rheo-HPDC, G = Samples from near the gate, V = Samples from near the vent, P = Thin wall samples in ground surface condition, B = Samples from the thick bottom plate ground to the half of thickness (Bulk samples).

Polarization tests and electrochemical impedance spectroscopy (EIS) were performed with a 3-electrode cell and a potentiostat (Parstat 2273, Ametek, Berwyn, PA, USA). The aluminum alloy with an exposure area of 1 cm^2 was connected as working electrode, platinum as a counter and silver/silver chloride (Ag/AgCl·3M·KCl) as reference electrodes. Due to the working environment of telecom components, diluted Harrison solution (0.5 g/L NaCl and 3.5 g/L (NH$_4$)$_2$SO$_4$) was used to simulate the electrochemical behavior of the alloys exposed to an acid rain [43,44].

Regarding the polarization test, the sweep rate was 0.166 mV/s, and the delay time before each the test was 600 s, to let the open circuit potential (OCP) reaches its stable value. For each sample, mainly anodic branch was collected (as cathodic branch did not exhibit any significant information), starting from OCP. Anodic polarization was stopped after the maximum current density of 9×10^{-4} A/cm^2 was reached. In order to highlight the effect of chloride ions, all the polarization experiments were also repeated using a solution of 3.5 g/L (NH$_4$)$_2$SO$_4$.

EIS measurements were collected for 24 h of immersion at the room temperature in the diluted Harrison solution from 100 kHz to 10 MHz with 36 points and 10 mV of amplitude of the sinusoidal potential.

EIS measurements were conducted on as-cast and ground surfaces of thin wall samples, and also on ground surface of the thick bottom plate samples, to investigate the effect of transverse macrosegregation and as-cast condition.

To distinguish the results of these different experiments, the letter P is added to the name of the samples of thin walls, which were ground before the electrochemical test. These samples were wet ground by silicon carbide abrasive papers from P600 to P4000 to the extent that the skin layer was completely removed. In the case of bulk samples (of the thick bottom plate), the surfaces were ground until the middle of each sample was reached. In this way, the electrochemical behavior of the bulk of each alloy can be investigated. Repeatability of the results was tested by conducting each experiment on at least three specimens. ZsimpWin^{TM} software (3.5, Echem software, Ann Arbor, MI, USA, 2013) was used to fit the EIS spectra. After the corrosion tests, corroded surfaces were examined using SEM/EDS (JSM-IT300) (Jeol, Akishima, Japan).

For metallographic analyses, samples were wet ground, followed by polishing using diamond paste (3 μm and 1 μm). NaOH solution (10 wt %) was used to clean the surface and to reveal the constituents of the microstructure. The microstructure of the surfaces and the bulk of components were studied using a light optical microscope (LOM) (Zeiss, Oberkochen, Germany) and scanning electron microscopy (SEM) (Jeol, Akishima, Japan). Energy/wave-dispersive X-ray spectroscopy (EDS/WDS) (EDAX, Mahwah, NJ, USA) was used to measure the composition of the different phases.

3. Results and Discussion

3.1. Microstructural Features

Generally, the microstructure of rheocast low silicon content aluminum alloys exhibit the presence of α-Al phase together with Al–Si eutectic mixture and some intermetallic particles [45]. Based on the presence of Fe in the two alloys (Table 1), and since its solubility in Al is very low [46], the presence of Fe-rich intermetallic particles is expected. The sequence of phase formation (aluminum phase and eutectic reaction) was calculated using ThermoCalc^{TM} (2015b, Solna, Sweden, 2015) software [47,48]. The results are shown in Figure 2. As it is predicted by the thermodynamic model, the needle shape β-AlFeSi intermetallic particle was the most favored intermetallic phase for precipitation, and it is formed before the eutectic silicon.

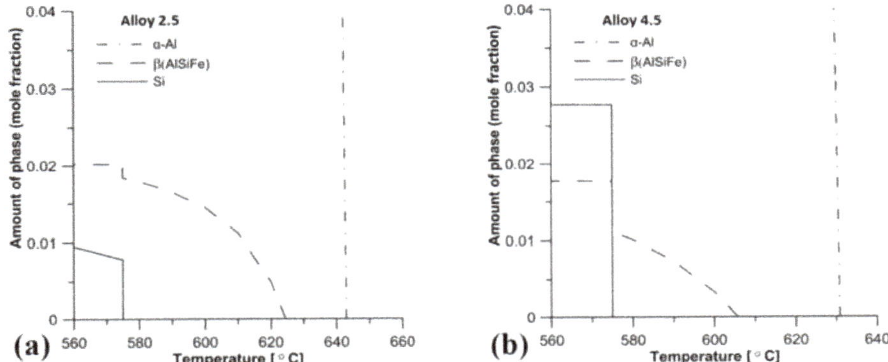

Figure 2. Sequence of formation of different phases in the (**a**) Alloy 2.5 and (**b**) Alloy 4.5.

The microstructural features of the polished surfaces of thin walls from different positions in the cavity, with different percentage of silicon, are illustrated in LOM images in Figure 3. From microstructure images in Figure 3, two different range sizes of α-Al phase can be observed: a coarse globular α-Al phase ($α_1$-Al) and a finer α-Al phase ($α_2$-Al).

The formation of $α_1$-Al and $α_2$-Al is related to the multi-stage solidification in the semisolid metal process. $α_1$-Al grains are nucleated in the ladle, under shear forces, due to stirring at a higher temperature due to contact with the EEM and form the slurry, while $α_2$-Al grains are mostly formed in the solidification stage inside the die, at a higher cooling rate [49].

Regarding the effect of the position, the microstructure near the gate (Figure 3a,c) consists of a higher amount of $α_1$-Al particles compared to the region near to the vent (Figure 3b,d). However, this difference is not significant in alloy 2.5.

This microstructure heterogeneity is attributed to the distribution among the die of the liquid and solid fraction of the slurry during the injection stage: the liquid part squeezes out and leaves the solid fraction behind near to the gate. Easton et al. [50] show this behavior in a SSM-HPDC component and defined this type of separation of the slurry as sponge effect.

$α_2$-Al particles in thin wall samples (of both alloys) are finer near the vent in comparison to the region near the gate, due to the higher undercooling in this region of the cavity. This effect is more evident for alloy 4.5. It seems that by increasing the amount of silicon, aluminum phase is refined by undercooling, which can be due to more nucleation of the eutectic silicon [51].

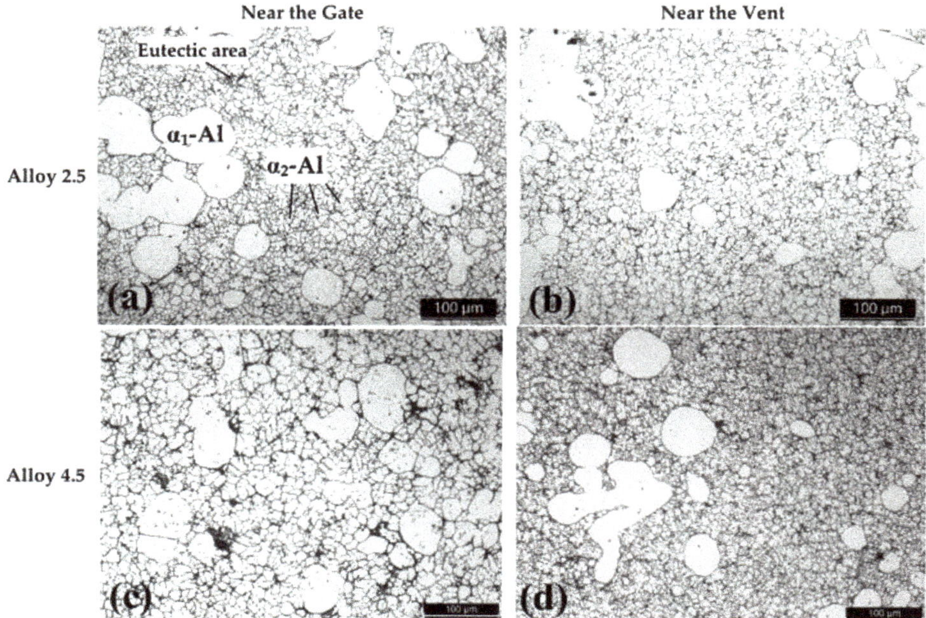

Figure 3. Light optical microscope (LOM) images of sample (**a**) 2.5 RGP; (**b**) 2.5 RVP; (**c**) 4.5 RGP and (**d**) 4.5 RVP.

LOM images of sample 2.5 RGB and 4.5 RGB from the thick plate are presented in Figure 4. In comparison to the microstructure of thin walls, the bulk microstructure contains more α_1-Al particles in both of the alloys, which is expected. In fact, due to the high viscosity of semisolid slurry, thin walls are mostly filled by liquid, while the relatively thicker parts are filled by the solid fraction of the slurry [15].

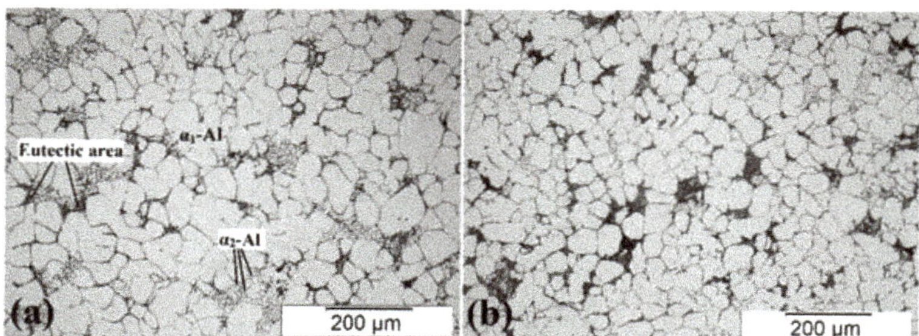

Figure 4. LOM image of sample (**a**) 2.5 RGB and (**b**) 4.5 RGB.

LOM image of cross-section of the thick plate of Alloy 2.5 in the region near the gate is reported in Figure 5. On the surface, there is a segregation of the liquid fraction to the surface and α_1-Al solid fraction, which tend to aggregate in the center of samples. This phenomenon is considered as transverse macrosegregation. Three different phenomena lead to transverse macrosegregation: skin effect [12], sponge effect [13,50] and shearing band during melt flow that lead to porosity or eutectic-rich segregation band [52]. Study of transverse macrosegregation in different positions of the same Rheo-HPDC component performed by Payandeh et al. [53] showed that the thickness of the surface segregation layer increases by increasing the liquid faction. Govender et al. [54] also reported the existence of a surface layer consisting of mainly liquid or eutectic phase in SSM-HPDC A356 Alloy.

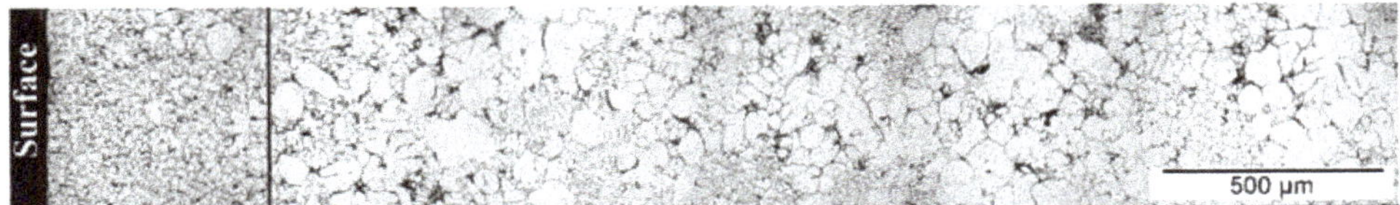

Figure 5. LOM images of cross-sectional view of sample 2.5 RGB.

Figure 6 compares the as-cast and polished surfaces of sample 2.5RG/P of the thin wall section. It is noticeable how the as-cast surface is enriched with eutectic phase and intermetallic particles. Moreover, localized defects, such as porosities and/or voids among the grains and inclusions, are visible.

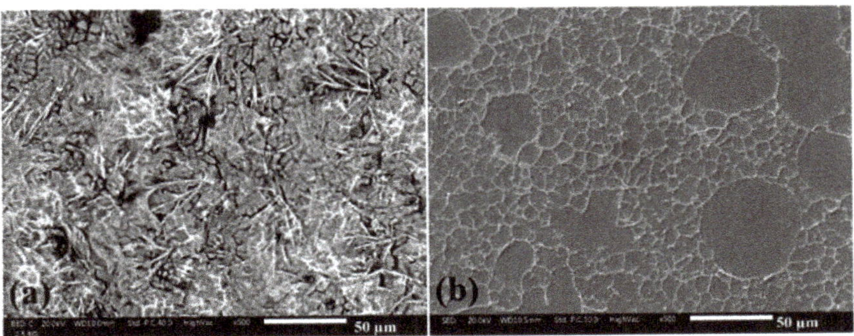

Figure 6. Scanning electron microscopy (SEM) image of sample (**a**) 2.5 RG (as-cast condition) and (**b**) 2.5 RGP (polished surface).

SEM image of sample 4.5 RGP at higher magnification in Figure 7a depicts the microstructure of the eutectic and the intermetallic particles. The Si particles have a flake shape and form a continuous network [55].

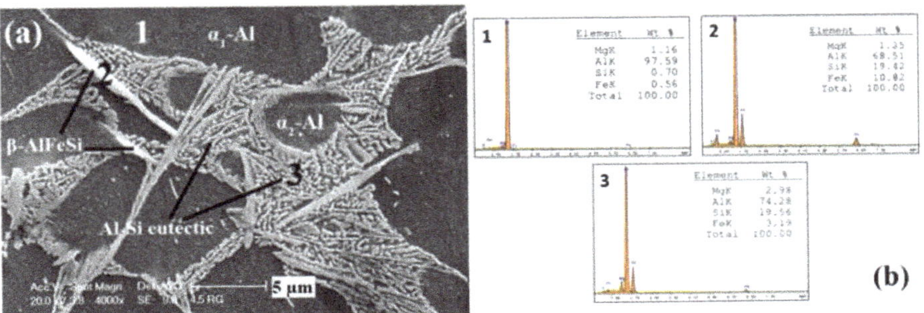

Figure 7. SEM image (**a**) and spot EDS analysis (**b**) of polished surface microstructure of the sample 4.5 RGP.

According to the EDS results in Figure 7b, in this figure, intermetallic particles are rich in iron, and as is expected in hypoeutectic alloys, the intermetallic particles are β-AlFeSi. These intermetallic particles usually have a needle shape [34], and a platelet morphology in 3D tomographic volume [56].

Figure 8 represents the concentration of silicon at the center of α_1-Al and α_2-Al grains in alloys 2.5 and 4.5. α_2-Al particles, nucleated from the liquid portion of the slurry [57], have higher amounts of silicon, which leads to smaller potential differences between them and the silicon eutectic phase and/or iron-rich intermetallic particles, and results in less severe corrosion [34].

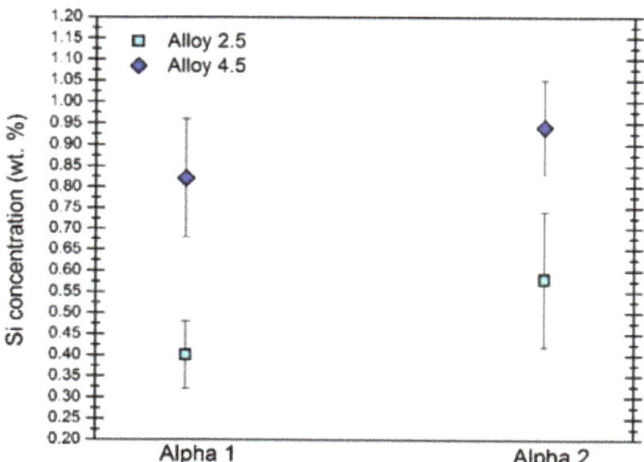

Figure 8. Concentration of Si in α_1 and α_2-Al phases in alloy 2.5 and 4.5.

3.2. Corrosion Studies

3.2.1. Potentiodynamic Polarization Curves

Potentiodynamic polarization curves of as-cast surface of the thin wall samples (2.5 RG, 2.5 RV, 4.5 RG, and 4.5 RV) in the diluted Harrison solution and in the solution of 3.5 g/L $(NH_4)_2SO_4$ are reported in Figure 9.

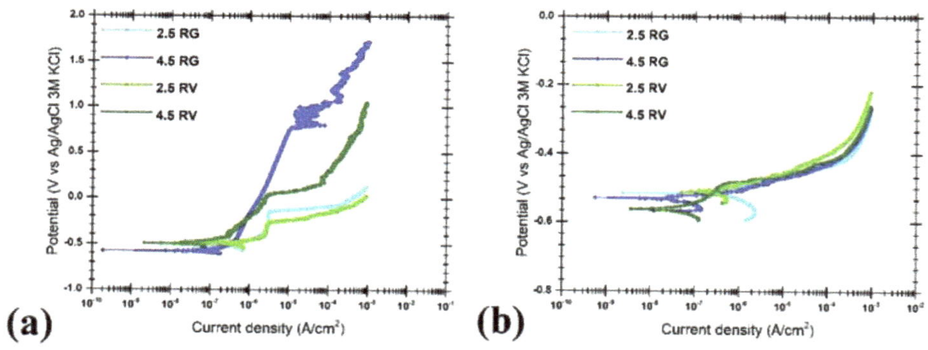

Figure 9. Potentiodynamic polarization curves in (**a**) the solution of 3.5 g/L $(NH_4)_2SO_4$ and (**b**) the diluted Harrison solution.

In both of the solutions, values of corrosion potential are similar for different samples regardless of the position and the silicon content. Regarding the effect of the position, samples from near the gate

(2.5 RG and 4.5 RG) possess higher pitting potentials, in the solution of 3.5 g/L $(NH_4)_2SO_4$, compared to the samples taken from near the vent (2.5 RV and 4.5 RV). In the diluted Harrison solution, all of the samples (except for 4.5 RV) show pitting from the OCP, which is due to the chloride ions and the higher amount of iron-rich intermetallic particles, as well as the defective condition of the as-cast surfaces.

Considering the bulk microstructure, the results of potentiodynamic polarization test in the diluted Harrison solution and in the solution of 3.5 g/L $(NH_4)_2SO_4$, of samples of 2.5 RGB and 4.5 RGB, are presented in Figure 10. In both of the solutions, the corrosion potentials of the two samples shows no significant difference.

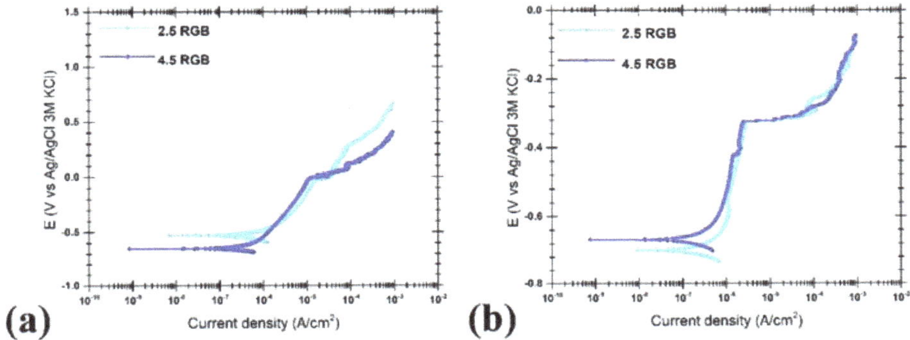

Figure 10. Potentiodynamic polarization curves in (**a**) the solution of 3.5 g/L $(NH_4)_2SO_4$ and (**b**) the diluted Harrison solution.

The stability of the passive oxide layer is higher for the bulk samples compared to the as-cast surface, especially in the chloride containing solution. This is due to the surface condition, since the bulk samples have higher α_1-Al fraction, less eutectic fraction, and are ground while the thin wall samples are tested in as-cast condition.

SEM images of the corroded surfaces of samples 2.5 RG and 4.5 RG after the polarization test in the diluted Harrison solution are presented in Figure 11. For both samples, corrosion is localized, and a galvanic couple between the eutectic silicon phase and/or iron-rich intermetallic particles and aluminum matrix can be observed.

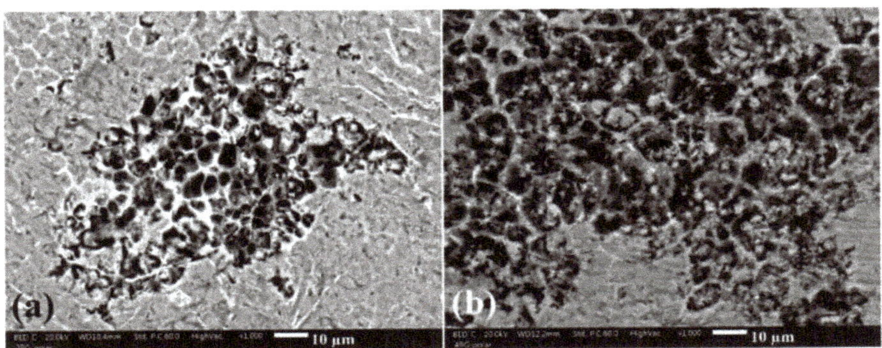

Figure 11. Corroded surfaces of sample (**a**) 2.5 RG and (**b**) 4.5 RG after polarization test in the diluted Harrison solution.

3.2.2. Electrochemical Impedance Spectroscopy

EIS spectra were obtained during 24 h of immersion in the diluted Harrison solution. The Bode plots of EIS spectra of thin wall samples in as-cast condition are reported in Figure 12. The figure also compares the effect of different positions in the cavity and the silicon content in the alloy.

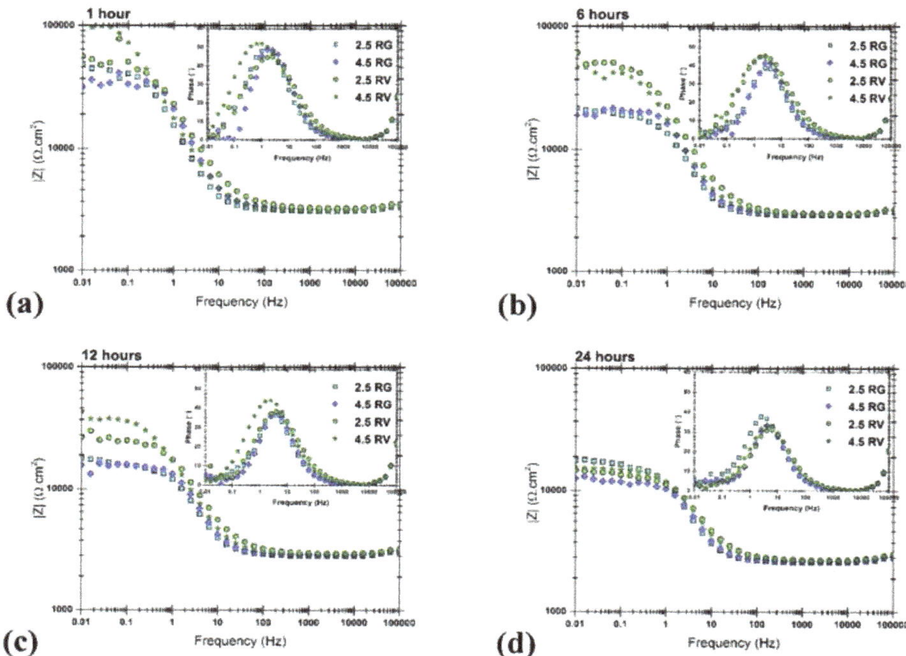

Figure 12. Bode plots of EIS spectra of samples 2.5 RG, 4.5 RG, 2.5 RV, and 4.5 RV after (**a**) 1 h; (**b**) 6 h; (**c**) 12 h; and (**d**) 24 h of immersion in the diluted Harrison solution.

For all of the samples, the impedance values at low frequencies decrease with the immersion time, which indicates the progressive corrosion process on the surface. In addition, the phase angle peak depresses through the immersion time, suggesting that the pitting corrosion activity is increasing [58].

Regarding the effect of position, samples taken from near the vent almost always, during the 24 h of immersion, show slightly higher impedance values at low frequencies, compared to the samples taken near the gate, except for 24 h, when all the samples show the same values. As it was discussed before, according to LOM images (Figure 3), the samples, which are taken from near the vent, possess a higher fraction of α_2-Al particles, and they also have a finer microstructure compared to the samples which are taken from near the gate.

Concerning the effect of silicon, no noticeable difference is detectable. However, all the previous research has indicated the positive effect of silicon on the pitting resistance of aluminum alloys [29,59,60]. Nevertheless, it is worth mentioning that these studies are focused on silicon content higher than 6%, which is higher than the percentage of silicon in both of our alloys.

To remove the effect of as-cast surface quality, the thin wall samples were ground using SiC abrasive papers to the extent that the skin layer was totally removed. The samples were then monitored by EIS during 24 h of immersion in the same solution. Bode presentation of EIS spectra of these samples are presented in Figure 13. According to these spectra, the ground thin wall samples show one order of magnitude higher impedance values at low frequencies compared to the same samples in as-cast condition. In addition, the impendence values increased during 24 h of immersion for all of these samples. This can be related to the presence of a protective oxide, which is more stable to pitting. This protective oxide later is provided by better finishing quality on the ground surfaces. By grinding, the skin layer is removed, and a surface containing more α_1-Al particles and fewer intermetallic particles (Figure 6) is exposed to the corrosive solution.

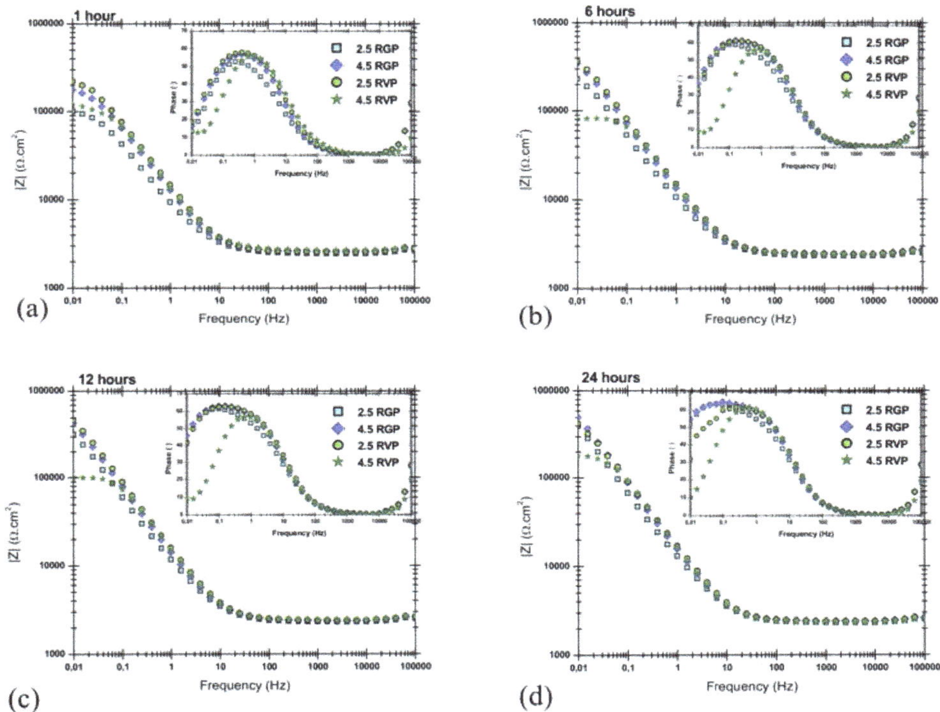

Figure 13. Bode plots of EIS spectra of samples 2.5 RGP, 4.5 RGP, 2.5 RVP, and 4.5 RVP after (**a**) 1 h; (**b**) 6 h; (**c**) 12 h and (**d**) 24 h of immersion in the diluted Harrison solution.

The difference between the corrosion performances of samples from different positions is negligible in this condition. Impedance tends slightly to increase after 6 h, and no pits are developed after 24 h. This proves that the poor behavior of as-cast surface is due to the poor surface quality and the higher amounts of intermetallic particles.

EIS spectra of the samples taken from the thicker plate, which are ground to the middle to expose the bulk microstructure, are reported in Figure 14. These results indicate the growth of a protective oxide layer on the surface.

The equivalent circuits used to describe the electrochemical responses of the samples 2.5 RG(P/B), 4.5 RG(P/B), 2.5 RVP, and 4.5 RVP are shown in Figure 15.

The electrochemical behavior of aluminum surface is affected by the presence of the passive oxide layer and the interface between the intermetallic particles and the aluminum. Although all these constituents are present, their time constants strongly overlap, or one dominates. Therefore, only one peak in the phase diagram can be observed for all of them.

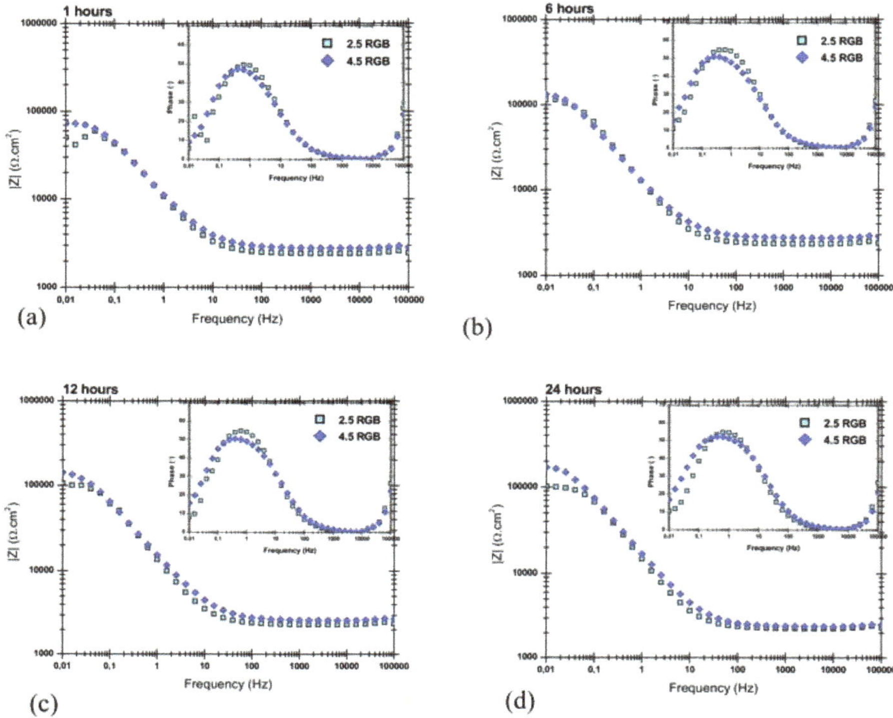

Figure 14. Bode plots of EIS spectra of samples 2.5 RGB, 4.5 RGB after (**a**) 1 h; (**b**) 6 h; (**c**) 12 h; and (**d**) 24 h of immersion in the diluted Harrison solution.

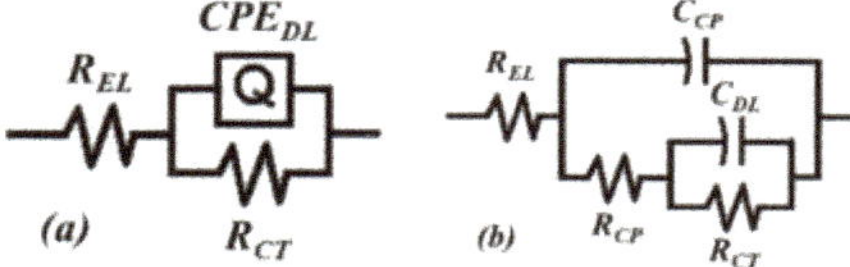

Figure 15. Equivalent circuits for the studied alloy after (**a**) a short time and (**b**) a longer time of immersion in the diluted Harrison solution.

The oxide layer is represented by a parallel circuit containing a resistor and a capacitor representing, respectively, the oxide ionic conduction and its dielectric properties [61].

The circuit in Figure 15a with one time constant has been used for the first 6 h of immersion, for almost all the samples, while the circuit in Figure 15b has been used for the later hours of immersion, when a second time constant was visible in the EIS spectrum. In the circuits depicted in Figure 15a,b, R_{EL} represents the resistance of the electrolyte. Q_{DL} (or C_{DL}) and R_{CT} stand for capacitive behavior of the electrical double layer at the interface between the surface and the solution, and for the resistance against the charge transfer (or polarization resistance), respectively. The second time constant at longer immersion time shows the oxidation of aluminum, and can represent the oxide layer (for ground surface of thin wall and thick plate samples). It can also stand for the corrosion products formed due to the localized corrosion attack in the case of thin wall samples in as-cast condition. R_{CP} and C_{CP} stand for the resistance and capacitive behavior of the oxide layer (or the corrosion products).

The fitting results of the two resistances (R_{CT} and R_{CP}) for thin wall samples, ground and in as-cast condition, and ground thick plate samples, are presented in Figures 16–18, respectively.

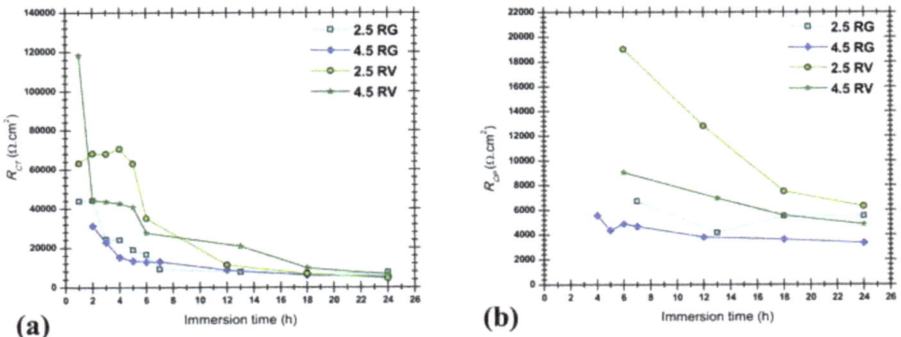

Figure 16. Fitted EIS parameters of as-cast thin wall samples: (**a**) R_{CT} and (**b**) R_{CP}.

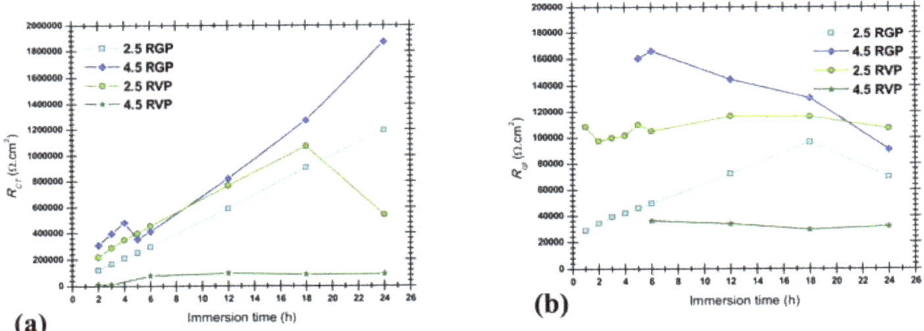

Figure 17. Fitted EIS parameters of ground thin wall samples: (**a**) R_{CT} and (**b**) R_{CP}.

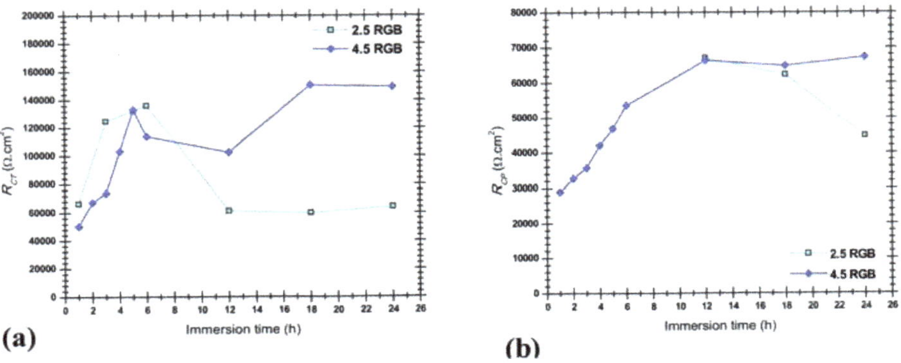

Figure 18. Fitted EIS parameters of ground thick plate (bulk) samples: (**a**) R_{CT} and (**b**) R_{CP}.

Values of R_{CT} of thin wall samples in as-cast condition (Figure 16a) decrease with the immersion time for all the samples, which indicates an increase of the activity and a higher corrosion rate in the localized attack [62]. It should be noted that this decrease is immediate for the samples taken from near the gate, while for the samples taken from near the vent, it starts after 5–6 h of immersion. These values are higher for the samples near the vent, compared to the samples near the gate. Regarding the effect of silicon content, samples with lower amounts of silicon show slightly higher R_{CT} values at the very first hours of immersion. The values of R_{CP} (Figure 16b), that for these samples represent the resistance of corrosion products in the pits, are very low, and slightly decrease for all the samples due to the corrosion attack.

Regarding the ground thin wall samples, R_{CT} values are generally one order of magnitude higher compared to the thin wall samples in as-cast condition. The values increase during immersion time for all the samples (Figure 17a). The values of R_{CT} for different samples, with regard to the position, and with different amounts of silicon, are close to each other. In the case of ground samples, values of R_{CP} represent the presence of the passive oxide layer. They remain almost constant for all of the samples, and show a similar trend.

In the case of ground thick plate (bulk) samples, reported in Figure 18, the behavior of 2.5 RGB and 4.5 RGB samples are very similar. After some initial fluctuation, values of R_{CT} remain almost constant for the two samples from 12 to 24 h of immersion, but never reach the high resistance showed by the ground thin wall samples. R_{CP} values increase from the first hour of immersion for sample 4.5 RGB.

SEM micrographs and map analysis of some selected corroded surfaces are shown in Figure 19. As clearly visible in Figure 19, corrosion is mainly localized in all the samples, and takes place especially in the eutectic region, at the interface between the silicon and aluminum and at the interface of aluminum grains and iron-rich intermetallic particles. These results are in accordance with results of other researchers about corrosion of Al–Si alloys in similar solutions [34,35].

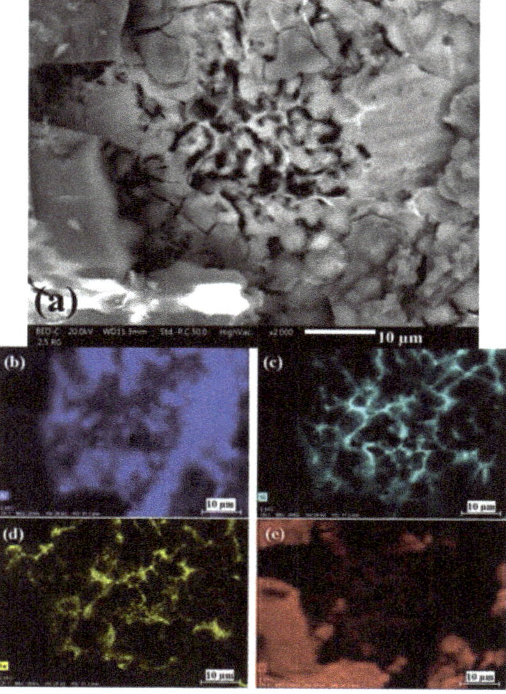

Figure 19. (**a**) SEM micrograph and elemental map analysis of corroded surface of sample 2.5 RG; (**b**) Al; (**c**) Si; (**d**) Fe; and (**e**) O.

4. Conclusions

In this paper, the corrosion behavior of the alloy in different parts of the component geometry were compared, and the microstructure macrosegregation due to Rheo-HPDC process was evaluated in terms of corrosion properties for two aluminum alloys containing 2.5 wt % and 4.5 wt % silicon. These alloys are prone to localized corrosion in the eutectic region at the interface between iron-rich intermetallic particles and the aluminum. Therefore, segregation of these phases influences the electrochemical behavior of the component. It was shown that the samples taken from different positions and different parts with various thicknesses differ in the amount of α_1-Al and α_2-Al particles. Samples also show a transverse macrosegregation. α_1-Al particles which are the solid fraction of the slurry, tend to segregate at the center of the samples, while the surface is richer in α_2-Al. Longitudinal segregation induces higher fraction of α_1-Al particles in the area nearer to the higher fraction of α_2-Al and the eutectic phase near the vent.

Both kinds of these segregations were shown to have influence on the corrosion behavior. A big change in the corrosion resistance was shown by grinding the samples. This is due to the as-cast surface morphology and composition, and to higher concentrations of intermetallic particles on the surface. In as-cast surface conditions, samples with higher amounts of α_2-Al and finer microstructure (near the vent) show slightly higher corrosion resistance.

Differently to most of the previous papers, the as-cast surface was tested. It was shown how the surface segregation and the increase of intermetallic particles on the very surface influence the pitting resistance of the component. Both ground thin wall and thick plate samples possess better corrosion resistance compared to the thin wall samples in as-cast condition. This improvement is due to the better surface condition. Nevertheless, relatively thinner samples show a higher corrosion resistance compared to the thicker samples. Surfaces of the rheocast component present a purely liquid microstructure due to the segregation, which makes them more resistant to corrosion when ground and therefore without surface defects.

Acknowledgments: This research work was partially supported by the KK-foundation (CompCast, Project No. 20100280) which is gratefully acknowledged. The authors would like to thank COMPtech AB, Sweden for the production of the component and technical support.

Author Contributions: Maryam Eslami and Caterina Zanella conceived and designed the experiments; Maryam Eslami performed the experiments; Maryam Eslami, Mostafa Payandeh, Flavio Deflorian, Anders E. W. Jarfors and Caterina Zanella analyzed the data; Flavio Deflorian contributed reagents/materials/ analysis tools; Maryam Eslami, Mostafa Payandeh and Caterina Zanella wrote the paper.

Conflicts of Interest: The authors declare no conflict of interest.

References

1. Bonollo, F.; Gramegna, N.; Timelli, G. High-pressure die-casting: Contradictions and challenges. *JOM* **2015**, *67*, 901–908. [CrossRef]
2. Qi, M.; Kang, Y.; Zhou, B.; Liao, W.; Zhu, G.; Li, Y.; Li, W. A forced convection stirring process for Rheo-HPDC aluminum and magnesium alloys. *J. Mater. Process. Technol.* **2016**, *234*, 353–367. [CrossRef]
3. Ji, S.; Zhen, Z.; Fan, Z. Effects of rheo-die casting process on the microstructure and mechanical properties of AM50 magnesium alloy. *J. Mater. Sci. Technol.* **2005**, *21*, 1019–1024. [CrossRef]
4. Ji, S.; Wang, Y.; Watson, D.; Fan, Z. Microstructural evolution and solidification behavior of Al-Mg-Si alloy in high-pressure die casting. *Metall. Mater. Trans. A* **2013**, *44*, 3185–3197. [CrossRef]
5. Jin, C.K.; Jang, C.H.; Kang, C.G. Die design method for thin plates by indirect rheo-casting process and effect of die cavity friction and punch speed on microstructures and mechanical properties. *J. Mater. Process. Technol.* **2015**, *224*, 156–168. [CrossRef]
6. Fan, Z.; Ji, S.; Liu, G. Development of the rheo-diecasting process for Mg-alloys. *Mater. Sci. Forum* **2005**, *488–489*, 405–412. [CrossRef]
7. Moller, H.; Stumpf, W.E.; Pistorius, P.C. Influence of elevated Fe, Ni and Cr levels on tensile properties of SSM-HPDC Al-Si-Mg alloy F357. *Trans. Nonferr. Metals Soc. China* **2010**, *20*, s842–s846. [CrossRef]

8. Flemings, M.C.; Riek, R.G.; Young, K.P. Rheocasting. *Mater. Sci. Eng.* **1976**, *25*, 103–117. [CrossRef]
9. Fan, Z. Semisolid metal processing. *Int. Mater. Rev.* **2002**, *47*, 49–85. [CrossRef]
10. Atkinson, H.V. Alloys for semi-solid processing. *Solid State Phenom.* **2013**, *192–193*, 16–27. [CrossRef]
11. Payandeh, M. Rheocasting of Aluminium Alloys: Slurry Formation, Microstructure, and Properties. Ph.D. Thesis, Jönköping University, Jönköping, Sweden, April 2015.
12. Chen, Z.W. Skin solidification during high pressure die casting of Al–11Si–2Cu–1Fe alloy. *Mater. Sci. Eng. A* **2003**, *348*, 145–153. [CrossRef]
13. Laukli, H.I.; Gourlay, C.M.; Dahle, A.K. Migration of crystals during the filling of semi-solid castings. *Metall. Mater. Trans. A* **2005**, *36*, 805–818. [CrossRef]
14. Kaufmann, H.; Fragner, W.; Galovsky, U.; Uggowitzer, P.J. Fluctuations of Alloy Composition and Their Influence on Sponge Effect and Fluidity of A356-NRC. In *Proceedings of the 2nd International Light Metals Technology ConferenceProceedings of the 2nd International Light Metals Technology Conference, St. Wolfgang, Austria, 8–10 June 2005*; Kaufmann, H., Ed.; LKR-Verlag: St. Wolfgang, Austria, 2005.
15. Zabler, S.; Ershov, A.; Rack, A.; Garcia-Moreno, F.; Baumbach, T.; Banhart, J. Particle and liquid motion in semi-solid aluminium alloys: A quantitative in situ microradioscopy study. *Acta Mater.* **2013**, *61*, 1244–1253. [CrossRef]
16. Park, C.; Kim, S.; Kwon, Y.; Lee, Y.; Lee, J. Effect of microstructure on tensile behavior of thixoformed 357-T5 semisolid Al alloy. *Metall. Mater. Trans. A* **2004**, *35*, 1407–1410. [CrossRef]
17. Möller, H.; Masuku, E.P. The influence of liquid surface segregation on the pitting corrosion behavior of semi-solid metal high pressure die cast alloy F357. *TOCORRJ* **2009**, *2*, 216–220. [CrossRef]
18. Masuku, E.P.; Moller, H.; Curle, U.A.; Pistorius, P.C.; Li, W. Influence of surface liquid segregation on corrosion behavior of semi-solid metal high pressure die cast aluminium alloys. *Trans. Nonferr. Metals Soc. China* **2010**, *20*, s837–s841. [CrossRef]
19. Möller, H.; Curle, U.A.; Masuku, E.P. Characterization of surface liquid segregation in SSM-HPDC aluminium alloys 7075, 2024, 6082 and A201. *Trans. Nonferr. Metals Soc. China* **2010**, *20*, s847–s851. [CrossRef]
20. Guillaumin, V.; Mankowski, G. Localized corrosion of 6056 T6 aluminium alloy in chloride media. *Corros. Sci.* **2000**, *42*, 105–125. [CrossRef]
21. Eckermann, F.; Suter, T.; Uggowitzer, P.J.; Afseth, A.; Schmutz, P. The influence of MgSi particle reactivity and dissolution processes on corrosion in Al–Mg–Si alloys. *Electrochim. Acta* **2008**, *54*, 844–855. [CrossRef]
22. Zeng, F.-L.; Wei, Z.-L.; Li, J.-F.; Li, C.-X.; Tan, X.; Zhang, Z.; Zheng, Z.-Q. Corrosion mechanism associated with Mg_2Si and Si particles in Al–Mg–Si alloys. *Trans. Nonferr. Metals Soc. China* **2011**, *21*, 2559–2567. [CrossRef]
23. Liang, W.J.; Rometsch, P.A.; Cao, L.F.; Birbilis, N. General aspects related to the corrosion of 6xxx series aluminium alloys: Exploring the influence of Mg/Si ratio and Cu. *Corros. Sci.* **2013**, *76*, 119–128. [CrossRef]
24. Li, H.; Zhao, P.; Wang, Z.; Mao, Q.; Fang, B.; Song, R.; Zheng, Z. The intergranular corrosion susceptibility of a heavily overaged Al-Mg-Si-Cu alloy. *Corros. Sci.* **2016**, *107*, 113–122. [CrossRef]
25. Brito, C.; Vida, T.; Freitas, E.; Cheung, N.; Spinelli, J.E.; Garcia, A. Cellular/dendritic arrays and intermetallic phases affecting corrosion and mechanical resistances of an Al–Mg–Si alloy. *J. Alloys Compd.* **2016**, *673*, 220–230. [CrossRef]
26. Li, C.; Sun, J.; Li, Z.; Gao, Z.; Liu, Y.; Yu, L.; Li, H. Microstructure and corrosion behavior of Al–10%Mg_2Si cast alloy after heat treatment. *Mater. Charact.* **2016**, *122*, 142–147. [CrossRef]
27. Kairy, S.K.; Rometsch, P.A.; Diao, K.; Nie, J.F.; Davies, C.H.J.; Birbilis, N. Exploring the electrochemistry of 6xxx series aluminium alloys as a function of Si to Mg ratio, Cu content, ageing conditions and microstructure. *Electrochim. Acta* **2016**, *190*, 92–103. [CrossRef]
28. Svenningsen, G.; Lein, J.E.; Bjørgum, A.; Nordlien, J.H.; Yu, Y.; Nisancioglu, K. Effect of low copper content and heat treatment on intergranular corrosion of model AlMgSi alloys. *Corros. Sci.* **2006**, *48*, 226–242. [CrossRef]
29. Rehim, S.S.A.; Hassan, H.H.; Amin, M.A. Chronoamperometric studies of pitting corrosion of Al and (Al–Si) alloys by halide ions in neutral sulphate solutions. *Corros. Sci.* **2004**, *46*, 1921–1938. [CrossRef]
30. Tahamtan, S.; Boostani, A.F. Evaluation of pitting corrosion of thixoformed A356 alloy using a simulation model. *Trans. Nonferr. Metals Soc. China* **2010**, *20*, 1602–1606. [CrossRef]
31. Park, C.; Kim, S.; Kwon, Y.; Lee, Y.; Lee, J. Mechanical and corrosion properties of rheocast and low-pressure cast A356-T6 alloy. *Mater. Sci. Eng. A* **2005**, *391*, 86–94. [CrossRef]

32. Bastidas, J.M.; Forn, A.; Baile, M.T.; Polo, J.L.; Torres, C.L. Pitting corrosion of A357 aluminium alloy obtained by semisolid processing. *Mater. Corros.* **2001**, *52*, 691–696. [CrossRef]

33. Pech-Canul, M.A.; Pech-Canul, M.I.; Bartolo-Pérez, P.; Echeverría, M. The role of silicon alloying addition on the pitting corrosion resistance of an Al-12 wt %Si alloy. *Electrochim. Acta* **2014**, *140*, 258–265. [CrossRef]

34. Arrabal, R.; Mingo, B.; Pardo, A.; Mohedano, M.; Matykina, E.; Rodríguez, I. Pitting corrosion of rheocast A356 aluminium alloy in 3.5 wt % NaCl solution. *Corros. Sci.* **2013**, *73*, 342–355. [CrossRef]

35. Tahamtan, S.; Boostani, A.F. Quantitative analysis of pitting corrosion behavior of thixoformed A356 alloy in chloride medium using electrochemical techniques. *Mater. Des.* **2009**, *30*, 2483–2489. [CrossRef]

36. Larsen, M.H.; Walmsley, J.C.; Lunder, O.; Mathiesen, R.H.; Nisancioglu, K. Intergranular corrosion of copper-containing AA6xxx AlMgSi aluminum alloys. *J. Electrochem. Soc.* **2008**, *155*, C550–C556. [CrossRef]

37. Zhan, H.; Mol, J.M.C.; Hannour, F.; Zhuang, L.; Terryn, H.; de Wit, J.H.W. The influence of copper content on intergranular corrosion of model AlMgSi(Cu) alloys. *Mater. Corros.* **2008**, *59*, 670–675. [CrossRef]

38. Svenningsen, G.; Larsen, M.H.; Walmsley, J.C.; Nordlien, J.H.; Nisancioglu, K. Effect of artificial aging on intergranular corrosion of extruded almgsi alloy with small Cu content. *Corros. Sci.* **2006**, *48*, 1528–1543. [CrossRef]

39. Qian, M.; Li, D.; Liu, S.B.; Gong, S.L. Corrosion performance of laser-remelted Al–Si coating on magnesium alloy AZ91D. *Corros. Sci.* **2010**, *52*, 3554–3560. [CrossRef]

40. Yu, Y.; Kim, S.; Lee, Y.; lee, J. Phenomenological observations on mechanical and corrosion properties of thixoformed 357 alloys: A comparison with permanent mold cast 357 alloys. *Metall. Mater. Trans. A* **2002**, *33*, 1399–1412. [CrossRef]

41. Eslami, M.; Fedel, M.; Speranza, G.; Deflorian, F.; Zanella, C. Deposition and characterization of cerium-based conversion coating on HPDC low Si content aluminum alloy. *J. Electrochem. Soc.* **2017**, *164*, C581–C590. [CrossRef]

42. Wessén, M.; Cao, H. The RSF Technology: A Possible Breakthrough for Semi-Solid Casting Processes. In Proceedings of the International Conference of High Tech Die Casting, Vicenza, Italy, 21–22 September 2006.

43. Cano, E.; Lafuente, D.; Bastidas, D.M. Use of EIS for the evaluation of the protective properties of coatings for metallic cultural heritage: A review. *J. Solid State Electrochem.* **2010**, *14*, 381–391. [CrossRef]

44. Letardi, P. 7—Electrochemical measurements in the conservation of metallic heritage artefacts: An overview. In *Corrosion and Conservation of Cultural Heritage Metallic Artefacts*; Watkinson, D., Angelini, E., Adriaens, A., Eds.; Woodhead Publishing: Cambridge, UK, 2013; pp. 126–148.

45. Kirkwood, D.H.; Suery, M.; Kapranos, P.; Atkinson, H.V.; Young, K.P. *Semi-Solid Processing of Alloys*; Springer: New York, NY, USA, 2010.

46. Belov, N.A.; Aksenov, A.A. *Iron in Aluminum Alloys: Impurity and Alloying Element*; Taylor & Francis Inc.: New York, NY, USA, 2002.

47. Sundman, B.; Jansson, B.; Andersson, J.-O. The thermo-calc databank system. *Calphad* **1985**, *9*, 153–190. [CrossRef]

48. Andersson, J.-O.; Helander, T.; Höglund, L.; Shi, P.; Sundman, B. Thermo-Calc & DICTRA, computational tools for materials science. *Calphad* **2002**, *26*, 273–312.

49. Payandeh, M.; Jarfors, A.E.W.; Wessén, M. Solidification sequence and evolution of microstructure during rheocasting of four Al-Si-Mg-Fe alloys with low Si content. *Metall. Mater. Trans. A* **2016**, *47*, 1215–1228. [CrossRef]

50. Easton, M.; Kaufmann, H.; Fragner, W. The effect of chemical grain refinement and low superheat pouring on the structure of NRC castings of aluminium alloy Al–7Si–0.4 Mg. *Mater. Sci. Eng. A* **2006**, *420*, 135–143. [CrossRef]

51. Zhu, B.; Seifeddine, S.; Persson, P.O.Å.; Jarfors, A.E.W.; Leisner, P.; Zanella, C. A study of formation and growth of the anodised surface layer on cast Al-Si alloys based on different analytical techniques. *Mater. Des.* **2016**, *101*, 254–262. [CrossRef]

52. Gourlay, C.; Dahle, A.; Nagira, T.; Nakatsuka, N.; Nogita, K.; Uesugi, K. Granular deformation mechanisms in semi-solid alloys. *Acta Mater.* **2011**, *59*, 4933–4943. [CrossRef]

53. Payandeh, M.; Jarfors, A.E.W.; Wessén, M. Influence of microstructural inhomogeneity on fracture behaviour in SSM-HPDC Al-Si-Cu-Fe component with low Si content. *Solid State Phenom.* **2015**, *217–218*, 67–74. [CrossRef]

54. Govender, G.; Möller, H. Evaluation of surface chemical segregation of semi-solid cast aluminium alloy A356. *Solid State Phenom.* **2008**, *141–143*, 433–438. [CrossRef]

55. Dinnis, C.M.; Dahle, A.K.; Taylor, J.A. Three-dimensional analysis of eutectic grains in hypoeutectic Al–Si alloys. *Mater. Sci. Eng. A* **2005**, *392*, 440–448. [CrossRef]

56. Mingo, B.; Arrabal, R.; Pardo, A.; Matykina, E.; Skeldon, P. 3D study of intermetallics and their effect on the corrosion morphology of rheocast aluminium alloy. *Mater. Charact.* **2016**, *112*, 122–128. [CrossRef]

57. Ji, S.; Das, A.; Fan, Z. Solidification behavior of the remnant liquid in the sheared semisolid slurry of Sn–15 wt % Pb alloy. *Scripta Mater.* **2002**, *46*, 205–210. [CrossRef]

58. Wang, X.-H.; Wang, J.-H.; Fu, C.-W. Characterization of pitting corrosion of 7A60 aluminum alloy by EN and EIS techniques. *Trans. Nonferr. Metals Soc. China* **2014**, *24*, 3907–3916. [CrossRef]

59. Amin, M.A. Uniform and pitting corrosion events induced by SCN^- anions on Al alloys surfaces and the effect of UV light. *Electrochim. Acta* **2011**, *56*, 2518–2531. [CrossRef]

60. Amin, M.A.; Hassan, H.H.; Hazzazi, O.A.; Qhatani, M.M. Role of alloyed silicon and some inorganic inhibitors in the inhibition of meta-stable and stable pitting of Al in perchlorate solutions. *J. Appl. Electrochem.* **2008**, *38*, 1589–1598. [CrossRef]

61. Despić, A.; Parkhutik, V.P. Electrochemistry of aluminum in aqueous solutions and physics of its anodic oxide. In *Modern Aspects of Electrochemistry No. 20*; Bockris, J.O.M., White, R.E., Conway, B.E., Eds.; Springer: Boston, MA, USA, 1989; pp. 401–503.

62. Moreto, J.A.; Marino, C.E.B.; Filho, W.W.B.; Rocha, L.A.; Fernandes, J.C.S. SVET, SKP and EIS study of the corrosion behaviour of high strength Al and Al–Li alloys used in aircraft fabrication. *Corros. Sci.* **2014**, *84*, 30–41. [CrossRef]

Article

Tribological Behavior of Nano-Sized SiCp/7075 Composite Parts Formed by Semisolid Processing

Jufu Jiang [1,*], **Guanfei Xiao** [1], **Ying Wang** [2] and **Yingze Liu** [1]

[1] School of Materials Science and Engineering, Harbin Institute of Technology, Harbin 150001, China;
guanfeixiao@163.com (G.X.); liuyingze1995@foxmail.com (Y.L.)
[2] School of Mechatronics Engineering, Harbin Institute of Technology, Harbin 150001, China;
wangying1002@hit.edu.cn
*   Correspondence: jiangjufu@hit.edu.cn; Tel.: +86-187-4601-3176

Received: 26 December 2017; Accepted: 17 February 2018; Published: 25 February 2018

Abstract: The tribological behavior of the rheoformed and thixoformed nano-sized SiC_p/7075 composite parts is investigated. The semisolid stirring temperature has a little influence on the friction coefficient and wear resistance of the rheoformed composite parts. As for the thixoformed composite parts, the average value of the steady-state coefficient of friction increases firstly and then decreases with increasing reheating temperature. Higher wear resistance is achieved at a reheating temperature of 580 °C. The average value of the steady-state friction coefficient of the rheoformed composite parts varies from 0.37 to 0.45 upon applied loads of from 20 to 50 N. Weight loss increases slightly upon an increase of applied load from 20 to 40 N. An applied load of 50 N leads to a significant increase of the weight loss. The wear rate decreases firstly and then increases with increasing applied load. As for the thixoformed composite part, the average value of the steady-state friction coefficient and the weight loss decreased with an increasing applied load. However, the wear rate decreases firstly with increasing applied load and then increases. As for the rheoformed composite part, the average value of the steady-state friction coefficient decreases firstly and then increases a little with increasing sliding velocity. Weight loss and wear rate show a first increase and a followed decrease with increasing sliding velocity. As for the thixoformed composite part, the average value of the steady-state friction coefficient shows a decrease with increasing sliding velocity. Weight loss and wear rate exhibit, at first, an increase, and then a decrease with increasing sliding velocity. The average friction coefficient varies from 0.4 to 0.44 with increasing volume fraction of SiC. Weight loss and wear rate decrease with increasing volume fraction of SiC. An increase in dislocation density around the nano-sized SiC particles and the mismatch of the coefficient of thermal expansion (CTE) between 7075 matrix and nano-sized SiC particles during solidification improve the wear resistance of the composite. The dominant wear mechanisms of the rheoformed and thixoformed composite parts are adhesive wear, abrasive wear and delamination wear.

Keywords: nano-sized SiC particle; wear rate; friction coefficient; rheoformed; thixoformed

---

1. Introduction

Particle reinforced aluminum matrix composite (PRAMC) has received much attention because of their improved specific strength and modulus, good wear resistance, and modified thermal properties [1–4]. Wear behavior is an important evaluation parameter of the PRAMC. A large amount of research has been focused on tribological behavior (or wear behavior) of the PRAMC. Abdollahi et al. [5] investigated dry sliding tribological behavior of Al2024-5 wt. % B_4C nanocomposite fabricated by mechanical milling and hot extrusion and found that mechanical milling and adding B_4C increased the wear resistance of the nanocomposite. Kumar et al. [6] reported dry sliding wear behavior of stir cast AA6061-T6/AlN_p composite and developed a regression model predicting wear rate. The results showed that wear rate of cast $AA6061/AlN_p$ composite decreased with an increase in the mass fraction of AlN particles, and the regression model could predict wear rate at a 95% confidence level. SiC_p/Al-Cu matrix composites were produced by the direct squeeze casting (i.e., liquid aluminum melt is infiltrated into a preform of SiC particles under pressure) method, and their dry sliding properties were examined [7]. It was concluded that the friction coefficient decreased with increasing applied load and sliding velocity. Abrasive wear properties of SiC reinforced aluminum matrix composite produced by compocasting (i.e., casting of a stirred mixture of liquid aluminum and SiC particles) were studied [8]. The results revealed that the matrix hardness had a strong influence on the dry sliding wear behavior of the composite, and the lowest wear rate occurred in the composite with the lowest matrix hardness. Tribological behavior of composites with different aluminum matrix fabricated by squeeze cast was reported [9]. The results revealed that the matrix alloy had no remarkable influence on the tribological performance of the composites at low test loads less than 3 N.

Semisolid processing (SSP) has been widely used in automotive and 3C fields since it was developed by M. C. Flemings and his coworker [10–12]. Two typical technical routes such as rheoforming and thixoforming are included in SSP [13–17]. Rheoforming involves direct forming of semisolid slurries with spheroidal solid grains and liquid phase. In thixoforming process, semisolid billet obtained via solidifying semisolid slurries undergoes reheating and forming. SSP has shown an apparent advantage in dispersing the ceramic reinforcements of composite [18,19]. It was illustrated that SSP was suitable for fabricating PRAMC [20,21]. Therefore, some research has focused on wear properties of the PRAMC fabricated by SSP. Mazahery and Shabani [22] investigated the wear behavior of the sol-gel coated B_4C particle reinforced A356 matrix composites and concluded that the wear rate of the composites reinforced with coated B_4C was less than that of the matrix alloy and decreased with increasing volume fraction of B_4C particles. The research results of wear behavior of the rheocasted SiC_p/Al metal matrix composites (MMC) showed that the wear rate of the 11% SiC MMC was higher than that of the 50% SiC MMC [23]. The $A356/Al_2O_3$ metal matrix composites were fabricated by conventional stirring and semisolid processing [24]. It was concluded that the volume loss of the composites fabricated by semisolid processing was lower than that of the composites fabricated by conventional casting. The friction and wear of the aluminum alloy reinforced by TiO_2 particles fabricated by semisolid stirring was mentioned by Sarajan [25]. The results showed that accumulated volume loss was significantly higher when wear debris was removed by camel brush during dry sliding wear.

The aluminum matrix composite (AMC), reinforced with nano-sized ceramics, has received much attention due to high temperature creep resistance and fatigue life [26–29]. However, research on the wear behavior of the AMC reinforced with nano-sized ceramic particles was little reported. Therefore, the present study aims to investigate the wear behavior of nano-sized $SiC_p/7075$ composite parts formed by SSP and find the influence laws of the process parameters such as the volume fraction of SiC particles, and the applied load and sliding velocity on the wear behavior of the nanocomposite parts.

2. Materials and Methods

2.1. Fabrication of the Rheoformed and Thixoformed Nanocomposite Parts

Wrought 7075 aluminum alloys are used as matrix material of the composite. Its chemical composition content contained 6.0 wt. % Zn, 2.3 wt. % Mg, 1.56 wt. % Cu, 0.26 Si wt. %, 0.27 wt. % Mn, 0.17 wt. % Cr, 0.03 wt. % Ti and balance of Al. Nano-sized SiC particles with an normalized average size of 80 nm supplied by Xuzhou Jiechuang New Materials Co. Ltd of China (Xuzhou, China) are used as reinforcement of the composite [30]. The solidus of 546 °C and liquidus of 637 °C temperatures were achieved by using the Differential Thermal Analyzer (DTA) (Mettler Toledo, Zurich, Switerland) [30]. DTA data were converted into DSC (differential scanning calorimetry) data and then gave a curve of solid fraction vs temperature by integrating the DSC data. The semisolid slurry of nano-sized SiC_p/7075 (this denotes 7075 aluminum matrix composite reinforced by nano-sized SiC particles) composite was fabricated by ultrasonic-assisted semisolid stirring (UASS) method [30,31]. 7075 aluminum alloy was melt and held for 10 min at 650 °C. The nano-sized SiC particles were added into the melt. The melt with added nano-sized SiC particles was treated for 10 min via a ultrasonic device with a 2 kw power at a frequency of 20 kHz. Then the 7075 melt was cooled down while stirring to the required semisolid temperatures and stirred isothermally for the required times in order to obtain semisolid slurries. The required temperatures and times were shown in Table 1. Some of the semisolid slurries were directly used as rheoforming the cylinder parts of the composite. The other semisolid slurries were rheoformed into a cylindrical semisolid billet with a diameter of 70 mm and height of 58 mm. These cylindrical semisolid billets were reheated to given semisolid temperatures, soaked and then used as thixoforming cylinder parts (Table 1).

Table 1. Experimental procedures of dry sliding wear tests of nano-sized SiCp/7075 composite parts fabricated by rheoforming and thixoforming.

Experimental Serial Number	Fraction of SiC Particles (%)	Wear Velocity (m/s)	Stirring Temperature (°C)	Remelting Temperature (°C)	Applied Load (N)	Status
1	1	0.8	615	-	30	Rheoformed
2	1	0.8	620	-	30	Rheoformed
3	1	0.8	625	-	30	Rheoformed
4	1	0.8	628	-	30	Rheoformed
5	1	0.8	620	580	30	Thixoformed
6	1	0.8	620	590	30	Thixoformed
7	1	0.8	620	600	30	Thixoformed
8	1	0.8	620	610	30	Thixoformed
9	1	0.8	620	-	20	Rheoformed
10	1	0.8	620	-	40	Rheoformed
11	1	0.8	620	-	50	Rheoformed
12	1	0.8	-	600	20	Thixoformed
13	1	0.8	-	600	40	Thixoformed
14	1	0.8	-	600	50	Thixoformed
15	1	1.2	620	-	30	Rheoformed
16	1	1.6	620	-	30	Rheoformed
17	1	1.2	620	600	30	Thixoformed
18	1	1.6	620	600	30	Thixoformed
19	0	0.8	620	600	30	Rheoformed
20	0.5	0.8	620	600	30	Rheoformed
21	1.5	0.8	620	600	30	Rheoformed
22	2	0.8	620	600	30	Rheoformed

In the rheoforming process, a die was firstly preheated to 300 °C and then the semisolid slurries were carried into the die cavity. The upper die moved downwards and kept closed with lower die. The semisolid slurries were filled into the die cavity and rheoformed into the final composite cylinder part under a pressure of 398 MPa. In the thixoforming process, the reheated semisolid billet was carried into the die cavity with a preheated termperature of 400 °C. Then the upper die moved down and kept closed with the lower die. The reheated semisolid billet was filled into the die cavity and thixoformed into the final composite cylinder part under a pressure of 398 MPa. For the rheoformed cylinder parts, the process parameters included the semisolid stirring temperatures of 615, 620, 625 and 628 °C (corresponding to solid fractions of 0.46, 0.35, 0.30, 0.23 [30]), the stirring time of 20 min, the ultrasonic treatment time of 10 min and the volume fractions of 0, 0.5, 1.0, 1.5 and 2%. For the thixoformed cylindrical parts, the process parameters involved the semisolid stirring of 620 °C, the stirring time of 20 min, the ultrasonic treatment time of 10 min, the soaking time of 20 min, the volume fractions of 1.5% and reheating temperatures of 580, 590, 600 and 610 °C. The micrograph of the rheoformed and thixoformed parts were presented in Figure 1. As shown in Figure 1, the solid grains of the rheoformed composite cylinder part exhibited smaller deformation along flowing direction as compared to the thixoformed composite part. The TEM (transmission electron microscope) images of the rheoformed and thixoformed parts exhibited a uniform distribution of nano-sized SiC particles in the 7075 alloy matrix. It was attributed to double effect of the acoustic and cavitation created by ultrasonic wave and controllable viscosity of semisolid slurry [31,32]. In addition, it can be noted that needlelike second phase η-MgZn$_2$ existed in the microstructure of the rheoformed and thixoformed composite parts.

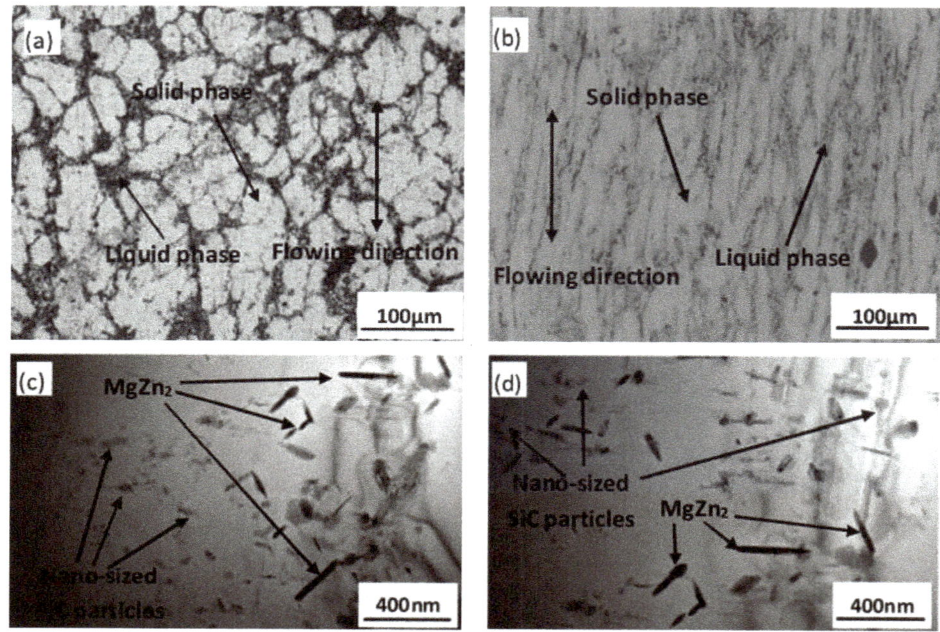

Figure 1. Micrographs of the rheoformed and thixoformed nanocomposite parts. (**a**) metallograph of the rheoformed at a stirring temperature of 620 °C (**b**) metallograph of the thixoformed at a remelting temperature of 590 °C. (**c**) TEM image of the rheoformed part (**d**) TEM image of the thixoformed part.

2.2. Dry Sliding Wear Tests of the Rheoformed and Thixoformed Parts

The dry sliding wear tests were carried out on a pin-on-disc wear-testing apparatus. The disc was made from GCr15 steel (Chinese National Standard GB/T18254-2016), which exhibited good quenching degree and high hardness (HRC 62) due to containing a large amount of chromium element [33]. Its chemical composition content contained 0.95–1.05 wt. % C, 0.25–0.45 wt. % Mn, 0.15–0.35 wt. % Si, 1.4–1.65 wt. % Cr, less than 0.02 wt. % S, less than 0.025 wt. % P, less than 0.1 wt. % Mo, less than 0.25 wt. % Ni, less than 0.25 wt. % Cu and balance of iron. After the rheoformed and thixoformed cylindrical parts were formed, they were machined into the samples with dimensions of Φ9 × 20 mm for the dry sliding wear tests. The diameter of the disc was 100 mm. The radius of the contact track of the dry slide samples used for their experiments was 27 mm. The *Ra* values of the sample and disc were 3.2 μm. Seven dry sliding wear tests were performed in these experiments, as shown in Table 1.The sample has the same section area (i.e., same contact area between pin sample and wear disc) so that applying force has the same effect of applying force per unit area. All the samples were performed under a wear distance of 1000 m. All dry sliding tests were performed under unidirectional sliding. 2.7 g/cm^3 was used as the density value of the 7075 aluminum alloy samples in order to calculate the wear volume. The density of the composite sample was determined by Archimedes drainage in order to calculate the wear volume of the composite sample. The density values of the composite sample with 0.5, 1.0, 1.5 and 2% SiC were 2.712, 2.716, 2.721 and 2.726 g/cm^3.

At the end of each wear test, the surface of the disc was washed by alcohol. The samples were weighed carefully and the weight loss was recorded and used as calculating wear rate of the composite parts. In order to investigate the surface morphology of worn samples and the wear mechanism, the worn surface was examined by scanning electron microscopy (SEM) (FEI, Hillsboro, OR, USA) with an energy dispersive X-ray spectrometer (EDS) (FEI, Hillsboro, OR, USA).

3. Results and Discussion

3.1. Influence of Semisolid Stirring Temperature on Tribological Behavior of Rheoformed Composite Parts

Figure 2 shows friction coefficient, weight loss, and wear rate of the rheoformed composite parts at different semisolid stirring temperatures. As indicated in Figure 2a–d, friction coefficient of the rheoformed composite parts at different semisolid stirring temperatures firstly increased significantly and then kept a fluctuation with increasing wear time. After the running-in period, the average value of the steady-state coefficient of friction was calculated as can be seen in Figure 2e. It is noticed that average value of friction coefficient varies from 0.42 to 0.46. It illustrates that semisolid stirring temperature has a little influence on friction of the rheoformed composite parts. As shown in Figure 2f, the weight loss varies from 2.84×10^{-3} g to 3.15×10^{-3} g. The wear rate was determined as defined by Equation (1):

$$K = \frac{W}{F_N \cdot S} \tag{1}$$

in which W is the wear volume (mm^3), F_N is the applied load (N), and S is the sliding distance (m), as reported by Zhang and Wang [34]. Wear rate is determined according to the data of weight loss (Figure 2g).

The wear rate of the rheoformed composite part at different semisolid stirring temperatures varies from 3.48×10^{-5} to 3.86×10^{-5} mm^3/m·N. Stirring temperature affects the solid fraction of semisolid slurries and further determines the deformation degree of solid grains rheoforming process. However, the deformation degree of solid grain during the rheoforming is obviously lower than that of thixoformed parts. Hence, the results of Figure 2f,g illustrate that semisolid stirring temperature has a little effect on the wear resistance of the rheoformed composite parts.

Figure 3 gives secondary electron (SE) images of the rheoformed composite parts at different semisolid stirring temperatures. As indicated in Figure 3, the delamination and shallow grooves are found in the microstructure of worn surface. Cracks are propagated in both transverse and longitudinal directions due to higher shear force on the sliding surfaces. It led to the loss of material from the worn surface in the form of flakes, as reported by Kumar [6]. In addition, some wear debris was found in the surface microstrucutre of the sample when semisolid stirring temperature was 628 °C (Figure 3d). It illustrates that abrasive wear also plays a role in the wear process of the composite.

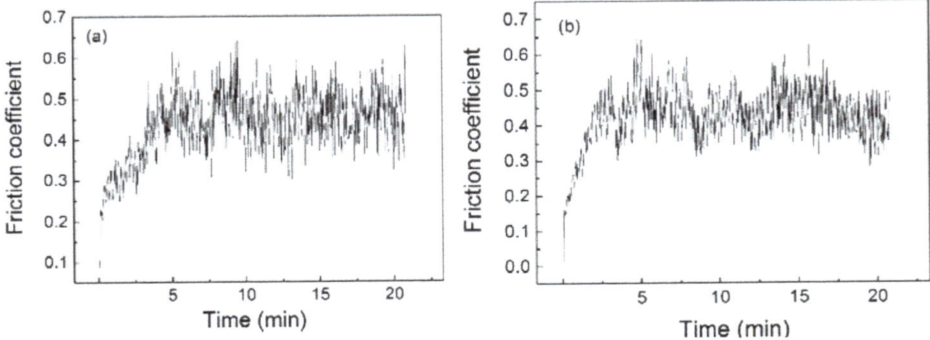

Figure 2. *Cont.*

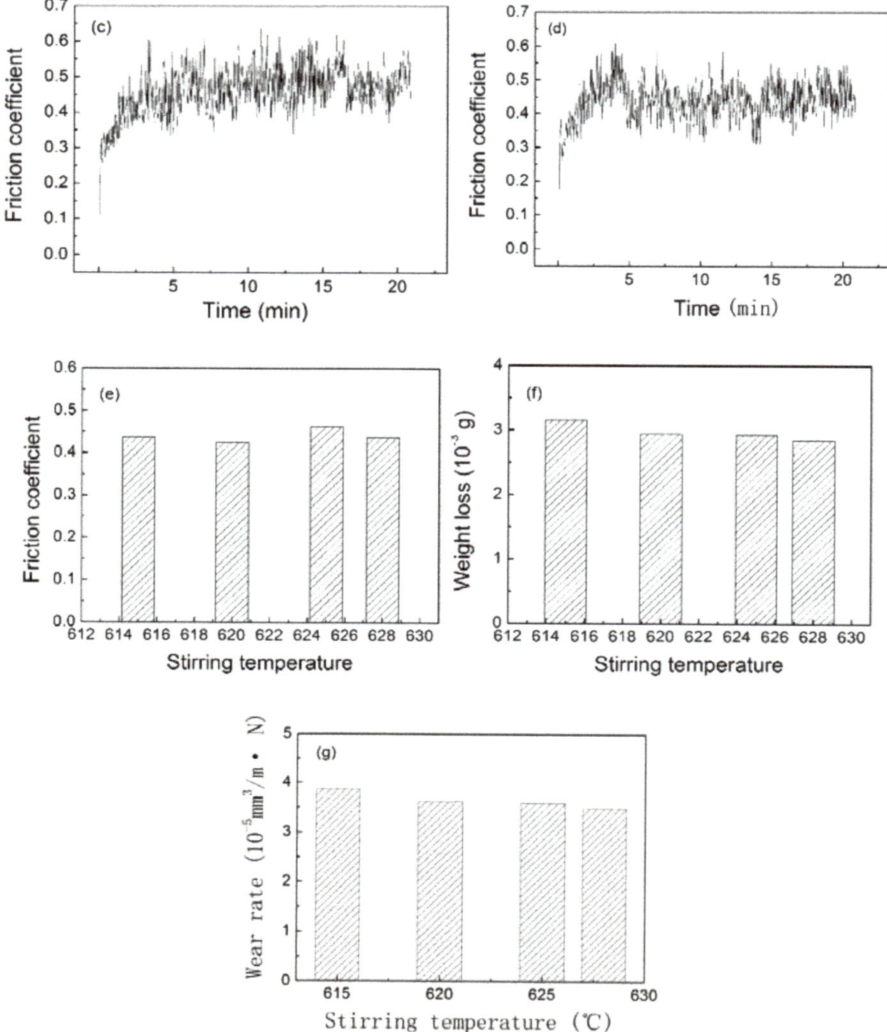

Figure 2. Friction coefficient, weight loss and wear rate of the rheoformed composite parts at different semisolid stirring temperatures: (**a**) friction coefficient at 615 °C, (**b**) friction coefficient at 620 °C, (**c**) friction coefficient at 625 °C, (**d**) friction coefficient at 628 °C, (**e**) average friction coefficient, (**f**) weight loss and (**g**) wear rate.

Energy dispersive X-ray (EDX) analysis revealed that some Fe and Cr elements occurred in the microstructure of the sample's surface (Figure 4). It illustrates also that wear debris of the disc made from GCr15 steel was retained on the surface of the composite sample, indicating the occurrence of abrasive wear.

Present oxygen element is due to the fact that wear test was done in an air environment, as reported by Sarajan [25]. Hence, the dominant wear mechanisms of the rheoformed composite parts at different semisolid stirring temperatures involve adhesive wear, abrasive wear and delamination wear.

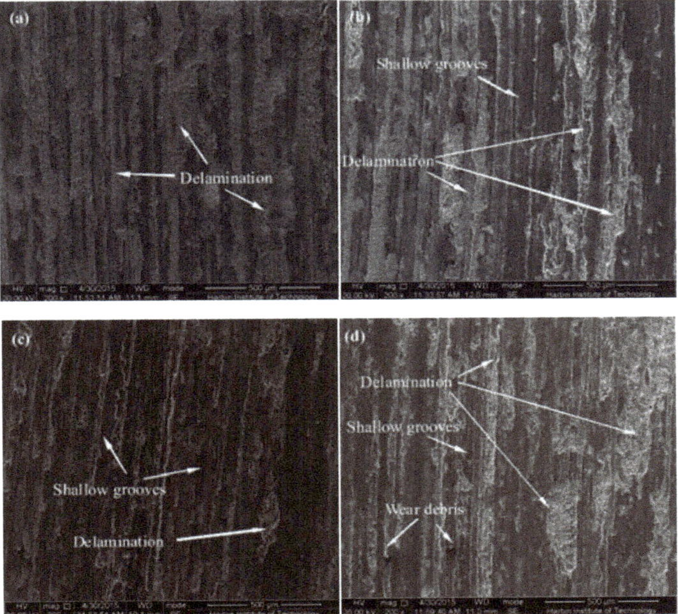

Figure 3. SEM images showing the worn surface morphology of the rheoformed composite parts at different semisolid stirring temperatures: (**a**) 615 °C, (**b**) 620 °C, (**c**) 625 °C and (**d**) 628 °C.

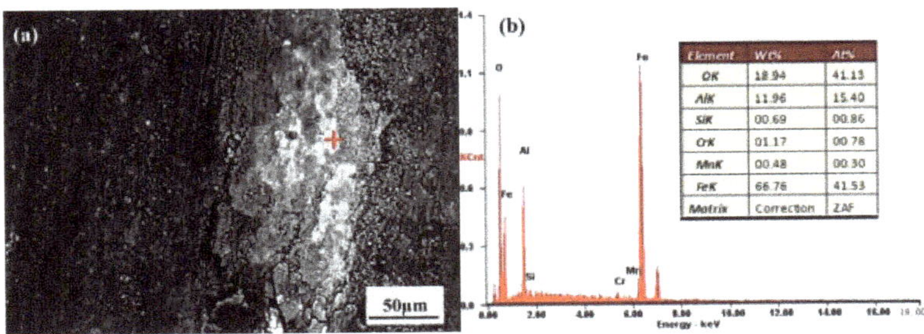

Figure 4. Energy dispersive X-ray (EDX) analysis of the rheoformed composite parts at a semisolid stirring temperature of 620 °C: (**a**) SEM micrograph and (**b**) distribution of elements.

3.2. Influence of Reheating Temperature on Tribological Behavior of Thixoformed Composite parts

Figure 5 exhibits friction coefficient, weight loss and wear rate of the thixoformed composite parts at different reheating temperatures. The curves of friction coefficient vs time reveal that friction coefficient increases firstly and then keeps a fluctuation with increasing time (Figure 5a–d). The average value of friction coefficient increases firstly and then decreases with increasing reheating temperature. When reheating temperature increases from 580 to 600 °C, the average friction coefficient increases from 0.41 to 0.54. As can be seen in Figure 5e, the average value of the steady state coefficient of friction decreases to from 0.54 to 0.50 upon a further increase from 600 to 610 °C. Weight loss and wear rate of the thixoformed composite part at reheating temperature of 580 °C are lower than those of the thixoformed composite parts at reheating temperatures of 590, 600 and 610 °C (Figure 5f,g). It indicates highest wear resistance was achieved in the thixoformed composite part at the reheating temperature of 580 °C.

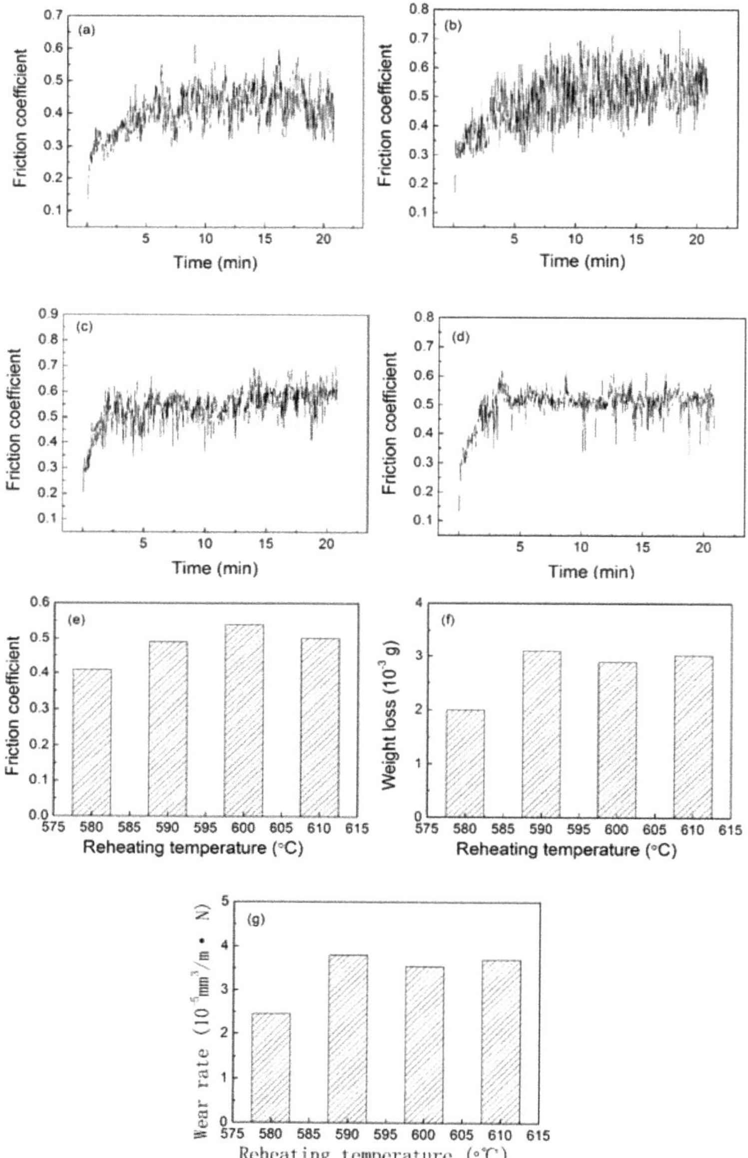

Figure 5. Friction coefficient, weight loss and wear rate of the thixoformed composite parts at different reheating temperatures: (**a**) friction coefficient at 580 °C, (**b**) friction coefficient at 590 °C, (**c**) friction coefficient at 600 °C, (**d**) friction coefficient at 610 °C, (**e**) average friction coefficient, (**f**) weight loss and (**g**) wear rate.

Plastic deformation of solid grains occurs severely because of low liquid phase fraction. It led to occurrence of more dislocations in solid grains. These dislocations can improve wear resistance of the thixoformed composite parts at 580 °C. When reheating temperature is elevated to a reheating temperature above 590 °C, the plastic deformation of solid grains decreases due to more liquid phase, leading to decreased dislocations. Therefore, the weight loss and wear rate change slightly when reheating temperatures are 590, 600 and 610 °C. In addition, it can be noticed that the weight loss and wear rate of the thixoformed composite parts at 590, 600 and 610 °C are close to those of the rheoformed composite parts (Figures 2 and 5).

Delamination, shallow grooves, craters and wear debris were found in the microstructure of worn surface (Figure 6). It illustrates that wear mechanisms of the thixoformed composite parts belong to delamination wear, abrasive wear and adhesive wear.

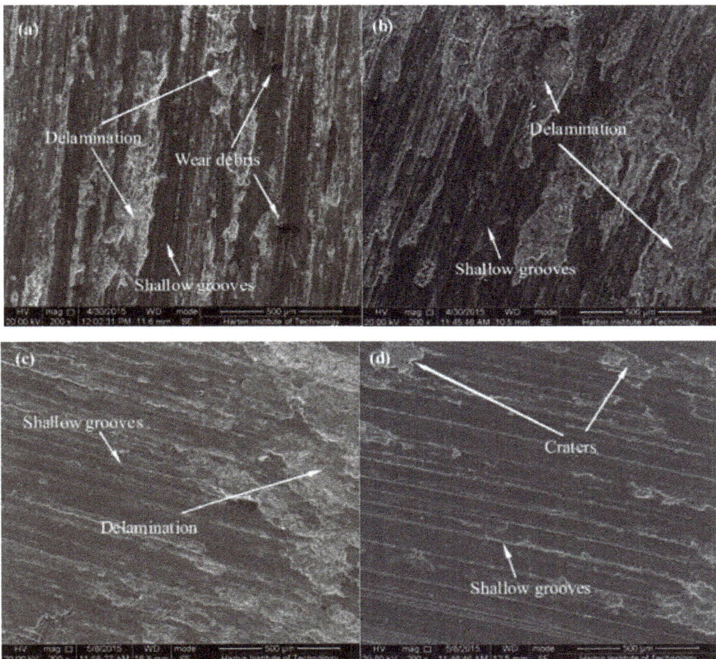

Figure 6. Worn surface morphology of the thixoformed composite parts at different reheating temperatures: (**a**) 580 °C, (**b**) 590 °C, (**c**) 600 °C and (**d**) 610 °C.

3.3. Influence of Applied Load on Tribological Behavior of the Rheoformed and Thixoformed Composite Parts

Friction coefficient, weight loss and wear rate of the rheoformed composite parts presented in Figure 7. Friction coefficient exhibits firstly a significant increase and then a fluctuation with increasing time (Figure 7a–c). Before 5min, friction coefficient shows firstly a significant increase and then fluctuation with increasing time. After 5 min, the friction coefficient exhibits a fluctuation again. The friction coefficient shows a significant increase and fluctuation with increasing time before 5 min. However, it is noted that friction coefficient increase slightly with increasing time after 5 min (Figure 7d). The average friction coefficient varies from 0.37 to 0.45 upon applied loads of from 20 to 50 N. As for applied load of 50 N, the average friction coefficient exhibited the lowest value of 0.37 (Figure 7e). The results achieved by Onat [7] revealed that friction coefficient decreased with increasing applied load. This result is almost consistent with the results achieved by Onat [7], except for the 40 N applied load. This work shows a decrease of friction coefficient upon an increase of applied load from 20 to 30 N. However, average friction coefficient increases with an increase of applied load from 30 to 40 N. Average friction coefficient exhibited a decrease with an increase of applied load from

40 to 50 N again. This may be due to different selected range of applied load. For research of Onat [7], the selected range of applied load is a range from 5 to 15 N. However, this selected range of applied load is from 20 to 50 N. Weight loss increases a little upon an increase of applied load from 20 to 40 N. However, when applied load reached to 50 N, the weight loss increased significantly (Figure 7f).

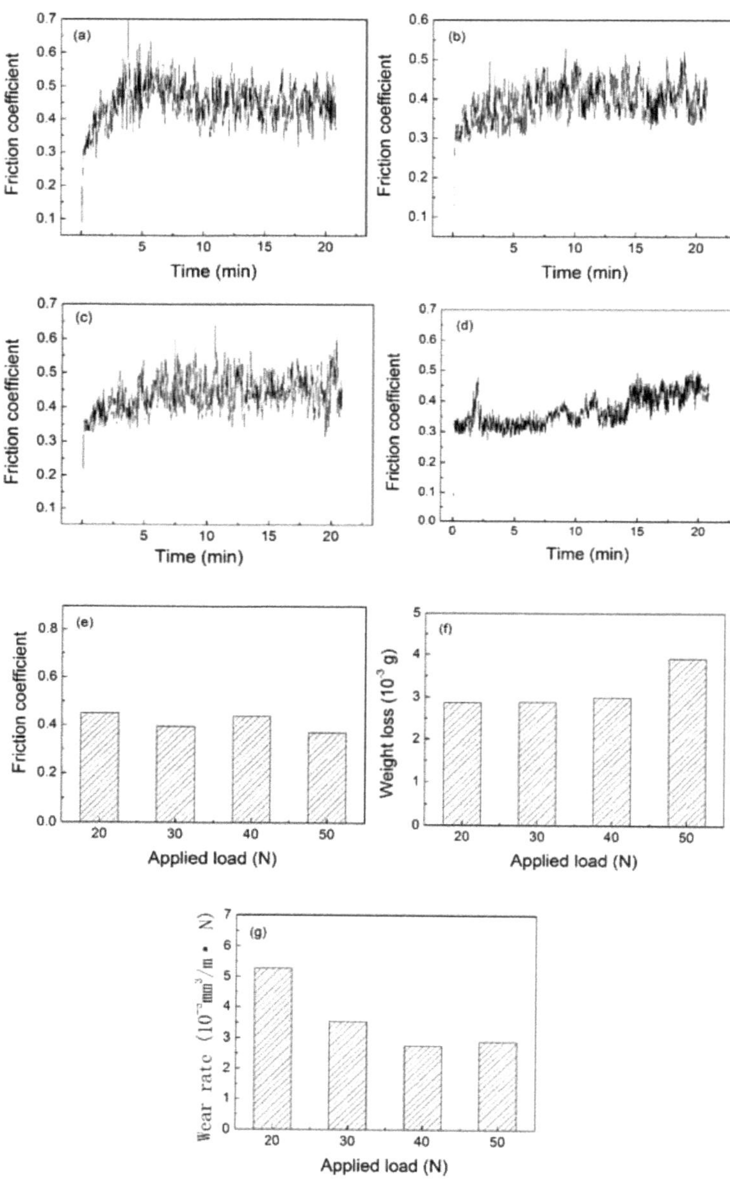

Figure 7. Friction coefficient, weight loss and wear rate of the rheoformed composite parts at different applied load: (**a**) friction coefficient at 20 N, (**b**) friction coefficient at 30 N, (**c**) friction coefficient at 40 N, (**d**) friction coefficient at 50 N, (**e**) average friction coefficient, (**f**) weight loss and (**g**) wear rate.

Wear rate exhibited a first decrease and then a slight increase with increasing applied load (Figure 7g). The surface microstructure is characterized by delamination, wear debris, and shallow grooves (Figure 8). It indicates that wear mechanisms depend on delamination wear, abrasive wear and adhesive wear. Especially, a large area of delamination was noted in the microstructure, indicating a dominant delamination wear.

Figure 8. Worn surface morphology of the rheoformed composite parts at different applied load: (**a**) 20 N, (**b**) 30 N, (**c**) 40 N and (**d**) 50 N.

Friction coefficient, weight loss and wear rate of the thixoformed composite part are present in Figure 9. A similar law to the rheoformed composite parts that friction coefficient increases significantly and keeps fluctuation is found in the curves of friction coefficient vs time (Figure 9a–d). The average friction coefficient decreased with increasing applied load (Figure 9e). This result is in agreement with the results achieved by Onat [7] and Zhang and Wang [34]. Weight loss increases with increasing applied load as shown in Figure 9f. It is an agreement with the results obtained by Nartarajan et al. [35].

However, the wear rate does not keep the similar law to weight loss with an increase of applied load. As to wear rate, it firstly keeps a decrease and then increases with increasing applied load. This change in trend is in agreement with the rheoformed composite parts as shown in Figure 9g. Delamination is a dominant characteristic of the surface morphology of the worn sample (Figure 10). It illustrates that wear mechanism of the thixoformed composite under different applied load belongs to delamination wear. In addition, it can be noticed that the delamination area firstly decreases and then significantly increases with increasing load. It is consistent with the change law of the wear rate.

Especially at an applied load of 50 N, the delamination area increases significantly, indicating a large weight loss. This is the main reason for increase of the wear rate.

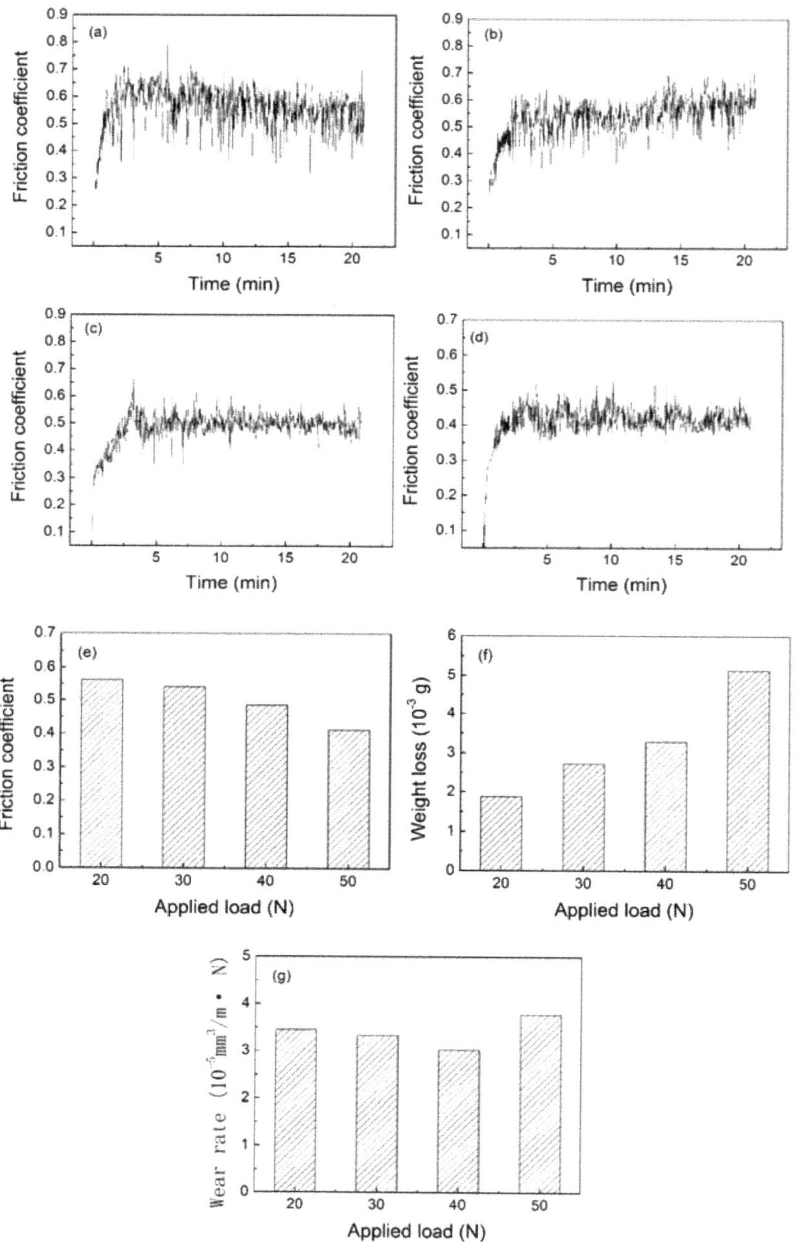

Figure 9. Friction coefficient, weight loss and wear rate of the thixoformed composite parts at different applied load: (**a**) friction coefficient at 20 N, (**b**) friction coefficient at 30 N, (**c**) friction coefficient at 40 N, (**d**) friction coefficient at 50 N, (**e**) average friction coefficient, (**f**) weight loss and (**g**) wear rate.

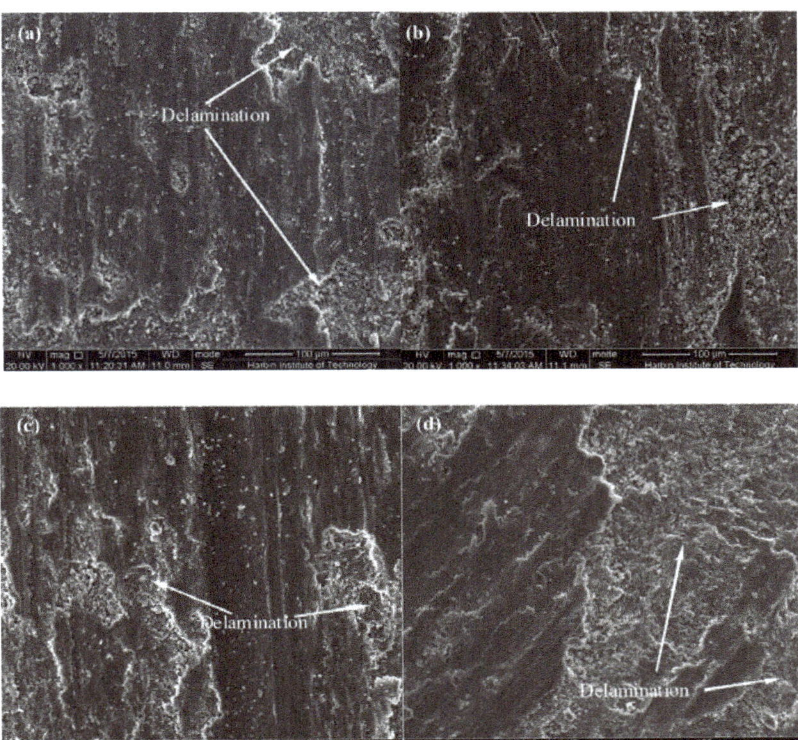

Figure 10. Worn surface morphology of the thixoformed composite parts at different applied load: (**a**) 20 N, (**b**) 30 N, (**c**) 40 N and (**d**) 50 N.

3.4. Influence of Sliding Velocity on Tribological Behavior of the Rheoformed and Thixoformed Composite Parts

Friction coefficient, weight loss, and wear rate of the rheoformed composite parts at different sliding velocities are displayed in Figure 11. When sliding velocity is 0.8 m/s, friction coefficient shows a significant increase and a followed fluctuation with increasing time (Figure 11a), which is similar to those of above mentioned rheoformed and thixoformed composite parts. However, an increase of from 0.8 to 1.2 m/s led to an obvious change. As shown in Figure 11b, three stages such as significant increase, significant decrease, and fluctuation were found in the curve of friction coefficient vs time.

A peak value of friction coefficient presented in the curve. The curve of friction coefficient vs time at a sliding velocity of 1.6 m/s exhibits a different change from those at 1.2 m/s (Figure 11c). Friction coefficient shows a significant increase and then enters a fluctuation stage with increasing time. A little difference is its fluctuation extent is lower as compared with the curve at 0.8 m/s. Average friction coefficient firstly decreases and then increases a little with increasing sliding velocity (Figure 11d). Weight loss shows a first increase and then a followed decrease with increasing sliding velocity (Figure 11e). Wear rate exhibits a similar law to weight loss upon an increase of sliding velocity (Figure 11f). Delamination and shallow grooves were also found in the worn surface microstructure, indicating an adhesive wear and delamination wear (Figure 12).

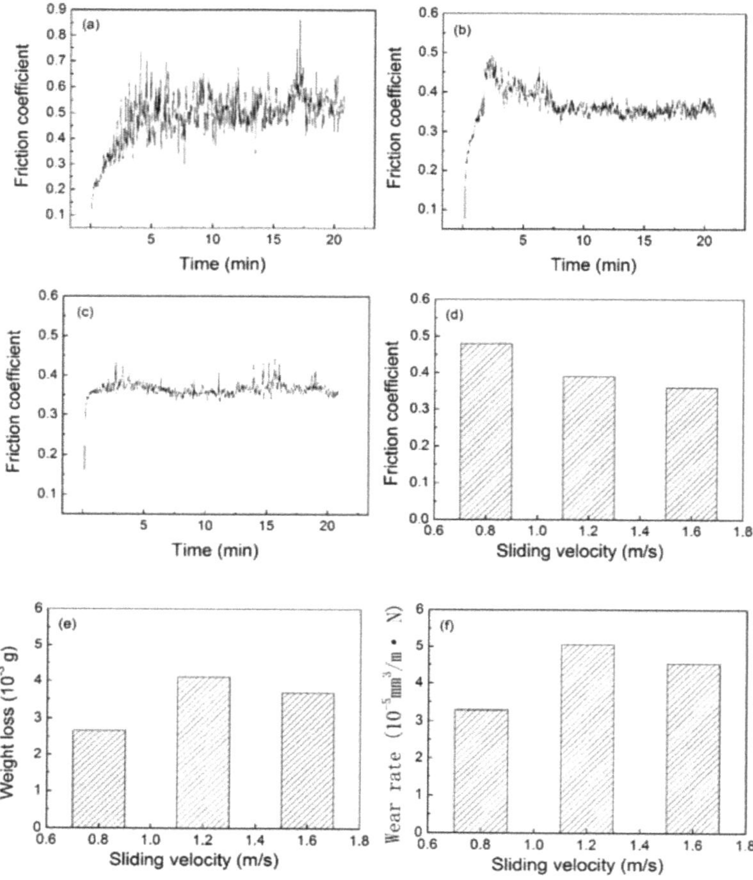

Figure 11. Friction coefficient, weight loss and wear rate of the rheoformed composite parts at different sliding velocity: (**a**) friction coefficient at 0.8 m/s, (**b**) friction coefficient at 1.2 m/s, (**c**) friction coefficient at 1.6 m/s, (**d**) average friction coefficient, (**e**) weight loss and (**f**) wear rate.

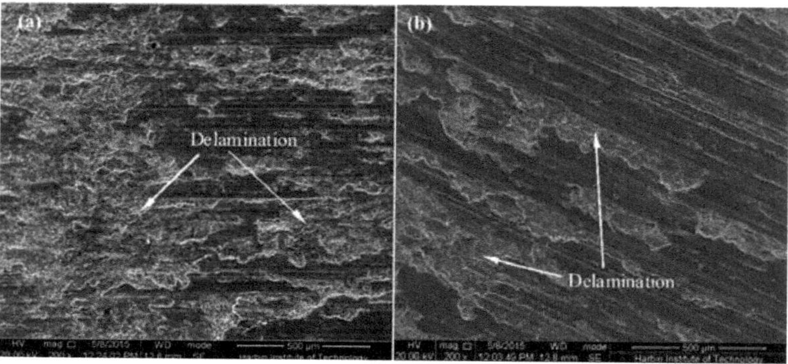

Figure 12. *Cont.*

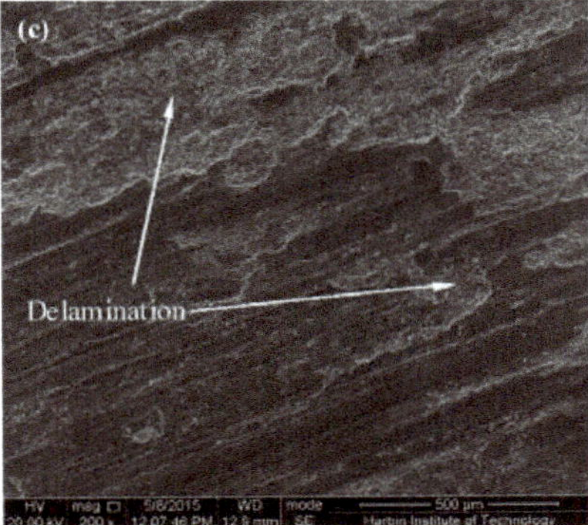

Figure 12. Worn surface morphology of the rheoformed composite parts at different sliding velocities: (**a**) 0.8 m/s, (**b**) 1.2 m/s and (**c**) 1.6 m/s.

Figure 13 depicts friction coefficient, weight loss, and wear rate of the thixoformed composite parts at different sliding velocities. A law of a significant increase and a fluctuation with increasing time was shown. A little difference among them is the fluctuation extents at 1.2 and 1.6 m/s are lower than that at 0.8 m/s. Average friction coefficient shows a decrease with increasing sliding velocity. This work shows a disagreement with the results obtained by Nartarajan et al. [35]. It may be due to effect of the different material sample and disc. Weight loss and wear rate exhibit a first increase and a decrease with increasing sliding velocity. Delamination and wear debris were found the worn surface microstructure (Figure 14), indicating delamination wear and adhesive wear play a role in the wear process of the thixoformed composite parts at different sliding velocities.

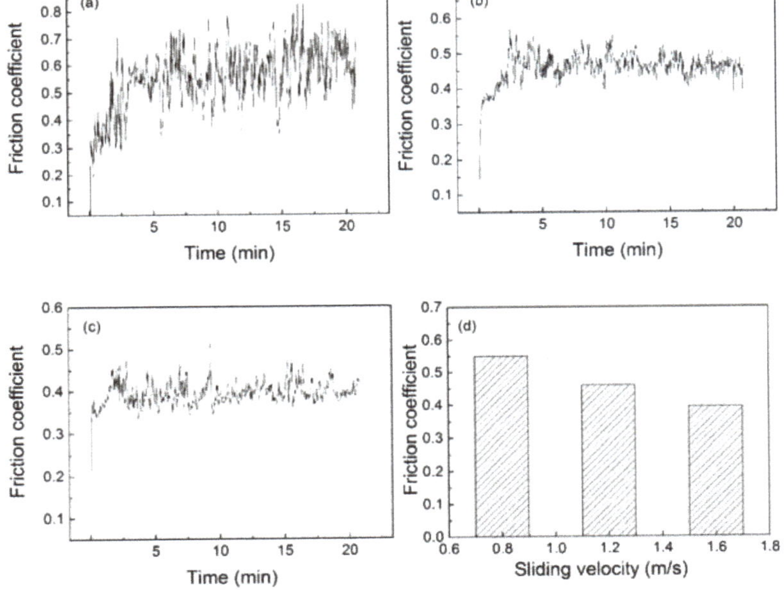

Figure 13. *Cont.*

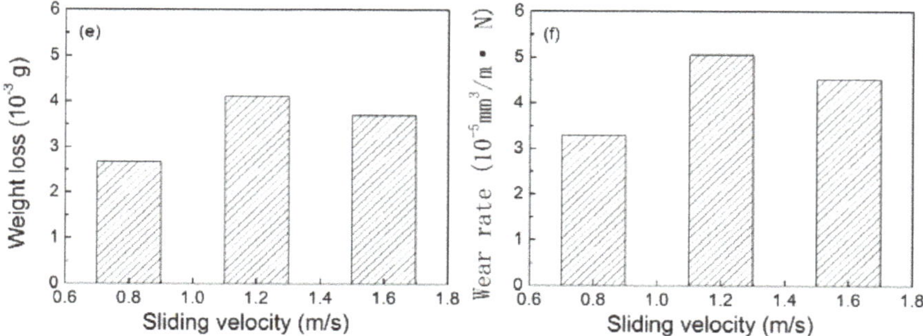

Figure 13. Friction coefficient, weight loss and wear rate of the thixoformed composite parts at different sliding velocities: (**a**) friction coefficient at 0.8 m/s, (**b**) friction coefficient at 1.2 m/s, (**c**) friction coefficient at 1.6 m/s, (**d**) average friction coefficient, (**e**) weight loss and (**f**) wear rate.

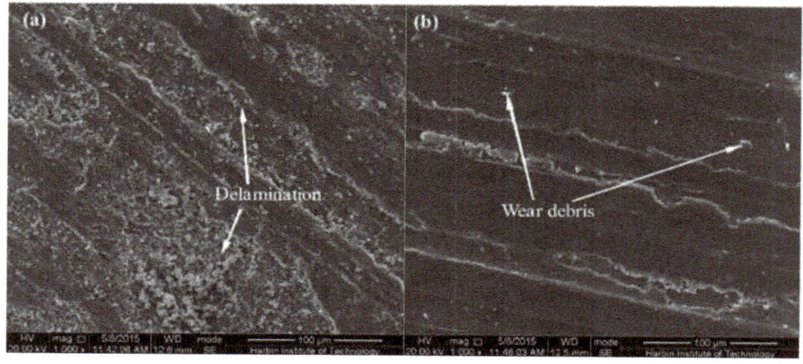

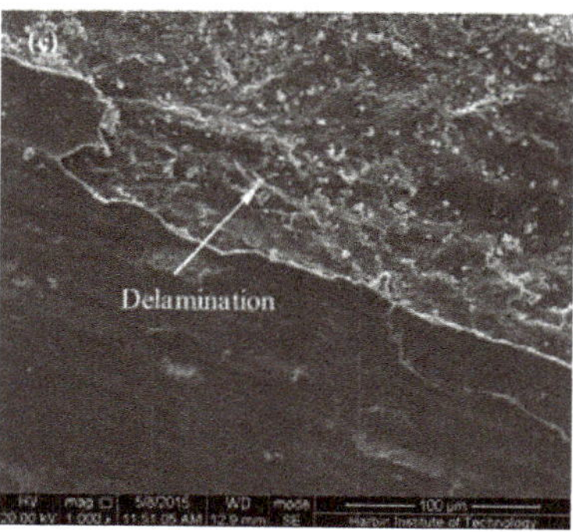

Figure 14. Worn surface morphology of the thixoformed composite parts at different sliding velocities: (**a**) 0.8 m/s, (**b**) 1.2 m/s and (**c**) 1.6 m/s.

3.5. Influence of Volume Fraction of SiC Particle on Tribological Behavior of the Rheoformed and Thixoformed Composite Parts

Figure 15 represents friction coefficient, weight loss, and wear rate of the rheoformed composite parts with different volume fraction of SiC particles. It is noted that friction coefficient also undergoes a significant increase and a fluctuation with increasing time (Figure 15a–e). It illustrates that the friction coefficient of the matrix 7075 aluminum alloy has a similar change law with those of the composite. The average friction coefficient varies from 0.4 to 0.44, and no obvious law is found in the curve of average friction coefficient with increasing volume fraction of SiC particles (Figure 15f). However, weight loss and wear rate decrease with increasing volume fraction of SiC particles (Figure 15g,h). The weight loss and wear rate of the matrix 7075 aluminum alloy are 3.78×10^{-3} g and 4.63×10^{-5} cm^3/m·N, respectively. Upon an addition of 2% nano-sized SiC particles, the weight loss and wear rate reach 2.16×10^{-3} g and 2.64×10^{-5} mm^3/m·N, respectively. The decrease extents are 42.8 and 43.0%, respectively. It illustrates that addition of nano-sized SiC particles can improve significantly the wear resistance of the composite parts. This work is consistent with the results presented in the A356/Al$_2$O$_3$ metal matrix composites by Alhawari et al. [24] and the stir cast AA6061-T6/AlN$_p$ composite, as reported by Kumar et al. [6]. The addition of nano-sized SiC particles with low coefficient of thermal expansion (CTE) into the alloy matrix with higher coefficient of thermal expansion led to mismatch of CTE of the composite, as mentioned by Kumar et al. [6] and Zhong et al. [36].

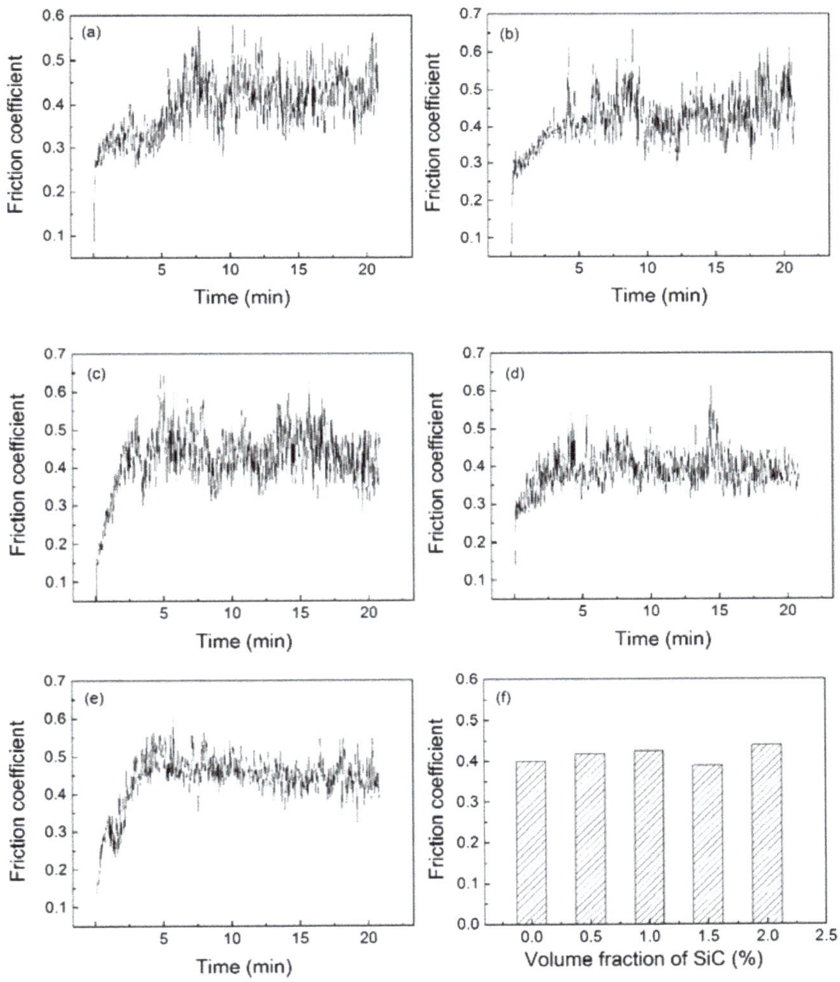

Figure 15. *Cont.*

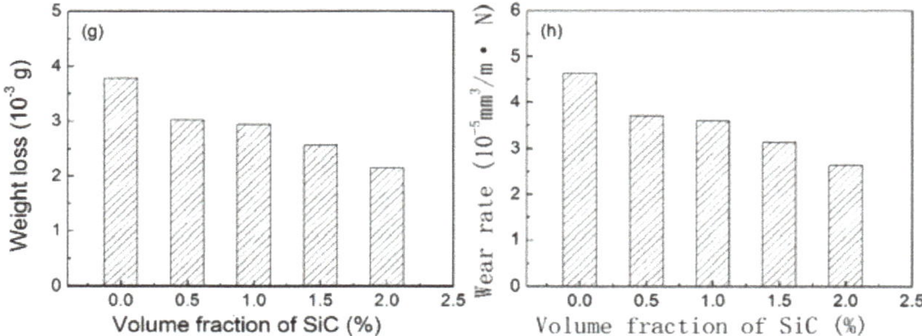

Figure 15. Friction coefficient, weight loss and wear rate of the rheoformed composite parts with different volume fraction of SiC particles: (**a**) 0, (**b**) 0.5%, (**c**) 1.0%, (**d**) 1.5%, (**e**) 2.0%, (**f**) average friction coefficient, (**g**) weight loss and (**h**) wear rate.

This contributes to the improved strength and wear resistance of the composite. An increase in the volume fraction of nano-sized SiC particles leads to an increase in dislocation density around the nano-sized SiC particles during solidification, as reported by Gutam and Srivastant [37]. The interaction between nano-sized SiC particles and dislocations also improves the wear resistance of the composite, as reported by Kumar et al. [6]. Delamination and shallow grooves were found in the worn surface microstructure (Figure 16). This shows that wear mechanisms depend mainly on delamination wear and adhesive wear.

Figure 16. *Cont.*

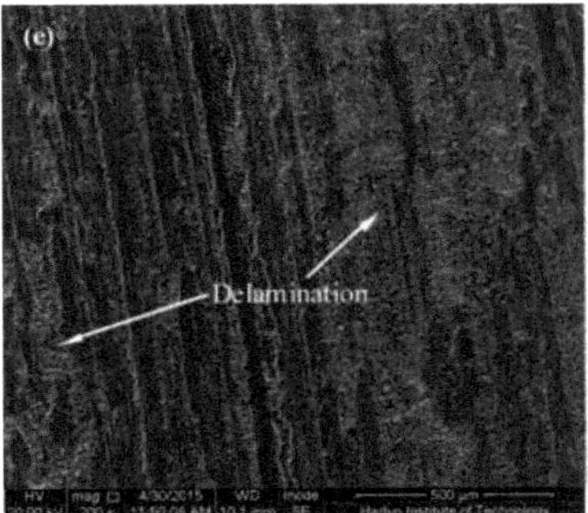

Figure 16. Worn surface morphology of the rheoformed composite parts with different volume fraction of SiC particles: (**a**) 0, (**b**) 0.5%, (**c**) 1.0%, (**d**) 1.5% and (**e**) 2.0%.

4. Conclusions

(1) The curves of the friction coefficient vs time exhibit a significant increase and a followed fluctuation with increasing time. As for the rheoformed composite part at the different semisolid stirring temperature, the average value of the steady-state friction coefficient varies from 0.42 to 0.46. This semisolid stirring temperature has a little influence on the friction and wear resistance of the rheoformed composite parts. As for the thixoformed composite parts, the average value of the steady-state friction coefficient increases firstly and then decreases with an increase in the reheating temperature. When the reheating temperature increases from 580 to 600 °C, the average value of steady-state friction coefficient increases from 0.41 to 0.54. The average value of the steady-state friction coefficient decreases to from 0.54 to 0.50 upon a further increase from 600 to 610 °C. The best wear resistance was achieved in the thixoformed composite part at a reheating temperature of 580 °C. The dominant wear mechanisms of the rheoformed and thixoformed composite parts involve adhesive wear, abrasive wear and delamination wear.

(2) As for the rheoformed composite part, the average value of the steady-state friction coefficient varies from 0.37 to 0.45 upon the applied loads of from 20 to 50 N. The weight loss increases slightly upon an increase of applied load from 20 to 40 N. However, it was noted that the weight loss increase significantly when the applied load reached 50 N. The wear rate decreases obviously and then slightly increases with the increasing applied load. Delamination, wear debris, and shallow grooves indicate wear mechanisms depend on delamination wear, abrasive wear, and adhesive wear. As for the thixoformed composite part, the average value of the steady-state friction coefficient decreased with the increasing applied load. Weight loss decreased with the increasing applied load. However, the wear rate firstly decreases with the increasing applied load and then increases.

(3) As for the rheoformed composite part, the average value of the steady-state friction coefficient firstly decreases and then increases a little with the increasing sliding velocity. Weight loss and wear rate show, at first, an increase and followed by a decrease with regard to the increasing sliding velocity. As for the thixoformed composite part, the average value of the steady-state friction coefficient shows a decrease with increasing sliding velocity. Weight loss and wear rate exhibit, at first, an increase and a then decrease with the increasing sliding velocity. Adhesive wear and delamination wear are the dominant wear mechanisms of the rheoformed and thixoformed composite parts at different sliding velocities.

(4) The average value of the steady-state friction coefficient varies from 0.4 to 0.44 and no obvious law is found in the curve of the friction coefficient with the increasing volume fraction of the SiC particles. However, weight loss and wear rate decrease with the increasing volume fraction of the SiC particles. The wear resistance of the composite parts was improved significantly due to the addition of the nano-sized SiC particles. An increase in the dislocation density around the nano-sized SiC

particles and the mismatch of the coefficient of thermal expansion (CTE) between the 7075 matrix and the nano-sized SiC particles during solidification improved the wear resistance of the composite.

Acknowledgments: This work is supported by the Natural Science Foundation of China (NSFC) under Grant No.51375112.

Author Contributions: Jufu Jiang designed most experiments, analyzed the results and wrote this manuscript. Guanfei Xiao and Yingze Liu performed most experiments. Ying Wang helped analyze the experimental data and gave some constructive suggestions about how to write this manuscript.

Conflicts of Interest: The authors declare no conflicts of interest.

References

1. Lee, S.S.; Yeo, J.S.; Hong, S.H.; Yoon, D.J.; Na, K.H. The fabrication process and mechanical properties of SiC$_p$/Al-Si metal matrix composites for automobile air-conditioner compressor piston. *J. Mater. Process. Technol.* **2001**, *113*, 202–208. [CrossRef]
2. Tjong, S.C.; Ma, Z.Y. High-temperature creep behaviour of powder-metallurgy aluminium composites reinforced with SiC particles of various sizes. *Compos. Sci. Technol.* **1999**, *59*, 1117–1125. [CrossRef]
3. Tatar, C.; Özdemir, N. Investigation of thermal conductivity and microstructure of the α-Al$_2$O$_3$ particulate reinforced aluminum composites (Al/Al$_2$O$_3$-MMC) by powder metallurgy method. *Phys. B Condens. Matter* **2010**, *405*, 896–899. [CrossRef]
4. Zhang, Q.; Wu, G.H.; Chen, G.Q.; Jiang, L.T.; Luan, B.F. The thermal expansion and mechanical properties of high reinforcement content SiCp/Al composites fabricated by squeeze casting technology. *Compos. Part A* **2003**, *34*, 1023–1027. [CrossRef]
5. Abdollahi, A.; Alizadeh, A.; Baharvandi, H.R. Dry sliding tribological behavior and mechanical properties of Al2024–5 wt. % B$_4$C nanocomposite produced by mechanical milling and hot extrusion. *Mater. Des.* **2014**, *55*, 471–481. [CrossRef]
6. Kumar, B.A.; Murugan, N.; Dinaharan, I. Dry sliding wear behavior of stir cast AA6061-T6/AlNp composite. *Trans. Nonferr. Met. Soc. China* **2014**, *24*, 2785–2795. [CrossRef]
7. Onat, A. Mechanical and dry sliding wear properties of silicon carbide particulate reinforced aluminium-copper alloy matrix composites produced by direct squeeze casting method. *J. Alloy. Compd.* **2010**, *489*, 119–124. [CrossRef]
8. Mazahery, A.; Shabani, M.O. Microstructural and abrasive wear properties of SiC reinforced aluminum-based composite produced by compocasting. *Trans. Nonferr. Met. Soc. China* **2013**, *23*, 1905–1914. [CrossRef]
9. Mindivan, H.; Kayali, E.S.; Cimenoglu, H. Tribological behavior of squeeze cast aluminum matrix composites. *Wear* **2008**, *265*, 645–654. [CrossRef]
10. Spencer, D.B. Rheology of liquid-solid mixtures of lead-tin. Ph.D. Thesis, Massachusetts Institute of Technology, Cambridge, MA, USA, 1971.
11. Spencer, D.B.; Mehrabian, R.; Flemings, M.C. Rheological behavior of Sn-15 pct Pb in the crystallization range. *Metall. Trans. B* **1972**, *3*, 1925–1932. [CrossRef]
12. Flemings, M.C. Behavior of metal alloys in the semisolid state. *J. Metall. Trans. B* **1991**, *22A*, 957–981. [CrossRef]
13. Guan, R.G.; Zhaom, Z.Y.; Zhang, H.; Lian, C.; Lee, C.S.; Liu, C.M. Microstructure evolution and properties of Mg-3Sn-1Mn (wt. %) alloy strip processed by semisolid rheo-rolling. *J. Mater. Process. Technol.* **2012**, *212*, 1430–1436. [CrossRef]
14. Guan, R.G.; Zhao, Z.Y.; Lee, C.S.; Zhang, Q.S.; Liu, C.M. Effect of wavelike sloping plate rheocasting on microstructures of hypereutectic Al-18 pct Si-5 pct Fe alloys. *Metall. Trans. B* **2012**, *43*, 337–343. [CrossRef]
15. Bolouri, A.; Kang, C.G. Characteristics of thixoformed A356 aluminum thin plates with microchannels. *Mater. Charact.* **2013**, *82*, 86–96. [CrossRef]
16. Chayong, S.; Atkinson, H.V.; Kapranos, P. Thixoforming 7075 aluminium alloys. *Mater. Sci. Eng. A* **2005**, *390*, 3–12. [CrossRef]
17. Birol, Y. Comparison of thixoformability of AA6082 reheated from the as-cast and extruded states. *J. Alloy. Compd.* **2008**, *461*, 132–138. [CrossRef]

18. Guan, L.N.; Geng, L.; Zhang, H.W.; Huang, L.J. Effects of stirring parameters on microstructure and tensile properties of (ABOw + SiCp)/6061Al composites fabricated by semi-solid stirring technique. *Trans. Nonferr. Met. Soc. China* **2011**, *21*, s274–s279. [CrossRef]

19. Cheng, F.L.; Chen, T.J.; Qi, Y.S.; Zhang, S.Q.; Yao, P. Effects of solution treatment on microstructure and mechanical properties of thixoformed Mg_2Si_p/AM60B composite. *J. Alloy. Compd.* **2015**, *636*, 48–60. [CrossRef]

20. Kang, C.G.; Youn, S.W. Mechanical properties of particulate reinforced metal matrix composites by electromagnetic and mechanical stirring and reheating process for thixoforming. *J. Mater. Process. Technol.* **2004**, *147*, 10–22. [CrossRef]

21. Zhang, H.W.; Geng, L.; Guan, L.N.; Huang, L.J. Effects of SiC particle pretreatment and stirring parameters on the microstructure and mechanical properties of SiC_p/Al-6.8Mg composites fabricated by semi-solid stirring technique. *Mater. Sci. Eng. A* **2010**, *528*, 513–518. [CrossRef]

22. Mazahery, A.; Shabani, M.O. Tribological behaviour of semisolid–semisolid compocast Al-Si matrix composites reinforced with TiB_2 coated B_4C particulates. *Ceram. Int.* **2012**, *38*, 1887–1895. [CrossRef]

23. Curle, U.A.; Ivanchev, L. Wear of semi-solid rheocast SiC_p/Al metal matrix composites. *Trans. Nonferr. Met. Soc. China* **2010**, *20*, s852–s856. [CrossRef]

24. Alhawari, K.S.; Omar, M.Z.; Ghazali, M.J.; Salleh, M.S.; Mohammed, M.N. Wear properties of A356/Al_2O_3 metal matrix composites produced by semisolid processing. *Procedia Eng.* **2013**, *68*, 186–192. [CrossRef]

25. Sarajan, Z. Friction and wear of aluminum alloy reinforced by TiO_2 particles. *Strength Mater.* **2013**, *45*, 221–230. [CrossRef]

26. Koli, D.K.; Agnihotri, G.; Purohit, R. A review on properties, behaviour and processing methods for Al-nano Al_2O_3 composites. *Procedia Mater. Sci.* **2014**, *6*, 567–589. [CrossRef]

27. Mazahery, A.; Shabani, M.O. Characterization of cast A356 alloy reinforced with nano SiC composites. *Trans. Nonferr. Met. Soc. China* **2012**, *22*, 275–280. [CrossRef]

28. Mazahery, A.; Abdizadeh, H.; Baharvandi, H.R. Development of high-performance A356/nano-Al_2O_3 composites. *Mater. Sci. Eng. A* **2009**, *518*, 61–64. [CrossRef]

29. Sajjadi, S.A.; Parizi, M.T.; Ezatpour, H.R.; Sedghi, A. Fabrication of A356 composite reinforced with micro and nano Al_2O_3 particles by a developed compocasting method and study of its properties. *J. Alloy. Compd.* **2012**, *511*, 226–231. [CrossRef]

30. Jiang, J.F.; Wang, Y. Microstructure and mechanical properties of the semisolid slurries and rheoformed component of nano-sized SiC/7075 aluminum matrix composite prepared by ultrasonic-assisted semisolid stirring. *Mater. Sci. Eng. A* **2015**, *639*, 350–358. [CrossRef]

31. Jiang, J.F.; Wang, Y. Microstructure and mechanical properties of the rheoformed cylindrical part of 7075 aluminum matrix composite reinforced with nano-sized SiC particles. *Mater. Des.* **2015**, *79*, 32–41. [CrossRef]

32. Jiang, J.F.; Wang, Y.; Nie, X.; Xiao, G.F. Microstructure evolution of semisolid billet of nano-sized SiCp/7075 aluminum matrix composite during partial remelting process. *Mater. Des.* **2016**, *96*, 36–43. [CrossRef]

33. Standarization Administration of the People's Republic China. *High-Carbon Chromium Bearing Steel*; PRC National Standard GB/T18254-2016; Standarization Administration of the People's Republic China: Beijing, China, 2016.

34. Zhang, S.Y.; Wang, F.P. Comparison of friction and wear performances of brake material dry sliding against two aluminum matrix composites reinforced with different SiC particles. *J. Mater. Process. Technol.* **2007**, *182*, 122–127. [CrossRef]

35. Natarajan, N.; Vijayarangan, S.; Rajendran, I. Wear behaviour of A356/25SiC_p aluminium matrix composites sliding against automobile friction material. *Wear* **2006**, *261*, 812–822. [CrossRef]

36. Zhong, X.L.; Wong, W.L.E.; Gupta, M. Enhancing strength and ductility of magnesium by integrating it with aluminum nanoparticles. *Acta Mater.* **2007**, *55*, 6338–6344. [CrossRef]

37. Gutam, M.; Srivastant, T.S. Interrelationship between matrix microhardness and ultimate tensile strength of discontinuous particulate-reinforced aluminum alloy composites. *Mater. Lett.* **2001**, *51*, 255–261.

 *metals*

Article

An Experimental Evaluation of Electron Beam Welded Thixoformed 7075 Aluminum Alloy Plate Material

Ava Azadi Chegeni and Platon Kapranos *

Department of Materials Science & Engineering, University of Sheffield; Sir Robert Hadfield Building, Mappin Street, South Yorkshire, Sheffield S1 3JD, UK; avaazadi.922@gmail.com
* Correspondence: p.kapranos@sheffield.ac.uk; Tel.: +44-114-22-25509

Received: 8 November 2017; Accepted: 13 December 2017; Published: 15 December 2017

Abstract: Two plates of thixoformed 7075 aluminum alloy were joined using Electron Beam Welding (EBW). A post-welding-heat treatment (PWHT) was performed within the semi-solid temperature range of this alloy at three temperatures, 610, 617 and 628 °C, for 3 min. The microstructural evolution and mechanical properties of EB welded plates, as well as the heat-treated specimens, were investigated in the Base Metal (BM), Heat Affected Zone (HAZ), and Fusion Zone (FZ), using optical microscopy, Scanning Electron Microscopy (SEM), EDX (Energy Dispersive X-ray Analysis), and Vickers hardness test. Results indicated that after EBW, the grain size substantially decreased from 67 μm in both BM and HAZ to 7 μm in the FZ, and a hardness increment was observed in the FZ as compared to the BM and HAZ. Furthermore, the PWHT led to grain coarsening throughout the material, along with a further increase in hardness in the FZ.

Keywords: 7075 aluminum alloy; thixoforming; post-welding-heat treatment; electron beam welding (EBW)

---

1. Introduction

7075 wrought aluminium alloys are used for a wide variety of applications in aerospace and automotive industries due to the outstanding characteristics that they possess, such as high-strength-to-weight ratio, ductility, toughness, low density, and resistance to fatigue [1–4]. Promising fabrication techniques are required to produce high quality and integrity parts for such applications. Hence, semi-solid metal processing as a single step manufacturing method providing good quality near net shape products has been widely employed to aluminium alloys due to the advantages that this technology offers over the conventional casting techniques [5–8].

Weldability of the materials is another important factor in aerospace and automotive industries. However, although Al alloys have in the past been considered as difficult-to-weld materials through conventional arc welding techniques, improvements have removed these difficulties and quite few studies have focused on other technologies that offer improvements of weld performance, such as high-power density fusion joining, namely, laser beam welding (LBW) and electron beam welding [1,9, 10], and, of course, Friction Welding.

Electron-beam welding (EBW) is a fusion welding process, in which a beam of high-velocity electrons is applied to two materials to be joined. The workpieces melt and flow together as the kinetic energy of the electrons is transformed into heat upon impact. EBW is often performed under vacuum conditions to prevent dissipation of the electron beam. Electron beam welding provides high-quality welded joints for a wide range of thicknesses, and can be operated with high welding speeds [11,12]. Using EBW generates low distortions in the Fusion Zone (FZ), together with a narrow Heat Affected Zone (HAZ), and low residual stresses in comparison with conventional welding techniques [12,13]. To take advantage of these features, many studies centred around the investigation of the EBW on different aluminium alloys. Cam et al. [14] investigated the effects of EBW on mechanical properties and microstructural characterisation of 5005, 2024, and 6061 aluminium alloys, and concluded that a defect free weld line was observed in these alloys. Kocak et al. [12] also evaluated the impacts of EBW on aluminium 7020 alloy, and reported that a loss of hardness in FZ was observed due to the loss of strengthening phases. Currently, there is no reported data published on EBW of thixoformed aluminium 7075 alloy and the aim of this study is to investigate the effect of EBW on the microstructure and mechanical properties of two thixoformed aluminium plates of Al 7075 alloy. These results have been obtained through optical microscopy, Scanning Electron Microscopy (SEM), EDX (Energy Dispersive X-ray Analysis) on the electron beam welded plates, along with hardness measurements. In addition, a post welding heat treatment of the weld zone was conducted at the semi-solid temperature range in order to investigate any microstructural and property changes in the weld material, as well as the parent material. The choice of the Semi-solid range was to replicate the conditions experienced during the thixoforming process that takes place within this temperature range at approximately 50% liquid content.

2. Materials and Methods

Two thixoformed plates of 7075 wrought aluminium alloy were used as starting materials. The chemical composition of the alloy is presented in Table 1 [1].

Table 1. Chemical composition of wrought 7075 Al alloy (wt %) [1].

Alloy	Cu	Zn	Mg	Mn	Cr	Ti	Si	Fe	Zr	B	Al
7075	0.94	4.52	2.24	<0.01	0.22	0.05	0.02	<0.01	<0.01	<0.01	Bal.

The plates were welded using electron beam welding with a speed of 1000 mm/min at TWI Ltd. Cambridge (Great Abington, Cambridge, UK). The accelerating voltage and beam current that were used were 130 kV and 21 mA, respectively. The microstructure of the welded plates was investigated using standard optical metallographic methods. Samples that were cut along the length of the weld line and from the parent material were ground with standard SiC grinding paper and polished with 6 and 1 µm monocrystalline diamond suspension and 0.05 µm silica suspension; the specimens were subsequently etched using sodium hydroxide (10 g NaOH diluted with 100 mL water). The microstructures of the Base Metal (BM), HAZ, and FZ were evaluated using a Nikon Eclipse LV150 optical microscope (Nikon, Tokyo, Japan) and TM3030Plus Tabletop Scanning Electron Microscope (Hitachi, Tokyo, Japan). The chemical analysis of the phases was performed using EDX (Energy Dispersive X-ray Analysis, Hitachi, Tokyo, Japan). Image J software (An open platform for scientific image analysis, https://imagej.net/Welcome) was used to measure the average grain size from optical images using the linear intercept method, the shape factor of the solid grains, calculated based on their perimeter and area, as well as the liquid fraction content of the specimens. The Vickers hardness measurements were conducted using a Zwick hardness tester (ZHU250CL, Ulm, Germany) with 10 kgf applied force for 10 s across different locations on the plates, as shown in Figure 1, and the average values were reported. To investigate the influence of post welding heat treatment on the microstructure and mechanical properties of the welded plates, specimens were heat treated by being kept for 3 min at 610, 617, and 628 °C, respectively (i.e., within the semi-solid temperature range of the alloy) and fast cooled in water.

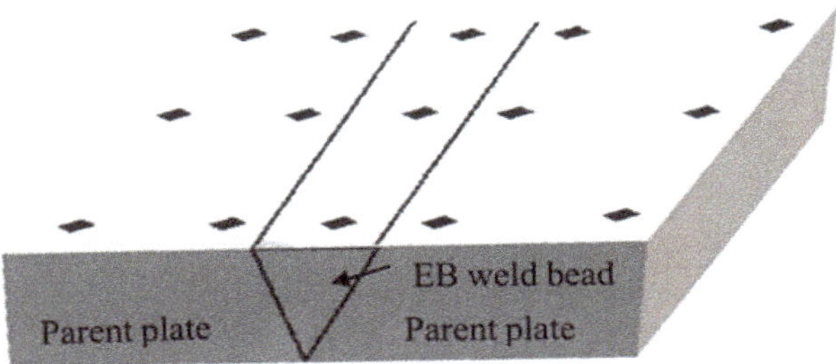

Figure 1. Illustration of the welded plates and the areas of hardness tests.

3. Results and Discussion

3.1. Microstructure of the Base Material

Micrographs of the as-received plate of 7075 thixoformed aluminum alloy are presented in Figure 2; the images are taken from different regions of two plates. When considering Figure 2 from the top left to the bottom right, the microstructure of the alloy is uniform throughout the thixoformed plates, and consists of globular, fine, non-dendritic grains in a solid matrix consisting of the last liquid to solidify.

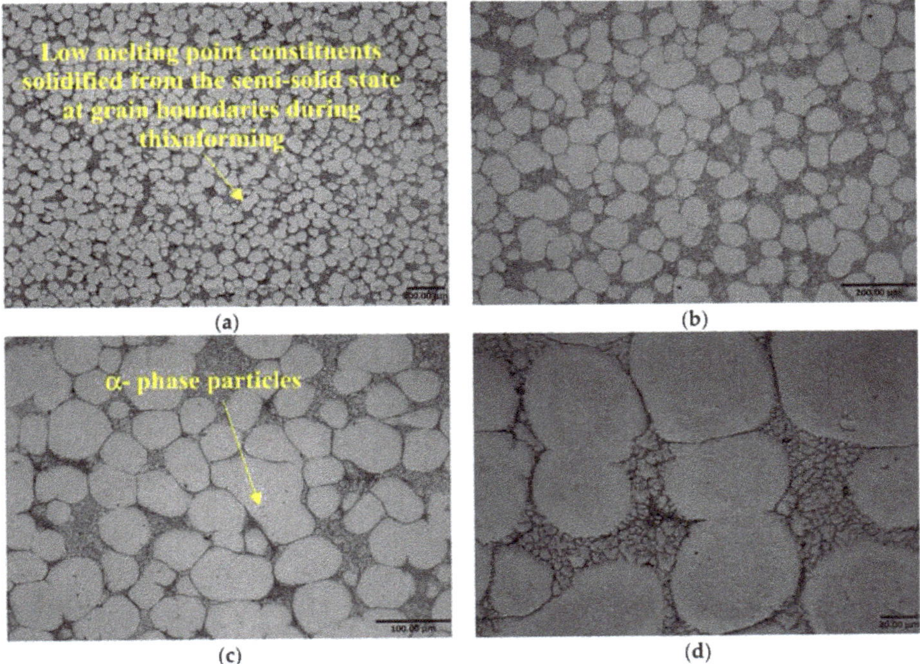

Figure 2. Optical micrographs showing the thixoformed base metal (BM) of 7075 Al Electron Beam (EB) welded plates (**a–d**) with typical near spheroidal microstructure at different magnifications.

Scanning Electron Microscopy (SEM) micrographs of the BM are shown in Figure 3a,b. It can be observed that the liquid phase specified with a square was formed at the grain boundaries during the thixoforming process. In addition, the last liquid to solidify from the semi-solid state around the grains is illustrated in Figure 3c. There are two ways by which the liquid can be entrapped inside the grains. First, by segregation of the alloying elements inside the solid grains, leading to the formation of fine liquid droplets during the partial re-melting, and, secondly, when grains are combined to reduce the solid-liquid interfacial energy during the heating stage of the thixoforming process, giving rise to the creation of relatively large liquid droplets contained within the sub-grains. Hence, the base metal consists of alpha-Al, eutectic liquid phase and occasional trapped liquid pools within sub-grains [5].

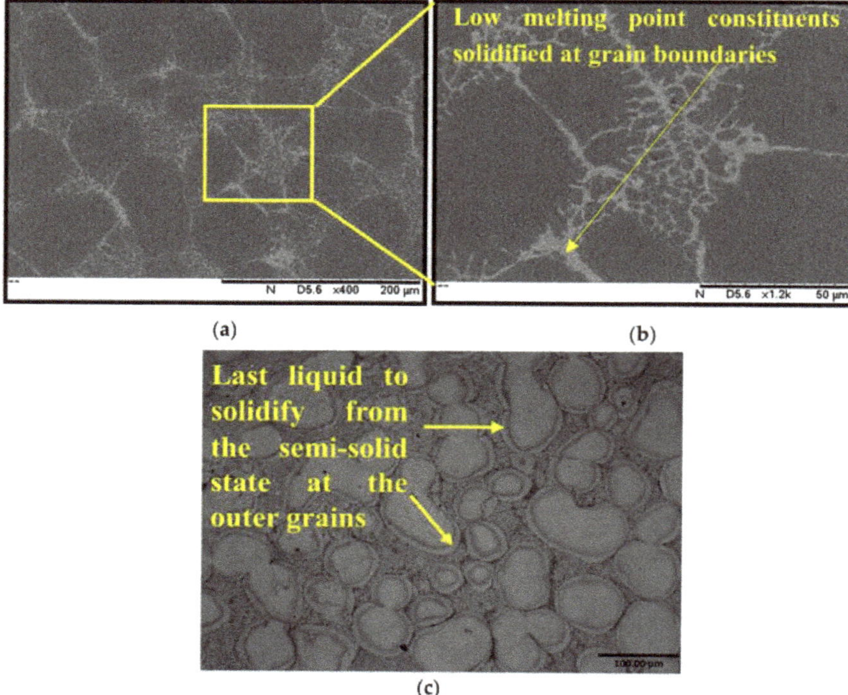

(a)

(b)

(c)

Figure 3. (**a**) Scanning Electron Microscopy image of Al 7075 alloy showing the last liquid phase to solidify at grain boundaries; (**b**) a higher magnification SEM image of the liquid phase; (**c**) optical microscopy image showing the last liquid solidified around the grains.

3.2. Microstructural Evolution after Welding

Figure 4 illustrates the micrographs of the EB welded plates. As can be seen, the microstructure contains three regions, namely, Heat-Affected Zone (HAZ), Fusion Zone (FZ), and Base Metal (BM), as presented in Figure 4a. Generally, high heat input and preheating are two factors that increase the width of the HAZ in precipitation hardenable aluminum alloys. However, it can be observed that the HAZ is relatively narrow due to the low heat input during the electron beam welding process [9,14,15]. In addition, it can be said that although aluminum 7075 alloy is prone to cracking, the EBW process did not pose any significant problems, and only few numbers of pores were formed in the FZ, as shown in Figure 4b. Possible reasons for the formation of these pores could be the high specific energy density and evaporation of metal that is associated with EBW and thermophysical features of aluminum, including its low melting point, high thermal conductivity, and surface oxide films, with high melting points [9,14,16]. Figure 4c,d show the weld area at different magnifications. As can be observed, the fusion zone consists of a fine grain microstructure with a significant reduction in the average grain size, as compared to the BM and HAZ. As the total heat input into the material during the EBW is lower than that of other fusion welding techniques due to the higher power density of the EBW process, a finer microstructure in the FZ can be typically obtained using EBW. The difference between the grain size in FZ and HAZ is presented in Figure 4e.

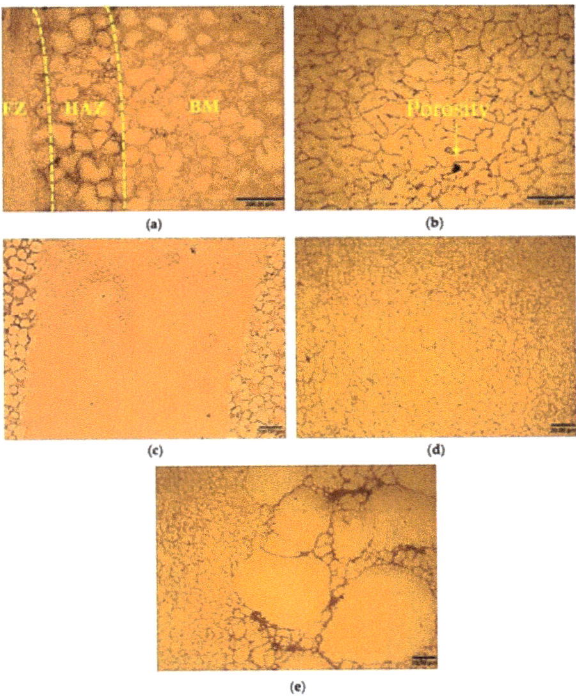

Figure 4. (**a**) Optical micrographs of Al 7075 EB welded plates, (**b–d**) show details of the Fusion Zone (FZ) and (**e**) detail of the plate weld boundary.

Figure 5a shows graphically the grain size as a function of the position from the weld zone. It can be seen that the mean grain size in the BM is around 67 µm, and this number is approximately the same in the HAZ, which shows a consistency of the microstructure in the thixoformed plate. However, the grain size in the FZ is significantly reduced to around 7 µm, which is considerably smaller than that of either the BM or HAZ. There are several reasons for the reduction of the grain size after welding. First, the presence of the alloying elements that precipitated out at the grain boundaries impedes the severe growth of the grains. In addition, the fast cooling of the joint after the welding process can also prevent grain growth in the FZ [17]. Furthermore, due to the high welding rates of the EB process that are caused by the high melting speeds of the focused heat source, the time that is required for the welding to be accomplished is also reduced so that the grains do not have enough time to grow during the EBW [12]. Figure 5b demonstrates a graph of the shape factor against the distance from the weld area. The shape factor represents the circularity of the grains, which has a maximum of one for a totally spherical grain and zero for a grain with a complex shape. As can be seen from the graph, more spherical grains can be observed in the BM when compared to the HAZ and FZ, since the material was exposed to the high temperatures during the EBW, leading to the deviation from the spherical grains that are present in the BM. Measurements were taken across various positions in the different zones, and there is clearly a deviation between the spheroidicity of the different grains, as expected as the FZ has undergone melting that destroyed the original non-dendritic, near spheroidal microstructure of the base material.

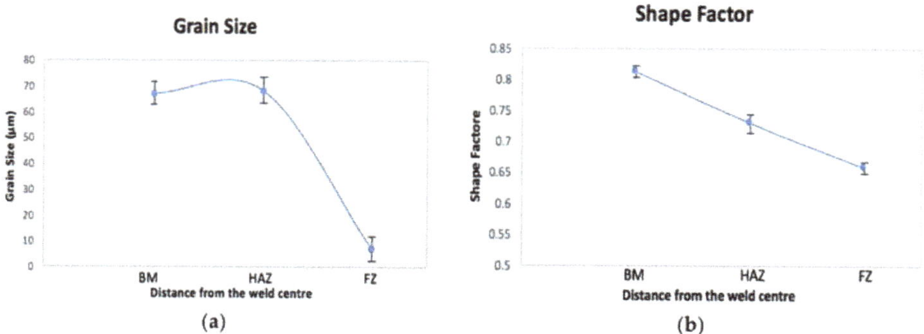

Figure 5. (**a**) Graph of grain size as a function of position, (**b**) Graph of shape factor against position.

Results of the EDX point analysis conducted for the last liquid to solidify phase at the boundaries and grains in the BM, as well as the grains in the FZ are shown in Figure 6 and Table 2. From the graphs, it can be said that the liquid phase at the boundaries mainly consisted of aluminum, magnesium, copper, and zinc, implying the presence of the alloying elements in these regions. In addition, it can be observed that aluminum was by far the highest constituent of the grains in the base metal, whilst the percentage of the other elements was low within the grains, which confirms that the alloying elements precipitated out at the boundaries. Moreover, aluminum is the main element in the FZ, with almost the same content as that of the BM. However, EDX point analysis reveals that the matrix phase in FZ contains less Zn than the BM matrix, whereas the Mg and Cu contents of the FZ are higher than that of the BM. Cam et al. [9] suggested that the heat input during the fusion welding processes may give rise to the evaporation of the solute atoms with low melting points in the fusion zone, hence, the lower amount of zinc in this region can be attributed to the evaporation mechanism due to the high amounts of heat applied during welding. The loss of Mg/Zn/Cu was also reported by other researchers [12,14]. Furthermore, rapid cooling after the EBW process leads to the presence of super saturated amount of Mg and Cu in the FZ. It is worth adding that the EDX data for the liquid phase at the boundaries in the FZ could not be measured due to the low amount of the last liquid to solidify phase that remained at the boundaries after EBW.

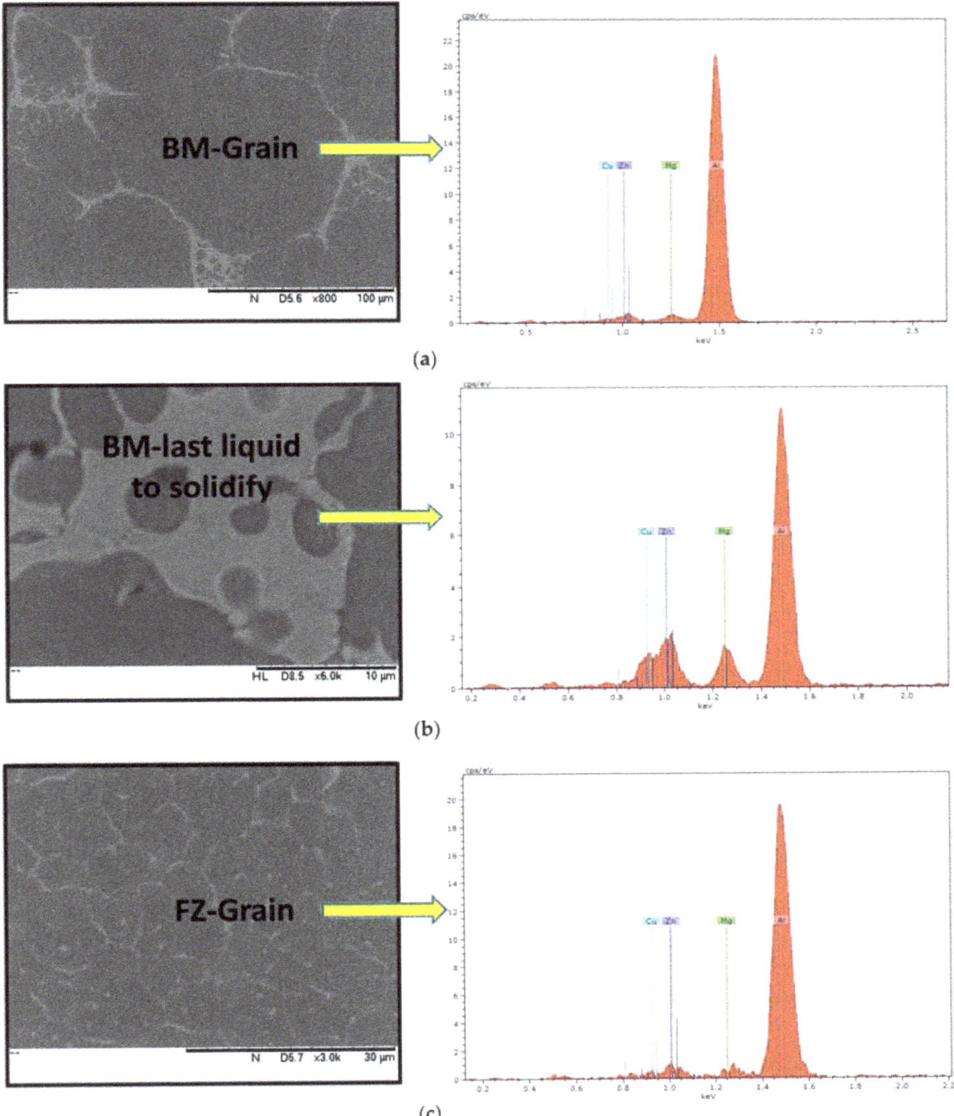

Figure 6. Results of Energy Dispersive X-ray Analysis (EDX) point analysis for: (**a**) grains in BM, (**b**) eutectic phase at boundaries, (**c**) grains in FZ.

Table 2. Weight percent of alloying elements at different points of microstructure. Reported numbers are the average of at least three attempts.

Locations	Al (wt %)	Zn (wt %)	Mg (wt %)	Cu (wt %)
BM-Grain	92.2825	4.5225	2.242	0.94075
BM-Boundary	65.91	15.453	10.39	8.216
FZ-Grain	91.105	4.2145	2.65	2.018

The amount of the last liquid to solidify phase on quenching was calculated using image J software, and a plot of liquid fraction versus distance from the weld centre is indicated in Figure 7. The graph represents a downward trend from the BM to FZ, which suggests that the fraction of eutectic, which is around 35% in the thixoformed plate, is higher than this amount in both HAZ and FZ, which is 22% and 16%, respectively. The higher percentage of the eutectic phase in the BM can be related to the nature of the semi solid metal processing technique, which typically contains between 30–50% liquid. On the other hand, the lower amount of the eutectic phase in the FZ can be attributed to the complete melting in the FZ during the EBW, followed by the fast cooling of this region. It is worth pointing out that the actual amount of the eutectic phase in the BM is higher than the above mentioned number, as can be seen from the micrographs, since some of the liquid is entrapped within the grains and the software could not measure it during the calculations. We will use the terms eutectic solid and last liquid to solidify interchangeably as the terminology used in the Semi-solid forming usually refers to the eutectic solid as liquid fraction.

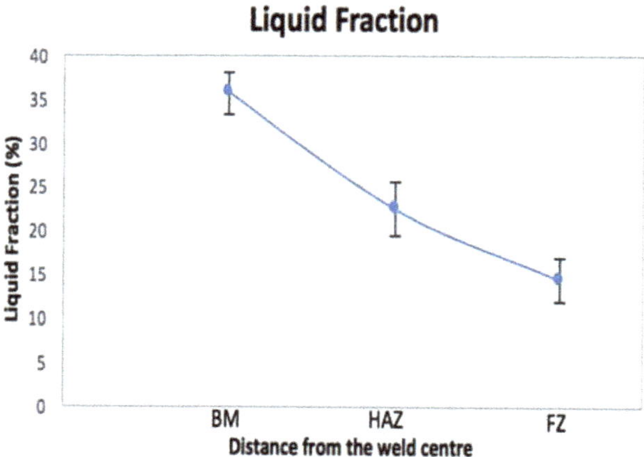

Figure 7. Liquid fraction against distance from the centre of the weld for Al 7075 alloy EB welded plate.

3.3. Post Weld Heat Treatment

Micrographs of EB welded specimens following heat treatment at 610, 617, and 628 °C are presented in Figure 8. According to Figure 8a, the microstructure of the base metal and the heat affected zone at different temperatures consisted of fine equiaxed solid grains that are uniformly distributed throughout the material; which is similar to that of the welded material before the heat treatment. The differences between the grain size of the FZ and HAZ can be observed from Figure 8b. Micrographs of the fusion zone for heat-treated samples, at the three temperatures, are displayed in Figure 8c, as compared to the microstructure of the joint before the post weld process. The grain structure in the FZ has coarsened after heat treatment at three temperatures. In addition, there is a noticeable change in the morphology in this region, in that grains are more spherical in the FZ when compared to the as-received grain structure. This can be explained due to the post weld heat treatment in the semi-solid temperature range to which the material was subjected.

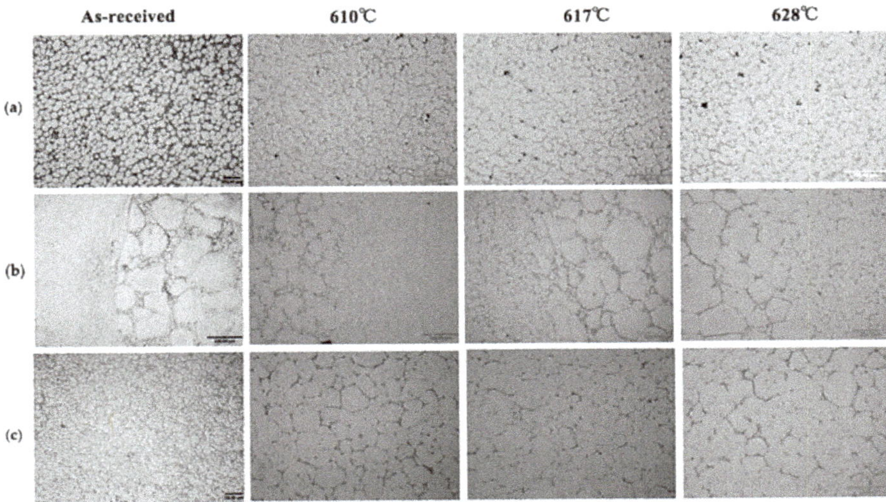

Figure 8. Optical microscopy micrographs of Al 7075 alloy EB welded plates after heat treatment at three temperatures: 610, 617 and 628 °C. (**a**) as received plate material, (**b**) HAZ and plate boundary and (**c**) FZ weld area. (Magnifications top from left to right in µm: 200, 500, 500, 500); (magnifications middle from left to right in µm: 100, 200, 100, 100); (magnifications bottom from left to right in µm: 20, 50, 50, 50).

A graph representing the grain size of the BM, HAZ, and FZ for the heat-treated materials at 610, 617, and 628 °C is shown in Figure 9a. From the graph, it can be seen that the mean grain size in the BM, HAZ, and FZ follows a similar trend, with a slight reduction in the grain diameter at 617 °C, followed by a rise in the average grain size at 628 °C. It is worth adding that grains started to grow in all three regions of specimens, including BM, HAZ, and FZ when compared to the grain structure before the heat treatment. This is due to the exposure of samples to high temperatures during the heat treatment and holding at these temperatures for three minutes, resulting in the grain coarsening. Moreover, the average grain size in the BM is slightly higher than that of the HAZ, but the mean grain diameter in the FZ is by far lower than those in both HAZ and BM, as shown in the micrographs of Figure 8b. Figure 9b shows the shape factor measurement for the heat-treated specimens. As can be seen, there is a significant change in the shape factor of particles in the FZ when compared to before heat treatment. During the heat treatment, solid grains in the FZ became more spherical as they were heated up to the thixoforming temperature range and held for three minutes at these temperatures. The shape factor of grains in the HAZ at the three temperatures is almost the same. In addition, the circularity of the particles in the BM has experienced an upward trend with temperature increase.

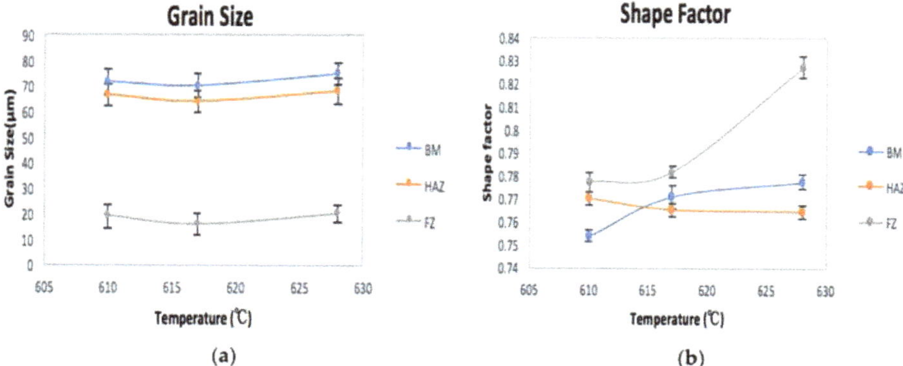

Figure 9. (**a**) Graph of grain size against temperature, (**b**) Plot of shape factor as a function of the three temperatures used, 610, 617, and 628 °C for Al 7075 alloy EB welded plates after heat treatment.

3.4. Vickers Hardness Test

Results of the Vickers hardness test performed on the EB welded plates of the thixoformed 7075 aluminium alloy are summarised in Figure 10. As can be observed, the graph shows a relatively symmetric shape, and the hardness number in the BM of two plates is roughly the same, since these regions belong to the parent material with similar properties and fabrication method, either side of the weld. In addition, a rise in hardness can be observed from both plates toward the FZ. As already mentioned, evaporation of zinc was observed after welding in the FZ, however, this amount was not considerable (0.307 wt % Zn depletion based on numerical values in Table 2) and can be eliminated, so it did not cause hardness reduction in this region. Moreover, based on the results of the EDX point analysis presented before, the weight percent of Mg and Cu increased in the FZ as compared to the BM as a result of rapid cooling that occurred after the welding process. Therefore, unlike the results provided by several researchers about some alloys of aluminium, including 5005 and 2024 series showing a hardness reduction in the fusion zone due to the loss of strengthening elements, such as Cu and Mg in this area [14,18], herein the hardness increment can be attributed to the surplus amount of these strengthening elements in the FZ, supported by researchers that have reported the hardness increase that is obtained in the FZ by using appropriate filler wire during the welding process to compensate for the evaporation of elements [9].

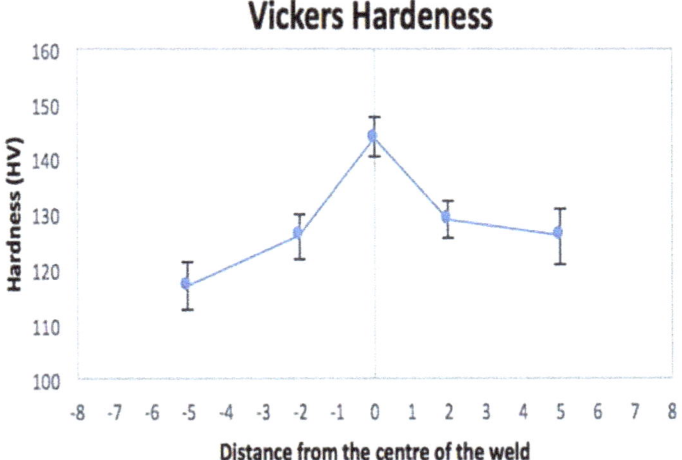

Figure 10. Graph of hardness versus position from the weld centre for Al 7075 alloy EB welded plates.

Figure 11 illustrates Vickers hardness over distance from the weld centre after heat treatment at 610, 617 and 628 °C. The overall behaviour of the three graphs is similar, demonstrating a peak in the FZ, and as the distance from the weld centre increases, a drop in the hardness can be observed, implying property improvement in the FZ. In comparison with the graph of the EB welded plate before heat treatment in the FZ, a rise in hardness can be seen after heat treatment at the three temperatures, with hardness increasing from 144 HV before to 156, 184, and 203 HV after heat treatment at 628, 610, and 617 °C, respectively, showing a maximum at 617 °C. After heat treatment, grain growth has occurred in the FZ of the material, and, hence, a drop in hardness was expected due to grain coarsening. However, the hardness increase can be attributed to the refined grain size, as well as chemical changes in the FZ after heat treatment as different diffusion coefficients of the critical elements in the semi-solid state locally can explain the movement of atoms from the Al-lattice at these high temperatures. In addition, the hardness number at the edges of both heat-treated plates is approximately 125 HV, which is roughly the same as that in the parent material before heat treatment. It can be concluded that although heat treatment improved the hardness in the FZ, it did not change the properties of the base metal.

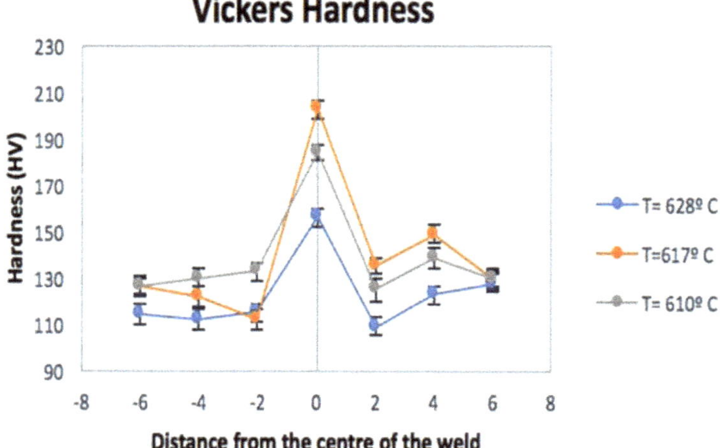

Figure 11. Graph of hardness versus position from the weld centre for Al 7075 alloy EB welded plates after heat treatment at three temperatures: 610, 617 and 628 °C.

Work done by various researchers on Laser hardening, [19] has shown that the most relevant parameter in the hardened layer depth is the scanning speed, followed by the hardened track width. Another group of researchers have given a good account of modelling such processes by modelling the Added Layers by Coaxial Laser Cladding [20].

4. Conclusions

Electron beam welding was employed on thixoformed plates of 7075 aluminium alloy, along with a post welding heat treatment in semi-solid temperature range in order to evaluate the resultant microstructures and mechanical properties in the weld area. The following conclusions were inferred from the work:

1. A fine microstructure with extremely smaller grain size than that in the parent material and HAZ was obtained in FZ of the electron beam welded plates, due to complete melting, followed by rapid cooling of the weld area.
2. Post weld heat treatment in the semi-solid range, 610, 617, and 628 °C, resulted in the growth of recrystallized grains in the fusion zone of the EB welded plates, with grains becoming more spherical in the FZ.
3. The results of EDX point analysis revealed that the weight percent of Mg and Cu increased in the FZ after electron beam welding leading to a rise in hardness in this area when compared to the BM and HAZ.
4. A further increase of hardness was obtained in FZ after post weld heat treatment in the semi-solid temperature range.

Acknowledgments: Technical staff at the University of Sheffield for their support in various analytical techniques used in this work.

Author Contributions: For research articles with several authors, a short paragraph specifying their individual contributions must be provided. The following statements should be used: Ava Azadi Chegeni and Platon Kapranos conceived and designed the experiments; Avi Azadi Chegeni performed the experiments; Avi Azadi Chegeni and Platon Kapranos analyzed the data; Avi Azadi Chegeni and Platon Kapranos wrote the paper.

Conflicts of Interest: The authors declare no conflict of interest.

References

1. Rachmat, R.S.; Takano, H.; Ikeya, N.; Kamado, S.; Kojima, Y. Application of Semi-Solid Forming to 2024 and 7075 Wrought Aluminium Billets Fabricated by EMC Process. *Mater. Sci. Forum* **2000**, *329–330*, 487–492. [CrossRef]
2. Li, J.F.; Peng, Z.W.; Li, C.X.; Jia, Z.Q.; Chen, W.J.; Zheng, Z.Q. Mechanical properties, corrosion behaviors and microstructures of 7075 aluminum alloy with various aging treatments. *Trans. Nonferr. Met. Soc. China* **2008**, *18*, 755–762. [CrossRef]
3. Williams, J.C.; Starke, E.A. Progress in structural materials for aerospace systems. *Acta Mater.* **2003**, *51*, 5775–5799. [CrossRef]
4. Isadare, A.D.; Aremo, B.; Adeoye, M.O.; Olawale, O.J.; Shittu, M.D. Effect of heat treatment on some mechanical properties of 7075 aluminum alloy. *Mater. Res.* **2013**, *16*, 190–194. [CrossRef]
5. Binesh, B.; Aghaie-Khafri, M. Phase evolution and mechanical behaviour of the semi-solid SIMA processed 7075 aluminium alloy. *Metals* **2016**, *6*, 42. [CrossRef]
6. Kapranos, P.; Ridgway, K.; Jirattiticharoean, W.; Haga, T.; Thomas, W. Friction Stir Welding (FSW) of Thixoformed and Rheocast Plates. In Proceedings of the 8th SSM Conference, Limassol, Cyprus, 21–23 September 2004.
7. Nakato, H.; Oka, M.; Itoyama, S.; Urata, M.; Kawasaki, T.; Hashiguchi, K.; Okano, S. Continuous semi-solid casting process for aluminium alloy billets. *Mater. Trans.* **2002**, *43*, 24–29. [CrossRef]
8. Chayong, S.; Atkinson, H.V.; Kapranos, P. Thixoforming 7075 aluminum alloys. *Mater. Sci. Eng. A* **2005**, *390*, 3–12. [CrossRef]
9. Cam, G.; Ipekoglu, G. Recent developments in joining of aluminium alloys. *Int. J. Adv. Manuf. Technol.* **2017**, *91*, 1851–1866. [CrossRef]
10. Cam, G.; Mistikoglu, S. Recent Developments in Friction Stir Welding of Al-alloys. *J. Mater. Eng. Perform.* **2014**, *23*, 1936–1953. [CrossRef]
11. Weglowski, M.S.; Blacha, S.; Phillips, A. Electron beam welding-techniques and trends-Review. In Proceedings of the 9th Symposium on Vacuum Based Science and Technology, Kolobrzeg, Poland, 17–19 November 2015. Programon-Elsevier Science Ltd..
12. Cam, G.; Kocak, M. Microstructural and mechanical characterization of electron beam welded Al-alloy 7020. *J. Mater. Sci.* **2007**, *42*, 7154–7161. [CrossRef]
13. Ning, Y.; Yao, Z.; Gua, H.; Fu, M.W. Hot deformation behaviour and hot working characteristic of Nickel-base electron beam weldments. *J. Alloys Compd.* **2014**, *584*, 494–402. [CrossRef]
14. Cam, G.; Ventzke, V.; Dos Santos, J.F.; Kocak, M.; Jennequin, G.; Gonthier-Maurin, P. Characterisation of electron beam welded aluminium alloys. *Sci. Technol. Weld. Join.* **2013**, *4*, 317–323. [CrossRef]
15. Brungraber, R.J.; Nelson, F.G. Effect of welding variables on aluminum-alloy weldments. *Weld. J.* **1973**, *52*, 97S–103S.
16. Olshanaskaya, T.V.; Salomatova, E.S.; Belenkiy, V.Y.; Trushnikov, D.N.; Permyakov, G.L. Electron Beam welding of Aluminium alloy AlMg6 with a Dynamically Positioned Electron Beam. *Int. J. Adv. Manuf. Technol.* **2017**, *89*, 3439–3450. [CrossRef]
17. Chen, S.C.; Huang, J.C. Comparison of post-Weld Microstructures and mechanical Properties of Electron- and Laser-Beam Welded 8090 Al-Li Alloy Plates. *Mater. Trans.* **1999**, *40*, 1069–1078. [CrossRef]
18. Pakdil, M.; Cam, G.; Kocak, M.; Erim, S. Microstructural and mechanical characterization of laser beam welded AA 6056 Al-alloy. *Mater. Sci. Eng. A* **2011**, *528*, 7350–7356. [CrossRef]
19. Martínez, S.; Lamikiz, A.; Ukara, E.; Calleja, A.; Arrizubieta, J.A.; Lopez de Lacalle, L.N. Analysis of the regimes in the scanner-based laser hardening process. *Opt. Lasers Eng.* **2017**, *90*, 72–80. [CrossRef]
20. Tabernero, I.; Lamikiz, A.; Martínez, S.; Ukar, E.; López de Lacalle, L.N. Geometric Modelling of Added Layers by Coaxial Laser Cladding. *Phys. Procedia* **2012**, *39*, 913–920. [CrossRef]

MDPI

St. Alban-Anlage 66

4052 Basel

Switzerland

Tel. +41 61 683 77 34

Fax +41 61 302 89 18

www.mdpi.com

Metals Editorial Office

E-mail: metals@mdpi.com

www.mdpi.com/journal/metals

9 783039 289752